AllAN KiNTIGH

ELECTRICAL CIRCUIT ACTION

ELECTRICAL CIRCUIT ACTION

HENRY C. VEATCH

**Instructor and Electronics Coordinator, Ret.
San Leandro Adult School
San Leandro, California**

 SCIENCE RESEARCH ASSOCIATES, INC.
Chicago, Palo Alto, Toronto, Henley-on-Thames, Sydney, Paris, Stuttgart
A Subsidiary of IBM

Acquisition Editor	Alan W. Lowe
Project Editor	James C. Budd
Technical Editor	Frank Meltzer
Compositor	Bi-Comp, Inc.
Illustrator	Basil Wood
Text Designer	SRA Staff
Cover Designer	Janet Bollow
Cover Photographer	Ernest Braun

Library of Congress Cataloging in Publication Data

Veatch, Henry C.
   Electrical circuit action.

   Includes index.
   1. Electric circuits.   I. Title.
TK454.V4   621.319'2   77-22049
ISBN 0-574-21510-7

© 1978, Science Research Associates, Inc. All rights reserved.

Printed in the United States of America.

10 9 8 7 6 5 4 3 2 1

CONTENTS

Chapter 1 INTRODUCTION   1
  About This Course of Study   1
  The Electronics Technician   2
  Electronics Application   3

Chapter 2 ELECTROSTATICS   22
  Matter   22
  Particles of Matter   23
  Atomic Structure   24
  Atomic Configuration   25
  Atomic Number and Weight   28
  Static Charge   28
  The Unit of Charge   29
  Basic Laws of Charges   30
  Electrical Circuit Units   31
  The Electrical Circuit   33
  Summary   37
  Questions   37

Chapter 3 BASIC CIRCUIT CHARACTERISTICS   39
  Basic Electrical Circuit Action   39
  Ohm's Law   40
  Unit Abbreviations   44
  Powers of Ten   44
  Practice Problems   46
  Practice Problems   47
  Practice Problems   48
  Practice Problems   49
  Multiple and Submultiple Circuit Problems   50
  Electrical Power   52
  Graphical Relationships   54
  Measuring Electrical Quantities   55
  Approximations   58
  Summary   60
  Questions   61
  Problems   62

Chapter 4 SERIES CIRCUITS — 63
Series Circuit Relationships — 63
Total Series Resistance — 64
Total Circuit Current — 64
Voltage Drops — 65
Voltage Drop Polarity — 68
Analyzing Series Circuits — 70
Multiple Power Sources — 71
Reference Voltage — 72
Ground — 73
Circuits with Various Unknowns — 75
Open and Short Circuits — 77
Practice Problems — 78
Summary — 80
Questions — 80
Problems — 81

Chapter 5 PARALLEL CIRCUITS — 86
Parallel Circuit Voltage — 86
Parallel Circuit Branch and Total Current — 86
Total Effective Resistance — 88
Conductances in Parallel — 90
Current Dividers — 91
Power in Parallel Circuits — 92
Parallel Circuit Problems — 93
Practice Problems — 94
Opens and Shorts in Parallel Circuits — 96
Summary — 97
Questions — 97
Problems — 98

Chapter 6 SERIES-PARALLEL CIRCUITS — 100
Total Series-Parallel Resistance — 100
Analytical Procedure — 102
Series-Connected Parallel Banks — 103
Parallel-Connected Series Strings — 104
Series-Parallel Circuit Examples — 105
Opens and Shorts in Series-Parallel Circuits — 112
Series-Parallel Ground Connections — 113
Bridge Circuits — 117
Practice Problems. — 118
Practice Problems — 119
Summary — 120
Questions — 120
Problems — 121

Chapter 7 DIRECT CURRENT METERS — 125
- The Basic Meter Movement — 126
- Meter Scales — 128
- Ammeters — 129
- Voltmeters — 134
- Ohmmeters — 140
- Multimeters — 148
- VOMs — 148
- Digital Multimeters — 150
- Summary — 151
- Questions — 152
- Problems — 153

Chapter 8 CONDUCTORS, INSULATORS, AND SEMICONDUCTORS — 155
- Characteristics of Conductors — 155
- Ion Current — 165
- Insulators — 168
- Basic Semiconductor Principles — 169
- Summary — 186
- Questions — 187
- Problems — 188

Chapter 9 RESISTORS — 190
- Types of Resistors — 190
- Color Coding — 196
- Design Principles — 198
- Nondiscrete Resistors — 200
- Testing Procedures — 201
- Potentiometers — 203
- Summary — 206
- Questions — 206
- Problems — 207

Chapter 10 CELLS AND BATTERIES — 208
- Basic Electrochemical Action — 208
- Battery and Cell Classification — 212
- Carbon-Zinc Characteristics — 214
- Alkaline-Manganese Characteristics — 218
- Mercury Cell Characteristics — 220
- Lead-Acid Characteristics — 221
- Nickel-Cadmium Characteristics — 224
- Other Systems — 226
- Internal Resistance of Sources — 227
- Summary — 233

Questions	234
Problems	235

Chapter 11 MAGNETISM — 236

Principles of Magnetism	236
Magnetic Fields	240
Magnetic Domains	242
Magnetic Quantities and Units	244
Electromagnetism	246
Magnetic Circuit Examples	254
B-H Curves	255
Systems of Measurement	260
Summary	262
Questions	263
Problems	263

Chapter 12 ELECTROMAGNETIC INDUCTION — 265

Motor Action	265
Induced Current	269
Induced Voltage	271
Induced Voltage Across a Coil	272
Self-Induction	274
Generator Action	278
Magnetic Devices	283
Summary	287
Questions	288
Problems	288

Chapter 13 ALTERNATING CURRENT AND VOLTAGE — 290

Transporting AC	291
The Sine Wave	291
Conversion Factors	295
Simple AC Circuit	296
Instantaneous Values	298
Signal AC	300
Phase Angles	303
Nonsinusoidal Waveforms	307
Summary	309
Questions	310
Problems	310

Chapter 14 INDUCTANCE AND INDUCTIVE DEVICES — 312

Inductance in AC Circuits	312
Circuit Examples	323

Inductors in Series and Parallel	325
Transformers	328
Inductance in DC Circuits	336
Summary	339
Questions	340
Problems	341

Chapter 15 INDUCTIVE CIRCUIT ANALYSIS 343

Complex Numbers	343
Power in Inductive Circuits	349
Circuit Examples	352
Types of Inductors	358
Trouble in Inductors	361
Summary	361
Questions	362
Problems	362

Chapter 16 CAPACITANCE AND CAPACITIVE DEVICES 364

The Elementary Capacitor	364
Types of Capacitors	369
Capacitor Characteristics	375
Capacitors in Series and Parallel	381
Opens and Shorts in Capacitors	384
Practice Problems	385
Summary	386
Questions	387
Problems	388

Chapter 17 CAPACITIVE CIRCUITS 389

Capacitance in AC Circuits	389
Capacitors in DC Circuits	402
Capacitive Voltage Dividers	407
Summary	409
Questions	409
Problems	410

Chapter 18 ALTERNATING-CURRENT CIRCUITS 413

AC Circuits	413
Series LC Circuits	414
Series RLC Circuits	418
Parallel RLC Circuits	420
Compound RLC Circuits	422
Conductance, Susceptance, Admittance	426
Measuring AC Circuits	427

Power in AC Circuits	437
Three-Phase Power	441
Rectified Power Supplies	444
Summary	449
Questions	449
Problems	450

Chapter 19 RESONANCE AND RESONANT CIRCUITS 453

Series-Resonant Circuits	453
Parallel-Resonant Circuits	460
Applications of Resonant Circuits	463
Passband of Resonant Circuits	465
Filter Circuits	467
Filter Traps	471
The Varactor Diode	473
Summary	476
Questions	476
Problems	477

Chapter 20 NONSINUSOIDAL CIRCUITS 479

Nonsinusoidal Waveforms	479
RC Networks	488
RL Networks	498
Summary	503
Questions	503
Problems	504

Chapter 21 CIRCUIT ANALYSIS TECHNIQUES 506

Current and Voltage Dividers	506
Kirchhoff's Laws	518
Thevenin's Theorem	522
Norton's Theorem	528
Superposition Theorem	529
Millman's Theorem	533
Delta-Wye Transformations	536
Summary	541
Problems	541

ANSWERS TO ODD-NUMBERED QUESTIONS	**546**
APPENDIX 1 STANDARD GRAPHIC SYMBOLS FOR ELECTRICAL AND ELECTRONICS DIAGRAMS	**561**
APPENDIX 2 COLOR CODES	**568**

APPENDIX 3	TRIGONOMETRIC FUNCTIONS	572
APPENDIX 4	COMMON LOGARITHMIC TABLES	578
APPENDIX 5	FREQUENCY BANDS AND ALLOCATIONS	583
APPENDIX 6	THE INTERNATIONAL SYSTEM OF UNITS	585
INDEX		589

PREFACE

This presentation was developed in response to instructor demand for an electrical fundamentals text that covers basics thoroughly, but without the extraneous topics and verbose explanations found in many similar offerings. This book is based on the premise that an electronics student's career survival is greatly enhanced by solid preparation in basic electrical concepts, and that this critical instructional phase should be carefully designed to ensure mastery of essential topics by the student.

In addition to avoiding extraneous "overkill," a deliberate effort has been made to keep the reading level as low as possible—without sacrificing accuracy and student comprehension. The material is primarily intended for service or industrial electronics technician programs in community colleges, technical institutes, adult-education programs, and area vocational-technical schools. Industrial education programs in four-year schools should also find the presentation ideal for their needs and interests.

The book, which reflects the author's 20 years' experience in teaching this subject, is organized in a manner that is not unlike many existing books, and many teachers may wish to rearrange some material to suit their own preferences or requirements. This can easily be done. Certain subjects have traditionally been given in a certain sequence, and these probably never change. Others, however, can be introduced at various times to suit particular needs. For example, some schools no longer cover the subject of dc power sources (batteries), and this can be eliminated altogether (or simply given as an extra reading assignment). Also, some teachers prefer to cover Kirchhoff's laws early in the course. In this book, these principles *are used* in the early chapters but are not identified as such. Then in a later chapter, the principles are restated and formally identified. The rationale for doing this is that an effort has been made in the early part of the course to minimize memory work and, in turn, reduce the student-attrition rate.

To meet a perceived trend, basic semiconductor and diode theory has been melded into the electrical theory (at the suggestion of potential users). Allowing students to experience some basic nonactive-device theory in the beginning course can serve as a motivator. Such coverage not only enlivens, to a degree, the first course, but also whets student appetites for subsequent electronic-circuit theory. Active-device theory, however, has been purposely excluded to avoid overload, and because such coverage most logically relates to, and belongs in, subsequent electronic device and circuit courses. In this connection, the subjects of voltage and current dividers, which are used so extensively in modern circuitry, are introduced early and are dealt with throughout the book as appropriate. This approach provides effective preparation for subsequent electronic circuit courses.

An attempt has been made to modernize the text, if such is possible to any extent. At the very least, many of the examples used are taken from equipment now available on the market. In the area of test equipment, digital multimeters and oscilloscopes are introduced and used to the extent possible. However, it should be noted that some books attempt to provide coverage of a very wide selection of topics and equipment. Much of this is inappropriate and tends to overwhelm the beginning student, in our opinion. Such practice needlessly increases the cost of the book, with little or no educational benefit. Thus the subject matter to be included (or excluded) must be carefully and intelligently selected. This book tends to be more selective and tries to give the student all that is needed to *thoroughly* understand the subject at the required level—but no more.

The book has been written with modern technology in mind; namely, the now widely utilized electronic calculator. Most of the problems in the later portion of the text are designed to be solved most easily by such a device. In order of importance, features beyond the four basic math functions are: 1. scientific notation (powers-of-ten); 2. square, square root, and reciprocal; 3. two or more addressable memories; 4. sin-cos-tan and the inverse of these; 5. polar and rectangular conversion; 6. nested parentheses; and 7. natural and common logs. Additional features beyond these are probably superfluous for this course. However, even if no such device is available, the solutions are readily accomplished by the use of the tabular material in the appendixes. (The student will find, of course, that it will take longer to work the problems without a calculator.)

In several areas, much effort has been exerted to remove confusion and conflicting standards or ideas. For example, when magnetism is introduced, many books thoroughly confuse the reader by simultaneously presenting both the cgs and mks systems, making memorization all but impossible. Herein, only the cgs system is used in the text so the student can become familiar with one system. Then, the mks system is introduced and explained in terms of the now-familiar cgs system. While it is true that the International System (SI) specifies the mks system at this level, the cgs system has proven easier to assimilate. Again, it is felt that student understanding is more important than other considerations.

Further, every effort has been made to introduce the student to standard practices. Thus all schematic symbols are taken from IEEE graphic standards, and all units of measure and their abbreviations also conform to current IEEE practice. This is a difficult period for the student—conventional units are being replaced by SI units, and the student must learn both systems. Every effort has been made to explain both systems as clearly and simply as possible.

To facilitate the learning process, questions and problems are provided at the end of each chapter. This is, of course, standard practice. A number of questions, either true or false, multiple choice, or fill-in-the blanks, are given to form a review of the content of the chapter that requires some recall. Then the problems provide exercises in numerical manipulation. Additionally, interspersed throughout the text are numerous examples—worked out in step-by-step detail—and practice problems, where such are thought to be helpful. Answers are given for all practice problems, so the student can have immediate reinforcement of the method of attack and the results. Further, there are approximately ten essay questions per chapter in the Instructor's Guide that are intended for supplemental use. (Some of these are similar to the questions in the textbook, but many are different.)

Finally, the author respectfully and actively solicits feedback from the classroom if problems of any kind (typographical errors or others) are encountered. These will be given full consideration and corrections made at the earliest possible moment. Please send your comments to the publisher, who will immediately forward them to me.

The author and publisher wish to extend their gratitude to the following for their thorough and constructive developmental reviewing:

Tom Bingham, St. Louis Community College at Florissant Valley
Richard Burchell, Riverside City College
Maurice Farris, Maricopa Technical Community College
Charles King, Fresno City College
George Knapp, Pennco Tech
Kenneth Moshier, California State University, Fresno
Edmund Turner, Modesto Junior College

The author wishes to express his appreciation to the many companies and firms who generously provided much of the content of the book. These are identified where their material appears. Also, thanks must be given to Carol Brodeur, who provided excellent typing of a difficult manuscript, and to the author's wife, Marilyn, for her secretarial and research assistance.

Finally, and far from least, Mr. Alan W. Lowe, SRA's Technical

Education Editor, provided many excellent suggestions throughout the creation of this book. His many years of experience and his dedication to technical education provided me with inspiration and guidance.

Henry C. Veatch

CHAPTER 1

INTRODUCTION

1-1 ABOUT THIS COURSE OF STUDY

The study of electronics must begin with the elementary concepts of electricity. The purpose of your present studies, then, is to build such a solid foundation in the basic principles of electricity that your future understanding of complex electronic equipment or systems is made easier. Mastering these fundamentals is not difficult, but as in any learning process, the more you apply yourself, the more you learn. Much of your early work will be mathematical in nature. This is unavoidable. To understand how electrical circuits work, the numerical interrelationships must be investigated and understood. Hence, a large number of practice problems is provided to aid and assist you in understanding these principles. (If you do not make full use of the practice problems, you will be cheating yourself.)

The study of basic electricity and electrical circuit action begins with a look into the structure of matter, since all electrical phenomena originate in the "stuff" of which our universe is made. A cursory glimpse into the atomic structure of matter is required before the movement of electrical charges can be visualized. Then, very simple electrical components are used to construct a circuit. Using this practical example, the characteristics of the fundamental electrical circuit are expressed in terms of very simple numerical relationships. Once the simplest of all electrical circuits is understood, the completion of this course of study consists in simply taking one more step at a time, progressing from the simple toward the more complex.

To make your progress as rapid as possible, this text is written in a simple, direct manner, completely avoiding the pedagogically accepted style so widely used by many writers. For example, consider the following statement: "It is certainly likely, in light of the recent history of the calculator market, that the frequency of utilization of these newer devices will significantly increase as their cost continues to decrease." Compare this with the following, which tells you the same information. "Pocket calculators will be used more and more as their cost decreases." This more direct approach to writing is used throughout this textbook.

To complete this course of study successfully, read the text carefully, if necessary more than once, and listen to your instructor; his experience is invaluable in helping you get over any rough spots. Study all practice problems and examples until you thoroughly understand the principles being explained. If you are diligent in your studies, you will complete this course most satisfactorily. Study at less than your best, and you will surely miss much of the valuable information that the first-class technician must know.

1-2 THE ELECTRONICS TECHNICIAN

An electronics technician is a person who has had *at least* one or two years of formal study (usually more) and who is capable of understanding, and therefore of working with, electronic equipment to produce useful results. The technician must be able to adjust, install, maintain, or repair the equipment. As opposed to an engineer, who has graduated from a four-year (or more) college, the technician often (although not always) has somewhat less training in mathematics, and fewer courses in design principles. He or she is often called the engineer's right hand.

While the above description is often true, many technicians do not work under the direction of an engineer. For instance, in a large manufacturing plant the technician often works side by side with the design engineer to implement the engineering designs in *prototype* form. That is, the technician works from a diagram that indicates the component parts to be used and the electrical connections to be made, and he is often responsible for the physical layout (the actual arrangement of parts). Thus, the technician is responsible for a very important part of the overall design, and as he develops experience, he also frequently assists the engineer in the design work itself. The service technician, on the other hand, will probably never meet a design engineer. In fact, depending on where he is employed, he may have little or no supervision of any kind. As he develops experience, the service technician can move upward, into a supervisory position that is on a par with engineers.

Because modern technology is advancing so rapidly, the conciencious technician is never through studying. New devices, new technologies, and new methods crop up with alarming frequency. New ideas must therefore be dealt with on nearly a daily basis. Hence, the study habits that you develop now will benefit you throughout your career in electronics.

The work done by electronics technicians is probably as diverse and challenging, or more so, as can be found in any other field of endeavor. You may find a job in any of the following branches of electronics, to name but a few.

- Microwave communications equipment
- Industrial electronics (automatic control and monitoring)
- Medical electronics
- Telephone communications
- Digital or analog computers
- Aerospace electronics
- Radio-TV broadcasting
- Home entertainment systems
- Automotive electronics
- Nuclear reactor instrumentation or research
- Electronic business machines
- Heavy-duty, automated manufacturing equipment

In any of the various branches of electronics, the technician may be asked to install, repair, adjust, operate, test, modify, calibrate, or log the results of his work upon any electronic equipment. He—or she—must have the ability to read and understand engineering instructions and schematic diagrams and to apply the information they impart. He must be able to extract significant data from manufacturers' literature and service manuals.

The technician must be able to operate various kinds of test equipment, including (1) volt-ohm-milliameters, (2) oscilloscopes, (3) audio- and radio-frequency signal generators, (4) spectrum analyzers, (5) transistor curve tracers, (6) level meters, (7) color-bar and dot generators, and (8) testers for a wide variety of electrical and electronic components.

The technician must therefore have a better than nodding acquaintance with all manner of electronic test equipment. In this course you will become familiar with such basic test equipment as the volt-ohm-milliameter and the oscilloscope. The actual kinds of equipment that you will be exposed to will depend largely upon the equipment that your school has. In any event, a reasonable amount of actual hands-on experience with such test instruments is essential to the successful completion of your studies. Additionally, many hand and power tools are used, ranging from simple side-cutting or long-nose pliers to a power brake, which is used for bending sheet metal to form metal-type chassis.

1-3 ELECTRONICS APPLICATIONS

The field of electronics covers such a vast array of equipment and so many subdisciplines that this entire book could be devoted to the subject. We have instead chosen a few examples, some mundane, some exotic, to provide a "feel" for several subdisciplines and types of equipment. Many other fields too numerous to mention exist (for example, medical electronics, or voice transmission using optical [glass] fibers in place of wires). Their exclusion here in no way diminishes their importance.

Telecommunications

Telecommunications Channel Unit

This plug-in device is widely used in telecommunications systems based on the *carrier* principle. A carrier system is one that can provide many two-way conversations over a single pair of wires. It does this by borrowing some techniques from radio broadcasting. Just as you can adjust the tuning dial of your radio to select one particular station from all stations broadcasting simultaneously in your area, the carrier telephone system literally broadcasts several conversations simultaneously over a pair of wires. A mechanism not unlike your tuning dial separates each conversation from all others and sends it to the previously selected *address* (the called number, but note that the broadcast is two-way once the connection is established).

The purpose of the illustrated channel unit (Figure 1-1) is twofold. First, it converts the signaling (dial-pulse address) and voice information generated at a telephone set to a radio-type signal on the transmit (send) side. Then, on the receive side, it converts the radio-type signal to the voice and signaling information as it originally existed. The unit can, therefore, be thought of as being two essentially separate entities; the *transmit* half, and the *receive* half. Both halves operate separately

Figure 1-1 Telecommunications channel unit. Courtesy Lynch Communications Systems, Inc.

and simultaneously so that full two-way (duplex) conversation can occur.

Shown in the illustration are a variety of electronic components, including resistors, capacitors, integrated circuits, transformers, and a piezo-electric (quartz) crystal that generates a highly accurate radio-frequency signal.

Telecommunications Channel Bank

A telecommunications (telephone) device that is helping to ease the traffic problem in your telephone service is shown in Figure 1-2. Traditional telephone circuits require a *dedicated pair* of wires (one pair per subscriber) between the exchange office and the telephone subscriber. These wire pairs are bundled together into a transmission cable, often of considerable cross-sectional area. Population growth has created a demand for telephone service that has overwhelmed existing transmission cables. The cost of new multipair transmission cables has become prohibitive, because of steeply rising material and labor costs. Some form of "cable relief" has become imperative. The Lynch B325 Channel Bank is a 24-channel PCM (pulse-code modulation) carrier system designed to provide the needed cable relief by transmitting and receiving 24 simultaneous telephone conversations over four wires (a 12-to-1 cable-pair improvement). This carrier system employs time-

Figure 1-2 Telecommunications channel bank. Courtesy Lynch Communications Systems, Inc.

6 Chapter 1

(a)

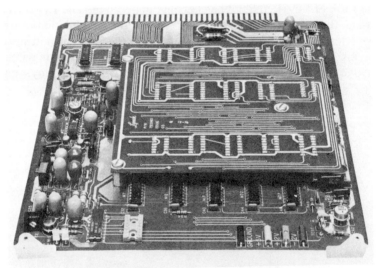

(b)

division multiplexing (TDM) and PCM techniques to simultaneously interleave 24 conversations. When first developed in 1962, the carrier system was quite expensive, but the cost has been continually reduced by such technical advances as multifunction medium- and large-scale integrated circuits (ICs), so that the system now costs considerably less than new cable that would provide the same communication capabilities.

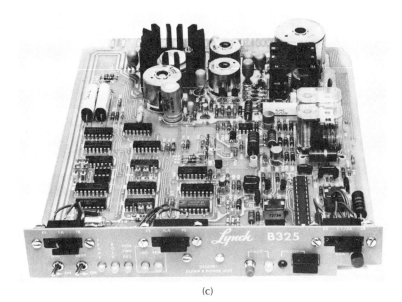

(c)

Figure 1-3 a, b, c,. Plug-in boards for the equipment in Fig. 1-2.
Courtesy Lynch Communications Systems, Inc.

Several of the printed-circuit (PC) plug-in boards that are a part of this system appear in Figure 1-3. Components shown that will be studied in this book are (1) resistors, (2) capacitors, (3) inductors, (4) transformers, (5) relays, and (6) diodes. Components shown that you will study in later courses are (1) transistors, (2) integrated circuits, (3) hybrid transformers, and (4) digital logic arrays.

Electronic Subscriber Switching System

The electronic subscriber switching system (ESSS) illustrated in Figure 1-4 is used in telephone communications as an extension of central-office equipment. Using a maximum of only 32 trunks (circuits) between itself and the central office, the ESSS automatically serves up to 128 subscribers. It uses computer-controlled logic to route and switch calls, and it stores in its "memory" the subscriber and trunk location information for each call made through the system. When the call is completed, the control unit requests the memory to provide the information required to return the appropriate electronic switches to their normal condition to await the next call.

One advantage of the ESSS is that it requires relatively few trunks between the office end and the subscriber end. The unit is especially

8 Chapter 1

Figure 1-4 Telecommunications electronic subscriber switching system ESSS. Courtesy Lynch Communications Systems, Inc.

useful when the number of subscribers increases in a given area after the telephone system is initially installed. A large number of new subscribers can easily be accommodated with *no* increase in cable (called "outside plant") facilities. Hence, existing cable can be used for a greatly increased number of users.

The unit illustrated uses the latest technology to achieve its purpose most economically. Modular design is used (see Figure 1-5) so that a PC board assembly can be instantly replaced if a failure occurs. A state-of-the-art microprocessor greatly increases the efficiency of the outside plant facilities.

Telecommunications Signaling Unit

In telecommunications systems it is necessary to transmit automatically what is called "supervisory and signaling" information, in addi-

Figure 1-5 One plug-in board for the ESSS. Courtesy Lynch Communications Systems, Inc.

tion to voice. A supervisory condition, for example, indicates to the office equipment that a subscriber's phone is either off hook (busy) or on hook (idle). An example of signaling information is a stream of pulses representing one digit of a dialed number. If a rotary-dial phone is to transmit to the central office a 9, for example, the dial is rotated nine spaces and released. As the dial mechanism is returning to its rest position, a contact (switch) is opened and closed nine times. The same thing happens for all dialed digits.

As long as the communication path is simply over wires, the signaling system just described is adequate. However, when the communication path is by way of radio, or if radio transmission over wires (carrier transmission) is used to increase the traffic (24 simultaneous conversations on just two wires is not uncommon as described above, while 4032 separate conversations are possible), the signaling and supervision information must be converted. The opening and closing of a switch is frequently converted to a high-pitched on-off tone which can be transmitted over the wires more efficiently than the switch-type signaling.

Such a unit is shown in Figure 1-6. This signaling unit is a versatile mechanism. It passes and modifies the voice signal. It also converts supervisory and signaling information to a complex tone that informs the far-end equipment of the dialed number. When the two-way connection is automatically completed, the signaling unit passes voice communication in two directions simultaneously. When one or both ends hang up, it signals this information to the opposite end and returns to the idle condition to await the next call.

Figure 1-6 Telecommunications signaling unit. Courtesy Lynch Communications Systems, Inc.

The equipment illustrated is made from a wide variety of electronic parts. Among those visible are resistors, capacitors, transformers, transistors, and potentiometers.

Data Processing

The Computer

The subject of data processing covers such a vast number of subjects and applications that it is impossible to cover everything in less space than a large book. Data processing includes business applications, scientific applications, military applications, and medical applications, to name but a few. Figures 1-7 and 1-8 illustrate medium-sized business data processing system and a somewhat larger one, respectively.

Basically, a data processing system consists of the following:

1. One or more *input* devices, which can be, for example, a punched card reader or an electric typewriter.

Introduction 11

Figure 1-7 A medium-sized data processing system. Courtesy National Cash Register Company.

Figure 1-8 A larger size data processing system. Courtesy IBM.

2. A *central processing unit* (CPU), which controls the operation of the entire computer system by executing the instructions it receives in the form of *programs*.
3. One or more *output* devices, often a card punch and a printer.

12 Chapter 1

(a)

These components are the *hardware* required to execute calculations. The programs, or *software,* are equally important—without programs, the computer would have no way to perform calculations in the correct order. In fact, an integral part of the CPU is *storage,* or memory, and it is the ability of a computer to store not only different programs, but also calculated results, that makes it so valuable a tool.

Data processors operate on a *digital,* rather than *analog,* basis. That is, the electrical energy that makes up the signals (as opposed to power) in your AM radio receiver are analog signals because they are con-

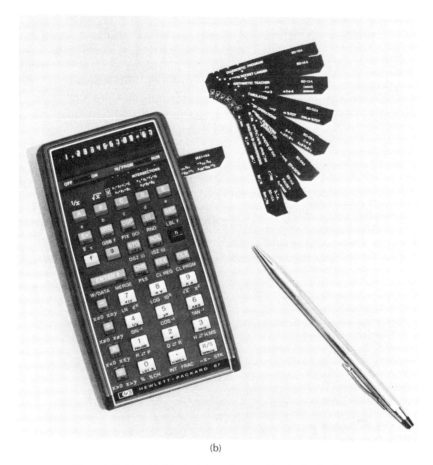

(b)

Figure 1-9 Hand-held calculators. Courtesy Hewlett-Packard Corp.

stantly changing. If plotted on a graph, such signals appear as smooth (analog) curves. A digital signal, on the other hand, appears as a two-level signal that is considered to be *on* or *off,* and as having *no other possible state*. This is why such circuits are often called *switching circuits,* since the results are as though thousands of tiny switches are simply turned on or off according to the program instructions.

Personal Computing Devices

While not as powerful as large computers, miniature hand-held calculators (Figure 1-9) have become so common and inexpensive that virtually anyone can afford one. In their present form, calculators can be considered small-scale data processors. Functions that only a few years ago added hundreds of dollars to the basic cost are now available for only a few dollars. The mathematical capability of the more complex calculators is nothing short of astounding, when compared to the earliest models. It is now virtually standard to have complete trigonometry functions, two separate logarithm systems ($\log_{10}$ and $\log_e$), plus scientific notation (powers of ten) from 1×10^{-99} through 9×10^{99}, rectangular and polar conversion, and other operations too numerous to mention. Add to these features the calculators that can be "programmed," and you can appreciate the fact that they are already taking over some functions of larger computers.

Space Communications

Shown in Figure 1-10 is a large radio-frequency antenna system used for space communications. Antennas similar to this are used for a variety of purposes, such as orbital tracking for satellites and for communicating with deep-space probes. Antennas such as this are also used to search for signals from other intelligent beings in the universe, and as radio-telescopes that "listen" to the radio energy emitted from certain types of galactic bodies. Such devices are used by astronomers in addition to conventional telescopes to increase our knowledge of the physics of the universe.

Figure 1-10 A space communications antenna. Courtesy SRI International.

The radio receivers used with such antennas are in many respects similar to the conventional AM or FM receivers that you are familiar with. They are, however, more highly refined and are capable of receiving much weaker signals. These sensitive receivers, plus the energy-gathering capability of so large an antenna, allow the reception of *very* weak signals. Generally, these antennas are situated far away from densely populated areas to minimize man-made radio interference.

Commercial Broadcasting

Commercial broadcasting includes regular AM, FM, and TV services. An extremely wide range of equipment is encountered in these fields. Voice-processing equipment includes microphones, audio amplifiers, audio tape recorders and players, audio mixers, and modulation equipment. Video (picture) processing includes TV cameras, video-tape recorders, video amplifiers, video mixers, and modulation equipment. And in all cases a radio-frequency transmitter and antenna system are required.

Television Camera

A broadcast-quality color TV camera is shown in Figure 1-11. Such equipment is extremely complex, being in effect three cameras in one. This instrument combines complex mechanics, electronics, and optics

Figure 1-11 Broadcast color TV camera. Courtesy International Video Corporation.

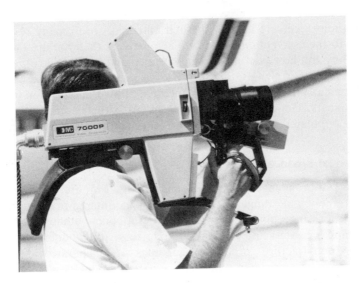

Figure 1-12 Portable color TV camera. Courtesy International Video Corporation.

to achieve the true-to-life color pictures that we are all accustomed to. The camera is mounted on a massive "dolly," which is often motorized so that the cameraman can easily and smoothly control camera movement. One standard lens system has a 30-to-1 zoom ratio, so the camera itself need not be moved for a wide range of indoor and outdoor scenes.

The camera is the mechanism that converts an optical scene to equivalent electrical signals. These signals, after processing, are transmitted and reprocessed to become the color picture on a TV receiver. A color TV transmission is, in effect, the equivalent of three pictures, one for each primary color. When combined in the proper proportions, the result is a sharp, accurate representation in full color of the original scene.

Figure 1-12 illustrates a portable color TV camera. It provides all of the basic functions of a studio-camera, but is designed to be carried as shown. It can be used for live coverage, or it might feed a video-tape recorder (VTR) for broadcast at a later time. Most TV stations now use such a "mini-cam" to provide on-the-spot coverage of news events.

Video-Tape Recorder

A broadcast studio-type video-tape recorder (VTR) is shown in Figure 1-13. This instrument records up to two-and-one-half hours on a single reel of 2-inch magnetic tape. The recording includes the picture information, two audio sound tracks, and an audio cue track. The VTR is

Figure 1-13 Studio broadcast video-tape recorder. Courtesy International Video Corp.

capable of recording the program and playing it back in *full color* with such fidelity that the viewer cannot tell that he is seeing a recording. In fact, a taped program is often better than a live performance since the tape can be edited to remove mistakes.

Because of such disturbances as tape stretch, tape flutter, and motor speed variations, a VTR must have, as an integral part of its electronics, a *time-base corrector*. This device (operating on either analog or digital principles) almost completely eliminates distortion caused by the problems mentioned above, which any tape and associated transport mechanism have. The net output of such a VTR with its time-base corrector is a reproduced picture that is not only rock-solid (perfectly synchronized), but is also one in which the colors do not vary. The slightest variation in reproducing the color part of the picture causes the colors to change, which is very noticeable and annoying to the viewer. Hence, in broadcast equipment, the time-base corrector is a necessary component.

Additionally, most broadcast VTRs have a device called a *drop-out compensator*. If, due to a malfunction, or a small blank spot on the tape, a portion of the program is missing, this remarkable device "remembers" the information recorded just previously and substitutes this information for the blank spot. Because the light and color values in adjacent parts of the picture are nearly the same, this produces a continuous, unbroken picture. Even a practiced viewer, knowing where the dropout will occur, cannot normally see any difference.

Finally, Figure 1-14 shows a more-or-less "portable" VTR. While not as sophisticated as the larger studio model, it nevertheless is capable of producing excellent color pictures. It has the advantage of being small enough to mount in a vehicle, serving as the VTR in a *remote* recording situation. As shown, the VTR is mounted in a console along with monitoring equipment to allow the operator to verify correct operation.

Home Entertainment Systems

Home entertainment systems include, but are certainly not restricted to, radio and TV receivers, stereophonic recorders and players (disc, cassette, and cartridge), and electronic organs. A recent addition to this area of electronics are the electronic games that are played on a TV

Figure 1-14 Color VTR. Courtesy International Video Corp.

screen. Such a wide variety of equipment is encompassed that it is impractical to even begin to provide adequate illustrations. Furthermore, a person who has not yet encountered at least one such device is probably not a candidate for this course of study anyway. Hence, it is assumed that you are familiar, at least to some degree, with home entertainment systems.

Test Equipment

As you will discover, electrical quantities cannot be seen directly (although they can often be *felt,* sometimes quite painfully). Therefore means are required to allow the technician to be able to measure them directly, and sometimes indirectly. Such devices as volt-ohm-milliameters (VOMs), Figure 1-15; FET VOMs, Figure 1-16; and oscil-

Figure 1-15 Volt-ohm-milliameter. Courtesy Simpson Electrical Co.

Figure 1-16 FET volt-ohm-milliameter. Courtesy Radio Shack.

loscopes, Figure 1-17; as well as other pieces of test equipment, are required to maintain, repair, and operate electronic devices. Many test instruments you will no doubt be asked to use rather frequently during this course, so that you can attain a degree of familiarity with them, and practice in applying them. Many others you will probably not encounter until you are out of school and are actually on the job. In such a case, you must make an effort to understand the operating characteristics of the instrument you are to use. The quickest and best way to accomplish this is to study the operator's manual carefully. Every manufacturer of test equipment provides slightly different features, and to become completely proficient in the operation of any equipment you must be aware of these and be able to apply them on the job.

The proper use of test instruments such as those shown is one goal worth achieving in this and subsequent courses. No matter what the

Introduction 21

Figure 1-17 An oscilloscope. Courtesy Simpson Electric Co.

cost of these devices, the results obtained by their use are only as good as the interpretation given the data by the operator. In other words, an expensive metering device cannot, of course, think. If the operator of the instrument is not careful and is not fully aware of *all* aspects of the particular measurements, the results are probably worse than no measurement at all. This book will give you the technical knowledge necessary to properly use the instruments; it is, however, up to you to ensure that you can actually perform such work effectively.

CHAPTER 2

ELECTROSTATICS

In order to understand electricity, you must begin by studying the principles of matter. In this chapter, you will make that beginning. First we look briefly at matter and at the particles of which it is made. Then we turn to atomic structure, and we describe the fundamental subatomic particles. Next, we investigate basic electrical phenomena. At the end of the chapter, we use your new knowledge of these phenomena to define the basic units of measurement relating to electricity—the volt, the ampere, the ohm, and the watt; and in doing so, we present and describe the basic electrical circuit.

The topics to be covered in this chapter are

2-1 Matter
2-2 Particles of Matter
2-3 Atomic Structure
2-4 Atomic Configuration
2-5 Atomic Number and Weight
2-6 Static Charge
2-7 The Unit of Charge
2-8 Basic Laws of Charges
2-9 Electrical Circuit Units
2-10 The Electrical Circuit

2-1 MATTER

Many of the fundamental characteristics of electricity can be explained in terms of the principles of matter. *Matter* is the "stuff," or material, that makes up our universe; in other words, matter is anything that occupies space and has weight. Generally speaking, matter can exist in any of three *states,* solid, liquid, or gas.

Solids have definite volume and shape. Solids yield to force, but when the force is removed tend to return to their original shapes. This tendency varies in different samples of solids. A strip of spring steel exhibits the tendency to a remarkable degree, but a strip of lead has virtually no such tendency.

Like solids, *liquids* have definite volume. Unlike solids, liquids have no definite shape—they assume the shape of their containers. If not contained, liquids flow from one point to another, due to such forces as gravity and air pressure.

Gases are similar to liquids in many ways but have neither definite volume nor definite shape. That is, gases expand without limit if not held in a completely closed container. Liquids do not compress easily, but gases are readily compressed.

The state in which a substance exists—solid, liquid, or gas—is usually a function of its temperature. As you know, water exists in all three states. At temperatures below 0°C, water occurs as a solid (ice); between 0°C and 100°C, it occurs as liquid water; and at temperatures above 100°C, it occurs as a gas (steam). The state in which a sample of water exists can be changed simply by heating or cooling it. Also, these changes are *physical* and *reversible;* that is, solid water can be changed to liquid (or gas) and then back to solid.

Other substances have widely varying melting and boiling points. Iron, for example, melts (becomes liquid) at a temperature of approximately 1525°C and becomes a gas at approximately 3000°C. Nitrogen is a solid *below* $-210°C$, is liquid between $-210°C$ and $-196°C$, and gaseous at temperatures above $-196°C$.

Still other substances have different characteristics. Wood, for instance, occurs *only* as a solid, and never as a liquid or a gas. When wood burns, its composition changes *irreversibly.* (The substances that are produced when wood burns do not become wood when cooled.) Burning wood is one example of a *chemical,* rather than a physical, change.

2-2 PARTICLES OF MATTER

All matter is made up of tiny particles. These *microparticles* are unbelievably tiny (less than one-billionth of an inch in diameter!). Chemists and physicists recognize the existence of thousands of microparticles, each with its own unique characteristics. Fortunately, to explain basic electrical phenomena, we need only consider a very few of these thousands of types.

Chemically, all samples of matter that we can see are made up of atoms or molecules. An *atom* is the smallest particle of an element that retains the chemical properties of the element. An *element* is a sample of matter that consists of atoms of only one kind. With very few exceptions, atoms do not exist in a free state. Instead, two or more atoms are joined together to form a *molecule,* a particle made up of the component parts of the two (or more) atoms. The molecule is in fact the basic structural unit of matter.

24 Chapter 2

A *compound* is a substance made up of two or more elements whose atoms are chemically bonded together. For instance, a molecule of water consists of two atoms of hydrogen and one atom of oxygen. From this fact, we can infer that a chemical compound has chemical characteristics that differ from those of the elements of which it is composed. That is, water is a liquid at room temperature, but hydrogen and oxygen are gases. Scientists have identified more than one hundred elements. The number of compounds that can be formed from these elements is infinite. Literally millions of compounds are known, and new ones are discovered daily.

2-3 ATOMIC STRUCTURE

The fundamental laws of electricity can be explained only in terms of *subatomic* microparticles; that is, the particles out of which atoms are formed. As we have said, physicists and chemists have discovered the existence of a great many types of these particles, each with its own special set of characteristics. For the time being, we are interested in only three of these particles. The electron and proton contribute directly to electrical phenomena, and the neutron is an integral part of the structure of atoms. Every atom contains at least an electron and a proton, and all atoms save those of hydrogen also contain neutrons. We shall describe atomic structure in a moment, but first we must describe these three basic particles.

The *electron* is a particle with a diameter of approximately 0.000 000 000 000 01 (one-hundred trillionth) inch. The *proton* is approximately 1836 times more massive than the electron, but is only about one-tenth its size. Electrons and protons possess an attractive force for each other; that is, these two types of particles tend to move toward each other, although normally other forces restrain them. The force of the electron is arbitrarily called *negative,* and the electron is said to carry a *negative* charge. The force of the proton is equal but opposite in sign to that of the electron. Thus, protons are called *positive* and are said to carry a *positive* charge. We represent the electron as e^- and the proton as p^+ to indicate these charges and their polarity.

The *neutron* is similar to the proton in size and mass. However, the neutron, as its name implies, carries no charge (is electrically neutral). We therefore represent the neutron with the symbol n.

Before we turn to the actual configuration of atoms, we want to emphasize the fact that atoms (and therefore the particles that make up atoms) are unbelievably small. Scientists have estimated that 1 gram (0.035 ounce) of helium contains 150,000,000,000,000,000,000,000 (150 sextillion) atoms. To give you some idea of the quantity represented by that number, it is approximately the number of grams of water in all the oceans of the earth.

2-4 ATOMIC CONFIGURATION

The typical atom would probably appear somewhat as depicted in Figure 2-1, if in fact we could see atoms. An atom has often been likened to the solar system, with the nucleus of the atom holding the same relative position as the sun, while the electrons orbit the nucleus as the planets orbit the sun. Electrons are thought to revolve around the nucleus at tremendous speeds. It is this rapid motion that keeps the electrons from "falling" into the nucleus due to the mutual attraction of electrons and protons.

Figure 2-1 illustrates the configuration of an atom of hydrogen. The nucleus consists of a single proton around which a single electron is orbiting. As might be inferred from this, hydrogen is the lightest of all known elements.

Figure 2-2 shows a pictorial diagram of a single atom of helium, also a lighter-than-air gas. This atom contains two protons, two neutrons, and two electrons. Helium is therefore more massive than hydrogen. A portion of the periodic table of the elements is shown in Table 2-1. Note that each successive addition is the result of the addition of more particles—electrons, protons, and neutrons. This is indicated by the increasing atomic numbers. Another important fact is that in a normal atom, *there are as many electrons as protons.* With few exceptions, this always holds true. As will be discussed in detail subsequently, the exceptions are the very cause of electrical phenomena.

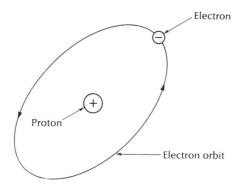

Figure 2-1 A representation of a hydrogen atom.

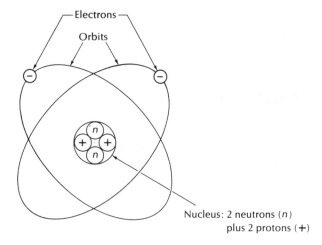

Figure 2-2 A representation of a helium atom.

The more massive and complex the atom, the more complex the configuration, particularly with respect to the electron paths, or orbits. Note the noncircular orbits of the helium atom in Figure 2-2. To avoid unnecessary complexity we shall use a more simple representation of the atom, an example of which is shown in Figure 2-3. For example, Figure 2-4 illustrates an atom of copper in this simplified manner. Copper is probably the most frequently used material in electronics.

The copper atom diagram can be used to illustrate the reason why copper is so useful to transport electricity. The outer electrons can be removed quite easily from any element having one, two, or three elec-

TABLE 2-1 A PORTION OF THE PERIODIC TABLE

Element and Symbol	Atomic Number	Atomic Weight
Hydrogen (H)	1	1.0080
Helium (He)	2	4.0026
Lithium (Li)	3	6.941
Beryllium (Be)	4	9.0122
Boron (B)	5	10.81
Carbon (C)	6	12.011
Nitrogen (N)	7	14.0067
Oxygen (O)	8	15.9994
Fluorine (F)	9	18.9984
Neon (Ne)	10	20.179
Sodium (Na)	11	22.9898

Electrostatics   27

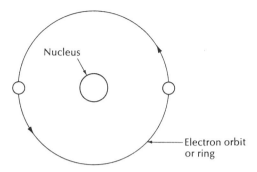

Figure 2-3   Simplified atomic representation.

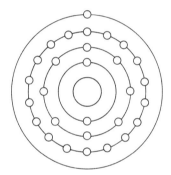

Figure 2-4   An atom of copper, drawn using the simplified method of representation.

trons in the outer orbit, or *shell*. Such materials are known as *conductors*. We have already said that copper is a conductor; other examples are gold, silver, and aluminum. As will be described later, the ease with which the electrons in the outer orbit can be removed determines the quality of the material for transporting electrical energy. On the other hand, elements having atoms with completely filled outer shells have electrical characteristics opposite from conductors. Because electrons are difficult to remove *if* the outer shells are full, these materials are known as *insulators,* and are equally useful in electrical circuitry. Some examples are certain plastics, hard rubber, porcelain, and glass.

Conductors are used to transport electricity from one point to another. Nearly everyone is familiar with the insulated copper wire used to provide electrical power to various appliances throughout the home. This plastic- or rubber-covered conductor, called zip-cord, is inexpensive and serves its purpose well. Anyone who has accidently touched the two wires together while the zip-cord is carrying electricity realizes the significance of the insulating cover. To be useful as well as safe, electrical energy must not only be transported, but must be limited to desired paths. Electricity is probably mankind's most useful servant; it can also be a deadly killer if not treated with respect and knowledge.

Between conductors and insulators are materials called semiconductors. These materials allow the transporting of electrical energy, but not so easily as in conductors. Semiconductors, such as carbon, germanium, and silicon, often serve to limit the amount of electrical energy carried by a conductor. We shall examine conductors, insulators, and semiconductors in some detail in Chapter 8.

2-5 ATOMIC NUMBER AND WEIGHT

The *atomic number* of an element is simply the number of protons in the nucleus. Hence, hydrogen is given the atomic number of 1, while helium is given number 2, copper is given number 29, and so on. *Atomic weight* is essentially determined by the number of protons and neutrons in the nucleus. For example, hydrogen has an atomic weight of 1 (actually 1.008) and is the reference for all weights. That is, all other elements are given an atomic weight relative to hydrogen. Helium has an atomic weight of 4.003, while that of copper is 63.54. That is, copper is approximately 63.5 times more massive than hydrogen.

Note above, that element number 1 has a relative atomic weight of 1.008. Similarly, neon (atomic number 10) has an atomic weight of 20.187. Nearly all elements exist in more than a single form. Variations in atomic weight for a single element are explained by the presence of *isotopes*. As an example, the element neon is known to have three isotopes. That is, three variations of the neon atom exist; all have the same chemical characteristics, but each has a different atomic weight. The atomic weight of neon then, arrived at by chemical means, is simply the average of all three isotopes in the proportions in which they exist. In fact, 90.5% of neon atoms have a weight of 20; 0.3% of neon atoms have a weight of 21; and 9.2% have a weight of 22. Hence, $[(20 \times 0.905) + (21 \times 0.003) + (22 \times 0.092)] = 20.187$, the atomic weight of neon.

2-6 STATIC CHARGE

Static electrical charge, or electricity at rest, can easily be demonstrated. A hard rubber comb run briskly through your hair will attract tiny bits of paper. The comb is giving evidence that it has received a static electrical charge. The *electrostatically* charged comb and the paper bits, which are neutral, attract one another. The paper bits move much more readily than the comb, which is relatively massive.

The friction of the comb rubbing against hair results in a *separation of charge* on the surfaces of the comb and hair. Because hard rubber is an insulator, the charge tends to remain on the comb surface. That is, frictional force removes some electrons from the hair and transfers them to the surface of the comb. The comb now carries an excess of negative charges, over and above the charges produced by the orbital electrons. However, orbital electrons do not yield any electrical effects that are noticeable *external to the material*. This is because each electron is "paired" with a proton in the nucleus, and each proton and electron cancels the other's charge. However, when a greater-than-normal quantity of electrons exists at a point, then *a negative charge exists*. This charge *is* noticeable and measurable, as evidenced by the attraction of the bits of paper to the comb.

Many other substances can be electrostatically charged. For example, a plastic-barrel pen rubbed briskly on a pad of paper will often generate enough charge to lift one corner of the paper. Experiments such as this work best in a very dry atmosphere. The more humid the atmosphere, the more rapid the discharge, because the moisture in the air attracts the free electrons on the charged surface.

Anyone living at a high altitude, where the air is typically very dry, experiences this separation of charge nearly every day. Plastic-soled shoes worn while walking on wool or nylon rugs can often cause the body to accumulate a charge large enough to draw a one-inch (or longer) spark when a large metal object is approached. This is an excellent, though often painful, demonstration of the accumulation of static charge. In instances such as this, the charge is much greater than the small amount required to lift a tiny bit of paper.

The spark mentioned above is a manifestation of the discharge that occurs when the insulating properties of the air are exceeded by an electrical pressure too great to be contained. Excessive charge flows from a point of high charge concentration to a conductor having a large mass. The flow of charge from one point to another is called *current*. For a brief instant, electrons flow and transfer the excessive charge until each surface has an equal number of electrons. Further discharge is then impossible until one surface again becomes charged relative to the other.

2-7 THE UNIT OF CHARGE

By international definition, the unit of static electrical charge is the coulomb (C). This unit represents a given quantity of free electrons, specifically 6,250,000,000,000,000,000 (6.25 quintillion) electrons. The coulomb is named for Charles A. Coulomb (1736–1806), a French physicist.

The basic unit of electrical charge is derived from the amount of charge on a single electron, which is calculated in terms of the forces of attraction and repulsion. This was first measured by an American physicist, Robert A. Millikan (1868–1953), in an ingenious experiment. In an enclosed container, he placed two metal plates horizontal to the earth's surface. Then he sprayed very fine drops of oil between the plates and applied known amounts of electrical energy to them. By precisely adjusting the electrical charge on the plates, he could exactly oppose the pull of gravity, thus suspending the oil droplets between the plates. Next, he bombarded the droplets with intense ultraviolet light. This dislodged electrons from the droplets. By carefully readjusting the charge on the plates, he could determine how much the charge on the droplets of oil was changed. He discovered that the change in charge was always a multiple of some minimum unit value. This indicated that if the charge changed by one unit, the droplet had lost one electron, if the charge changed by two units the droplet had lost two electrons, and so on. The basic electrical unit of charge was found to be 0.000 000 000 000 000 000 16 coulomb.

2-8 BASIC LAWS OF CHARGES

As has been mentioned, the two basic electrical charges are assigned polarities—negative for the electron and positive for the proton. In a solid material the nucleus of each atom is fixed in place, so the protons are immovable. Hence, to create an electrical charge, it is the electrons that must be moved. At first thought, it may seem that only negative charges can be caused to accumulate at a surface. However, this is not true. All matter contains *both* positive and negative charges. These charges normally exist in equal quantities; for every electron orbiting a nucleus, there is a proton in the nucleus. When a number of electrons are removed from a solid substance, perhaps by friction, protons are left locked in place by the very nature of solids. These protons are now "uncovered" charges (not canceled by an electron), and the substance exhibits a positive charge. Because the charge on electrons and protons is equal but of opposite sign, the magnitude of positive charge left behind equals the number of electrons removed.

An atom that exhibits an electrical charge, either an excess or deficiency of electrons, is known as an *ion*. A material exhibiting a charge, such as the comb we discussed previously, is said to be ionized. Thus, to produce separation of charges by any means is to ionize the materials used.

There are two basic laws regarding the behavior of electrical charges.

1. Unlike charges (+, −) attract.
2. Like charges (−, − or +, +) repel.

The first law was introduced earlier; electrons and protons tend to move toward (attract) each other. The second law, however, has not been discussed yet. If other forces are not considered, electrons tend to repel other electrons, and protons tend to repel other protons. However, because protons are normally locked in place in the nucleus, their repulsion effects are seldom noticed.

2-9 ELECTRICAL CIRCUIT UNITS

We found earlier that if a large enough charge has been accumulated on a surface, and if this surface is brought close to a large metallic surface or a surface with an opposite charge, then a brief static discharge will occur. The discharge, or spark, consists of a flow of electrons from an area of high concentration or ionization to one of lower concentration. While the spark exists (while electrons are flowing from one place to another), a current is said to flow. The cause of the current flow is the *electrical pressure* caused by the *difference in charge levels*.

Electrical pressure caused by a difference in charge levels is known as *electromotive force* or EMF, or more popularly, *voltage*. Voltage either appears between two or more points, or it does not. It can be caused to increase, or decrease. It *cannot,* however, move from one place to another. In certain respects voltage possesses the same properties as water pressure. A large tank of water at an elevated position is capable of producing a large flow of water. As the water in the tank is depleted, the water pressure decreases. It goes nowhere—it simply decreases as the volume of water decreases.

Volt

The unit of electrical pressure is the *volt* (V), named in honor of the Italian physicist Alessandro Volta (1745–1827). One volt is defined as the EMF required to cause 1 coulomb (6.25 quintillion electrons) to be moved in 1 second.

The amount of energy represented by the spark in the earlier example is much too small to do appreciable work. In fact, generating usable

Figure 2-5 An assortment of representative cells and batteries.

voltage by frictional means is not practical for most purposes. More practical means of generating voltage include electrochemical systems, a few of which are illustrated in Figure 2-5. These cells and batteries generate electrical energy by expending chemical energy. Electrical separation of charge occurs continuously until the chemicals are expended. Cells and batteries are described in detail in Chapter 10. Other means of producing electrical energy include (but are not limited to) huge generators operated by steam or water turbines, solar cells, and thermocouples.

Ampere

The unit of measure for current flow is the *ampere* (A), so named after André Marie Ampère (1775–1836), a French physicist. One ampere is defined as one coulomb flowing past one point in an electrical circuit, per second.

2-10 THE ELECTRICAL CIRCUIT

Basically, an electrical circuit must consist of *at least* three items:

1. A source of EMF
2. Conductors (wires) to carry current to and from the device using the energy
3. The device to be operated by electrical energy, called the *load*

Figure 2-6 illustrates such a circuit, both pictorially and by schematic diagram. The circuit consists of a battery, a switch, a fuse, wires, and a flashlight lamp. Note carefully the schematic representation of each of these parts. The battery symbol represents the internal plates of the battery; the wire symbol needs no explanation; the lamp symbol represents the internal filament of the lamp. The fuse is a protective device, and its symbol also represents the internal filament. The switch provides a convenient way of turning the lamp on and off, and its symbol shows this clearly. It is important to memorize graphic symbols as they are introduced so that you can read schematic diagrams. Common symbols are shown in Appendix A.

Fundamentally, such a circuit functions as follows. This action will be described in detail in later chapters; our purpose here is simply to obtain a general feeling for this circuit action. The battery, of course, is the source of EMF for the circuit. As mentioned earlier, the battery converts chemical energy to electrical energy. The wiring delivers the current to the lamp, which then produces light. In a circuit such as this, the battery voltage and the voltage rating of the lamp must agree. If the battery voltage is too low, the lamp cannot emit enough light to be useful; if the battery voltage is too high, the lamp will overheat and quickly burn out.

Under normal conditions the lamp will remain energized but will gradually become dimmer and dimmer until the battery exhausts its chemicals. If the circuit is broken, perhaps by removing a connection from the battery or by opening the switch, the lamp will of course go out. Note that there must be a complete metallic path from battery post through the load and back to the other post in order for current to flow and for the load to be energized.

The battery's chemical action strips electrons from one battery plate and deposits them on the other. Hence, one post is positive (+) and the other is negative (−). As long as a complete circuit exists, electrons flow *out* of the negative post, through the wire connected to the negative post, through the load (lamp), and back to the battery through the other wire. A circuit such as this is known as a *direct-current (dc) circuit*. Current (electrons) always flows *out* the negative battery post

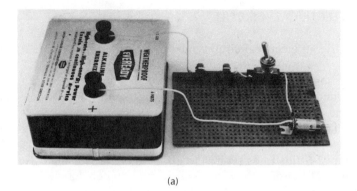

(a)

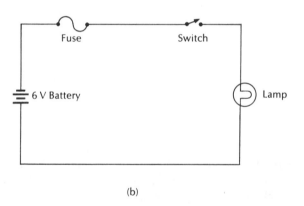

(b)

Figure 2-6 A basic electrical circuit: (a) a photograph of the actual circuit; (b) the schematic diagram.

and *into* the battery at the positive post. It should be noted here that prior to the discovery of electron theory it was thought that current flows from positive to negative (called *conventional* current flow). Hence, many older, tradition-bound concepts are still explained using conventional current flow. In recent years such positive- to-negative current has been found to exist in such devices as semiconductors and vacuum tubes. Current flow is therefore defined most accurately as the *concerted movement of charge carriers*. Because we are concerned for the most part in this book with electron flow, we will consider that current flows from negative to positive. We shall nevertheless remember that some people consider the reverse to be true, and that in the long run, it makes little difference as long as the end results are the same.

This circuit also serves to demonstrate another important circuit property. We could form a complete circuit by simply connecting a piece of wire between the two battery posts. This would, however, damage the battery by placing a *short circuit* on it. The short circuit is damaging because it contains nothing to limit the amount of current flow, so that all the current the battery can supply flows. Such unlimited flow quickly depletes the chemicals, so the battery will very so go "dead," or it will explode due to excessive internal heat. An exploding battery can spread toxic materials and metal fragments from the battery case over a wide area, and can cause severe burns and cuts. In a normal circuit—a circuit with a load—the battery does not go dead so quickly, nor does it explode. Obviously, the load has a property that *limits* current flow to reasonable values. This property is called *resistance,* and some amount of resistance is present in every electrical circuit.

Resistance

The unit of resistance is the *ohm,* named for the German physicist Georg Simon Ohm (1787–1854). The symbol for ohm is Ω, the Greek letter omega. All conductors possess resistance; the better the conductor, the lower its resistance. All nonconductors (insulators) also possess resistance; the better the insulator, the higher its resistance.

In general, wiring used in electrical circuits should have as low a resistance value as possible. As will be discussed in more detail later, the amount of resistance provided by a piece of wire depends upon three factors:

1. Length: the greater the length, the greater the resistance.
2. Diameter: the greater the cross-sectional area, the smaller the resistance.
3. Material used: various materials have different values of resistance for a given size and length.

The electronic component designed to introduce large amounts of resistance into a circuit in a very small package is called a *resistor.* Figure 2-7 illustrates a variety of resistors, most of which are made primarily from carbon. The larger ones, however, are made by coiling a special nickel-chromium wire around a core. Nickel-chromium wire has a relatively high resistance in reasonably small bulk.

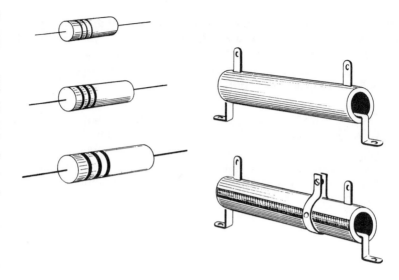

Figure 2-7 Typical resistors.

Electrical Power

A final electrical characteristic of interest is that of the power expended in an electrical circuit. In order to move electrons in a complete circuit, enough force must be generated to dislodge the electrons and to keep them moving. Relatively little force is required to move electrons in copper wire, but generally much more force is needed to push electrons through a load. This, of course, is related to the fact that copper wire has relatively low resistance, while most devices using electrical energy have moderate to high resistance.

This force of electrical work being done is called *power,* and its unit of measurement is the *watt* (W). One watt of power is expended when one ampere of current (6.25 quintillion electrons per second) is caused to flow with an applied EMF of one volt.

Electrical power is generally accompanied by the generation of heat. Electric stoves, heaters, and light bulbs are just a few familiar examples of the heat generated by electrical forces. Some applications, such as a small transistor radio, generate heat so slowly and in such small amounts that the human senses cannot detect it. Nevertheless, whenever current flows, power is expended and therefore heat is generated, even if the amount is very small.

SUMMARY

- Matter is the material from which the universe is made.
- Matter exists as solids, liquids, or gases.
- Matter is made up of atoms and molecules.
- An element is matter, the molecules of which cannot be subdivided by chemical means.
- An atom is the smallest particle of an element which retains the chemical properties of the element.
- A compound is a substance composed of two or more elements in a chemical bond.
- An electron is an elemental particle of matter having a negative electrical charge and very small mass.
- A proton is an elemental particle of matter having a positive electrical charge and 1836 times the mass of the electron.
- The neutron is similar to the proton in mass, but it has no electrical charge.
- An atom consists of a nucleus containing one or more protons and neutrons (except hydrogen); the nucleus is surrounded by spinning electrons in a number equal to the number of protons.
- Conductors, insulators, and semiconductors have distinctive electrical properties that are due to the structure of their outer, or valence, electron shells.
- A static discharge, or spark, is caused by an accumulation of charge being dissipated through the air to a large conducting surface.
- The unit of static charge is the coulomb, which represents 6.25 quintillion electrons.
- Unlike charges attract; like charges repel.
- Electrical pressure is called voltage and is measured in volts.
- The unit of measurement for current is the ampere.
- An electrical circuit must contain at least a source of voltage, conductors, and a load.
- The unit of electrical resistance is the ohm.
- The unit of electrical power is the watt.

QUESTIONS

1. True or false: All matter contains particles that possess an electrical charge.
2. True or false: All electrons, protons, and neutrons have an electrical charge.
3. The _____ carries the basic unit of negative charge.

4. True or false: The electron carries the basic unit of negative charge.
5. Comparing the charges on an electron and a proton, we can say that they are _____ and _____.
6. True or false: Charges on electrons and protons are equal and opposite.
7. True or false: Like charges repel; unlike charges attract.
8. True or false: Like charges attract; unlike charges repel.
9. A free electron will _____ another nearby electron.
10. True or false: Two adjacent protons in a piece of copper will repel each other until they are far apart.
11. True or false: Electrons tend to be mobile.
12. True or false: Protons tend to be mobile.
13. True or false: Neutrons tend to be mobile.
14. True or false: Static electrical charge is electrical energy at rest.
15. True or false: Current flow is the movement of voltage in a wire or other conductor.
16. True or false: Voltage is the electrical pressure that causes current to flow.
17. An element having 16 orbital electrons normally has _____ protons in the nucleus.

CHAPTER 3

BASIC CIRCUIT CHARACTERISTICS

In this chapter we introduce several basic ideas relating to simple electrical circuits. The precise numerical relationships between volts, amps, ohms, and watts are investigated. Basic circuit configuration is analyzed and many of its characteristics are discussed.

In order to prove to the student that the basic circuits used throughout this book come from real, commercially available equipment—and are not merely figments of imagination—we at times illustrate the equipment from which the examples have been taken. Not only is the equipment shown, but the circuit layout and method of construction are illustrated.

The first (and very simple) example is illustrated in Figure 3-1, which shows nothing more than an ordinary flashlight. Such a simple instrument exemplifies the basic electrical circuit that is the subject of this chapter. The electrical apparatus shown in Section 3-1 has the same general function as the flashlight (except the flashlight has no fuse), and this is the circuit arrangement that we shall use to introduce the fundamental relationships in all such circuits.

The topics covered in this chapter are:

3-1 Basic Electrical Circuit Action
3-2 Ohm's Law
3-3 Unit Abbreviations
3-4 Powers of Ten
3-5 Multiple and Submultiple Circuit Problems
3-6 Electrical Power
3-7 Graphical Relationships
3-8 Measuring Electrical Quantities

3-1 BASIC ELECTRICAL CIRCUIT ACTION

One kind of electrical circuit is called a *series* circuit. A series circuit is one in which there is one, and only one, path for current flow. Such a circuit is shown in Figure 3-2, both pictorially and schematically. The battery, of course, provides the power to energize the lamp. The picture shows that the battery voltage is 6 volts. Therefore, the lamp must

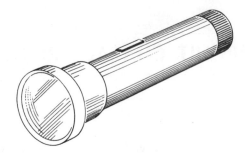

Figure 3-1 A flashlight is an example of the simplest possible electrical circuit.

also be rated at 6 volts. The unit labeled F_1 is a fuse. Its purpose is to protect the lamp from excessive current. While a fuse is not ordinarily included in such a simple circuit, it is shown here to introduce it as a separate component. The fuse is constructed of very fine wire that has a low melting point. If excessive current flows, the fuse wire melts and opens the circuit, thus protecting the load (the lamp) and the source. The switch (S_1) simply provides a convenient method of turning the lamp on or off.

To energize the lamp, the switch (S_1) is moved to the *on* position, closing or completing the circuit. The electrical action is as described in the previous chapter. By electrochemical action the battery provides a voltage of 6 volts. Electrons emerge from the negative post of the battery; flow around the circuit through the lamp, the switch, and the fuse; enter the battery at the positive post and flow through the battery, only to emerge from the negative post again.

If you could observe a single electron during this trip around the circuit, you would note that it moves rather slowly along the wire, actually only two or three centimeters per hour. However, the *electrical energy* is transmitted almost instantly down the wires. That is, a lamp situated at the end of a very long cable will energize at the same instant that the switch is thrown. This happens because the free electrons in a length of wire constitute a "cloud" of *current carriers:* Electrons repel each other, so that as one electron moves a given distance, impelled by the battery voltage, it moves adjacent electrons by the same amount and in the same direction. These electrons in turn repel *their* adjacent electrons, and so on around the entire circuit.

3-2 OHM'S LAW

In 1847, Georg Simon Ohm (1787–1854), a German physicist, formulated the precise relationship between voltage, current, resistance, and

(a)

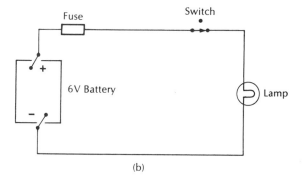

(b)

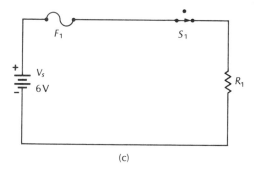

(c)

Figure 3-2 A simple electrical circuit: (a) a photograph of the apparatus; (b) a pictorial diagram; (c) the schematic.

power. This is the most basic and useful law to technicians and engineers and it must be learned thoroughly. In general terms, Ohm's law specifies that current in a circuit varies *directly* as the voltage and *inversely* as the resistance. That is, if the voltage is exactly doubled in a given circuit configuration, current will exactly double, if all other conditions remain the same. On the other hand, if voltage is kept constant and the resistance is doubled, the value of current will be halved. This general relationship can be restated in terms of an algebraic formula.

$$\text{Current} = \frac{\text{Voltage}}{\text{Resistance}}$$

or

$$I = \frac{E}{R}.$$

The terms I, E, and R are used to represent the words so as to shorten the written expression. You will encounter many other such abbreviations as your studies progress. I represents *intensity* of current; E represents *electromotive force* (EMF); R represents *resistance*.

Ohm's law is extremely useful in determining unknown values in virtually any circuit. The circuit shown in Figure 3-3a will be used to demonstrate these relationships. This circuit is essentially the same as the one shown in Figure 3-2, but without the fuse and switch. As in Figure 3-2c, the symbol for a resistor is used in place of the lamp; the symbol represents the resistance of the filament wire inside the lamp.

Note that both voltage and resistance are given, but that current is unknown. To determine the value of current, simply substitute the known values and compute.

$$I = \frac{E}{R} = \frac{6}{10} = 0.6 \text{ A}.$$

Knowing any two values allows you to find the third by use of one of the following relationships:

$$I = \frac{E}{R} \qquad E = I \times R \qquad R = \frac{E}{I}.$$

To provide several examples of these three formulas, different values will be substituted for the circuit shown in Figure 3-3.

Example 1. $E = 12$ V, $I = 1$ A; find R.

$$R = \frac{E}{I} = \frac{12}{1} = 12 \text{ }\Omega.$$

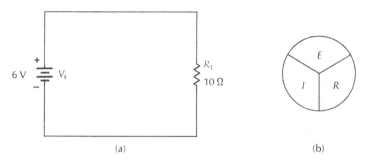

Figure 3-3 (a) A simple series circuit used to introduce Ohm's law, (b) a memory aid for Ohm's law relationship.

Example 2. $E = 12$ V, $R = 120$ Ω; find I.

$$I = \frac{E}{R} = \frac{12}{120} = 0.1 \text{ A}.$$

Example 3. $I = 0.12$ A, $R = 100$ Ω; find E.

$$E = I \times R = 0.12 \times 100 = 12 \text{ V}.$$

As a general rule, practical values of current and resistance are somewhat different from those in the preceding three examples. The following examples of circuit analysis (still using Figure 3-3) demonstrate this.

Example 4. $I = 0.001$ A; $R = 10,000$ Ω; find E.

$$E = I \times R = 0.001 \times 10,000 = 10 \text{ V}.$$

Example 5. $I = 0.001$ A; $E = 6$ V; find R.

$$R = \frac{E}{I} = \frac{6}{0.0001} = 60,000 \text{ Ω}.$$

Example 6. $R = 1,000,000$ Ω, $E = 24$ V; find I.

$$I = \frac{E}{R} = \frac{24}{1,000,000} = 0.000024 \text{ A}.$$

Figure 3-3*b* illustrates a simple memory aid for using Ohm's law. Simply cover the unknown quantity with your finger, and the remaining values are given as they appear in the formula. In other words, if R is unknown, cover R—the required operation is E over (divided by) I. To find E, cover E and you will see that you must multiply I and R.

3-3 UNIT ABBREVIATIONS

When using large values of resistance, which result in small values of current, it is not convenient to write the numbers in conventional style, as we did in the preceding examples. A shorthand method is universally used to shorten the writing and computing of large and small numbers. The basic electrical units used thus far—ampere, volt, and ohm—are used with multiple and submultiple units. Table 3-1 lists the most common prefixes together with their symbols and multiplying factors.

To use the table, simply replace the appropriate zeros, in groups of three, by the prefix:

$$1000 \text{ volts} = 1 \text{ kilovolt (1 kV)}$$
$$1{,}000{,}000 \text{ ohms} = 1 \text{ megohm (1 M}\Omega\text{)}$$
$$0.001 \text{ ampere} = 1 \text{ milliampere (1 mA)}$$
$$0.000001 \text{ ampere} = 1 \text{ microampere (1 }\mu\text{A)}$$
$$0.000000001 \text{ ampere} = 1 \text{ nanoampere (1 nA)}$$
$$0.000000000001 \text{ ampere} = 1 \text{ picoampere (1 pA)}$$

3-4 POWERS OF TEN

The use of powers of ten (exponents) allows the simple and quick handling of very large and very small numbers and greatly simplifies the attendant arithmetic. One of the problems encountered in working with large and small numbers is the correct placement of the decimal point and the chance of error in performing the calculations. Using powers of ten reduces the chance of error to essentially zero. Additionally, the display on many electronic calculators automatically switches to powers of ten when appropriate, so the mastering of this shorthand method of notation (often called *scientific notation*) is necessary.

TABLE 3-1 PREFIXES, SYMBOLS, AND MULTIPLYING FACTORS OF COMMON MULTIPLE AND SUBMULTIPLE UNITS

Prefix	Symbol	Multiplying Factor
mega	M	1,000,000
kilo	k	1,000
milli	m	0.001
micro	μ	0.000001
nano	n	0.000000001
pico	p	0.000000000001

To introduce this subject, suppose that it is necessary to multiply 0.000012 by 5,000,000. To perform this multiplication manually is not difficult, but it is tedious and the chance of error is considerable. Try it and see! The problem is simplified by first converting to powers of ten:

$$(1.2 \times 10^{-5})(5 \times 10^6) = 1.2 \times 5 \times 10^{-5} \times 10^6 = 6 \times 10^{-5+6}$$
$$= 6 \times 10^1 = 6 \times 10 = 60.$$

In this example, note that multiplying 1.2 by 5 is very easy. The remainder of the problem consists simply of algebraically *adding the exponents* (-5 and 6). Hence, powers of ten reduces much of the multiplication to simple addition.

The first step in learning this system is to be able to convert ordinary numbers to an equivalent power of ten, and to be able to perform the reverse operation. Table 3-2 lists several numbers and their corresponding power of ten. Note that the exponent of the power-of-ten expression is a measure of the number of places that the decimal point is moved in going from one form to the other.

For example, if 0.0001 is to be written as a power of ten, the decimal point is moved four places to the *right,* and the exponent is -4. Thus, $0.0001 = 1 \times 10^{-4}$. A negative exponent signifies that the original expression is smaller than the power-of-ten numbers. That is, 0.0001 is certainly smaller than 1. On the other hand, 1000 expressed as a power of ten is 1×10^3 and the positive exponent signifies that the original expression is greater than the power-of-ten numbers. That is, 1000 is greater than 1. The exponent is simply an indication of decimal-point movement.

Any number can be written as a power of ten. For example, the number 4700 can be expressed as 4.7×10^3. Again, the exponent simply tells the number of places to move the decimal point. To convert the

TABLE 3-2 SOME CONVENTIONAL NUMBERS AND THEIR CORRESPONDING POWERS OF TEN

Number	Power of Ten
0.0001	1×10^{-4}
0.001	1×10^{-3}
0.01	1×10^{-2}
0.1	1×10^{-1}
1.0	1×10^0
10	1×10^1
100	1×10^2
1000	1×10^3
10,000	1×10^4

power-of-ten expression 4.7×10^3 to conventional form, simply move the decimal to the *right* three places: $4.7 \times 10^3 = 4700$.

PRACTICE PROBLEMS

Perform the following conversions until you are certain that you can easily change any number to a power of ten and vice versa.

1. $123 =$
2. $4500 =$
3. $750,000 =$
4. $1,250,000 =$
5. $0.25 =$
6. $0.0056 =$
7. $0.0000002 =$
8. $1.23 \times 10^2 =$
9. $4.5 \times 10^3 =$
10. $7.5 \times 10^5 =$
11. $1.25 \times 10^6 =$
12. $2.5 \times 10^{-1} =$
13. $5.6 \times 10^{-3} =$
14. $2 \times 10^{-7} =$

Note that the first seven problems are the same as the second seven and that each group provides the answers for the other. Also note that there is more than one correct answer for these problems. For example, 123 can be expressed as 1.23×10^2 or as 12.3×10^1 or as 0.123×10^3. In electronics, it is often most useful to express these powers of ten in multiples of 1000 (10^3), so that values used in computations are compatible with multiple and submultiple units such as kilo, micro, and so on. Thus, 4700 Ω is best expressed as 4.7×10^3 Ω, which is the same as 4.7 kΩ. The number 18,000,000 is best expressed as 18×10^6 instead of, perhaps, 1.8×10^7 or 180×10^5.

Adding Powers of Ten

Rule: To add powers of ten, the expressions *must have the same exponent*. Simply add the numbers and bring down the common power of ten.

Example 7. Add the following:

$$\begin{array}{r} 4.5 \times 10^3 \\ + \; 6.8 \times 10^3 \\ \hline 11.3 \times 10^3. \end{array}$$

If the two numbers do not have the same exponent, one of the exponents must be changed.

Example 8. Add the following:

$$\begin{array}{r} 4.5 \times 10^3 \\ +3.4 \times 10^4. \end{array}$$

These *cannot* be added until the two exponents are the same. Therefore, choose one and change it to agree with the other. For example, $3.4 \times 10^4 = 34 \times 10^3$. Then

$$\begin{array}{r} 4.5 \times 10^3 \\ 34.0 \times 10^3 \\ \hline 38.5 \times 10^3. \end{array}$$

Note that when the decimal point is moved to the *right*, the exponent is *decreased* by the same number as the places moved. That is, if, the number is made *larger*, the exponent must be made *smaller*. If the number is made smaller, the exponent must be made larger.

PRACTICE PROBLEMS

Find the following sums:

1. $(4.68 \times 10^4) + (6.98 \times 10^4) =$ _____ .
2. $(3.1 \times 10^5) + (7.9 \times 10^4) =$ _____ .
3. $5.56 + (3.3 \times 10^2) =$ _____ .

Answers: (1) $11.66 \times 10^4 = 116.6 \times 10^3$; (2) 38.9×10^4 or $3.89 \times 10^5 = 389 \times 10^3$; (3) 335.56 or 0.33556×10^3. Note that $5.56 = 5.56 \times 10^0 = 5.56 \times 1 = 5.56$.

Subtracting Powers of Ten

Rule: To subtract powers of ten, the expressions *must have the same exponent*. Simply subtract the numbers and affix the common power-of-ten exponent.

$$\begin{array}{r} 6.8 \times 10^3 \\ -4.5 \times 10^3 \\ \hline 2.3 \times 10^3. \end{array}$$

As in adding powers of ten, if the numbers do not have the same exponent, change one of them to agree with the other.

Example 10. Find the remainder:

$$\begin{array}{r} 4.5 \times 10^3 \\ -2.3 \times 10^4. \\ \hline \end{array}$$

Change the exponent of one to agree with the other. Thus, $4.5 \times 10^3 = 0.45 \times 10^4$. Then

$$\begin{array}{r} 0.45 \times 10^4 \\ -2.3 \times 10^4 \\ \hline -1.85 \times 10^4. \end{array}$$

PRACTICE PROBLEMS

Perform the following subtractions:
1. $(7.98 \times 10^7) - (3.46 \times 10^7) = $ _____ .
2. $(4.79 \times 10^{-3}) - (2.39 \times 10^{-3}) = $ _____ .
3. $(4.9 \times 10^3) - (7.95 \times 10^2) = $ _____ .

Answers: (1) 4.53×10^7; (2) 2.4×10^{-3}; (3) 4.105×10^3 or 41.05×10^2.

Multiplying Powers of Ten

Rule: To multiply powers-of-ten expressions, multiply the numbers and affix the power-of-ten exponent which is the *sum* of the exponents. Expressed more simply, multiply the numbers and add the exponents.

Example 11. Find the following products:

$(3 \times 10^3)(5 \times 10^5) = 15 \times 10^8$.
$(7.9 \times 10^{-2})(4.3 \times 10^2) = 33.97 \times 10^0$ *or* 33.97.
$(4.5 \times 10^{-6})(2.22 \times 10^{-3}) = 9.99 \times 10^{-9}$.
$(8.0 \times 10^3)(4.0 \times 10^6)(3.13 \times 10^{-11}) = 100.16 \times 10^{-2}$ *or* 1.0016.

Dividing Powers of Ten

Rule: To divide powers of ten, divide the numbers and *subtract* exponents. That is, divide the numbers as usual, and subtract the exponent in the denominator from that in the numerator.

Example 12. Perform the following division problems:

$$\frac{35 \times 10^3}{7 \times 10^2} = 5 \times 10^{3-2} = 5 \times 10^1 = 50.$$

$$\frac{4.5 \times 10^{-3}}{5 \times 10^3} = 0.9 \times 10^{(-3)-(+3)} = 0.9 \times 10^{-6}.$$

$$\frac{4.5 \times 10^4}{0.05 \times 10^{-5}} = 90 \times 10^{4-(-5)} = 90 \times 10^9 = 9 \times 10^{10}.$$

$$\frac{4.5 \times 10^{-3}}{5 \times 10^{-3}} = 0.9 \times 10^{(-3)-(-3)} = 0.9 \times 10^0 = 0.9.$$

Note in these examples that the power of ten in the denominator is simply moved to the numerator, *its sign is changed,* and the exponents are then added. Using the last example,

$$\frac{4.5 \times 10^{-3}}{5 \times 10^{-3}} = \frac{4.5 \times 10^{-3+3}}{5} = 0.9 \times 10^0 = 0.9.$$

This is a useful and valid operation. The power of ten can be moved from numerator to denominator or vice versa as often as necessary *if its sign is changed each time.*

Raising to a Power

Rule: To raise a number expressed as a power of ten to a power, raise the number to the power and *multiply exponents.*

Example 13. Raise the expressions to the given power:

$$(3 \times 10^2)^3 = 3^3 \times 10^{2(3)} = 27 \times 10^6$$
$$(5 \times 10^{-3})^2 = 5^2 \times 10^{-3(2)} = 25 \times 10^{-6}.$$

Extracting a Root

Rule: To extract the root of a power-of-ten expression, extract the root of the number and divide the exponent of the power of ten by the root to be extracted.

Example 14. Extract the following roots:

$$\sqrt{9 \times 10^6} = \sqrt{9} \times \sqrt{10^6} = 3 \times 10^{6/2} = 3 \times 10^3$$
$$\sqrt{100 \times 10^{12}} = \sqrt{100} \times \sqrt{10^{12}} = 10 \times 10^{12/2} = 10 \times 10^6$$

PRACTICE PROBLEMS

The following practice problems are designed to increase your ability to work easily with powers of ten. Work each problem in detail so as to be sure that you fully understand each operation.

1. Convert 9.45×10^7 to a conventional number.
2. Convert 555,000 to a power of ten.
3. Convert 0.0000015 to a power of ten.
4. Convert 0.065×10^{-3} to a conventional number.
5. $768 + (0.123 \times 10^3) =$
6. $(4.7 \times 10^3) + (6.3 \times 10^4) =$
7. $6,800,000 + (0.2 \times 10^6) =$
8. $4,500,000 + (0.1 \times 10^{-3}) =$
9. $9000 - (4 \times 10^3) =$
10. $(4.5 \times 10^3) - (37.5 \times 10^2) =$

50 Chapter 3

11. $(4.35 \times 10^3) - 50 =$
12. $0.001 - (0.001 \times 10^{-3}) =$
13. $(45)(3.5 \times 10^3) =$
14. $(1 \times 10^4)(3.7 \times 10^{-2}) =$
15. $(2.5 \times 10^{-3})(34 \times 10^{-5}) =$
16. $(25 \times 10^6)(6.8 \times 10^{-12}) =$
17. $(1 \times 10^4) \div (3.7 \times 10^{-2}) =$
18. $(3.4 \times 10^{-4}) \div (2.5 \times 10^{-3}) =$
19. $68,000 \div (2.34 \times 10^7) =$
20. $(0.45 \times 10^{-7}) \div (2.2 \times 10^{-4}) =$
21. $(2.5 \times 10^3)^2 =$
22. $(4.6 \times 10^{-2})^3 =$
23. $(3.33 \times 10^{-4})^2 =$
24. $(4.5 \times 10^{-2})^2 =$
25. $\sqrt{3.35 \times 10^6} =$
26. $\sqrt{25 \times 10^{12}} =$
27. $\sqrt{100 \times 10^4} =$
28. $\sqrt{49 \times 10^2} =$

Answers: (1) 94,500,000; (2) 5.55×10^5 or 555×10^3; (3) 1.5×10^{-6}; (4) 0.000065; (5) 891 or 0.891×10^3; (6) 67,700 or 67.7×10^3; (7) 7,000,000 or 7×10^6; (8) 4,500,000.0001; (9) 5000 or 5×10^3; (10) 750 or 0.75×10^3; (11) 4300 or 4.3×10^3; (12) 9.99×10^{-4}; (13) 157,500 or 157.5×10^3; (14) 370 or 0.37×10^3; (15) 8.5×10^{-7} or 850×10^{-9}; (16) 1.7×10^{-4} or 170×10^{-6}; (17) 270,270.27 or 0.270×10^{-6}; (18) 0.136 or 136×10^{-3}; (19) 2.906×10^{-3}; (20) 2.045×10^{-4} or 0.2045×10^{-3}; (21) 6,250,000 or 6.25×10^6; (22) 9.73×10^{-5} or 97.3×10^{-6}; (23) 1.11×10^{-7} or 11.1×10^{-8} or 0.111×10^{-6}; (24) 20.25×10^{-4} or 2.025×10^{-3}; (25) 1830.3 or 1.8303×10^3; (26) 5,000,000 or 5×10^6; (27) 1000 or 1×10^3 or 10×10^2; (28) 70 or 7×10^1.

3-5 MULTIPLE AND SUBMULTIPLE CIRCUIT PROBLEMS

You should now be able to combine the use of multiple and submultiple units—and the use of powers of ten—with your ability to find unknown values in a circuit. The following examples show you how. The circuit in Figure 3-3 will be used again, but again the component values will be changed.

Example 15. $E = 12$ V, $R = 1$ MΩ; find I.

$$I = \frac{E}{R} = \frac{12}{1,000,000} = \frac{12}{1 \times 10^6} = 12 \times 10^{6} = 0.000012 \text{ A} = 12 \text{ } \mu\text{A}.$$

Basic Circuit Characteristics   51

Example 16. $E = 6$ V, $R = 470$ kΩ; find I.

$$I = \frac{E}{R} = \frac{6}{470,000} = \frac{6}{4.7 \times 10^5} = 0.000012765 \text{ A} = 12.765 \ \mu\text{A}.$$

Example 17. $E = 6$ V, $I = 120 \ \mu$A; find R.

$$R = \frac{E}{I} = \frac{6}{0.00012} = \frac{6}{1.2 \times 10^{-4}} = 50,000 \ \Omega = 50 \text{ k}\Omega.$$

Example 18. $E = 24$ V, $I = 240 \ \mu$A; find R.

$$R = \frac{E}{I} = \frac{24}{0.00024} = \frac{24}{24 \times 10^{-5}} = 1 \times 10^5 = 100,000 \ \Omega = 100 \text{ k}\Omega.$$

Example 19. $I = 100 \ \mu$A, $R = 100$ kΩ; find E.

$$E = I \times R = 0.0001 \times 100,000 = (1 \times 10^{-4})(1 \times 10^5) = 1 \times 10^1 = 10 \text{ V}.$$

Example 20. $R = 150$ kΩ, $I = 160 \ \mu$A; find E.

$$E = I \times R = 0.00016 \times 150,000 = (16 \times 10^{-5})(15 \times 10^4)$$
$$= 240 \times 10^{-1} = 24 \text{ V}.$$

Example 21. $R = 33.3 \times 10^3 \ \Omega$, $I = 1 \times 10^{-3}$ A; find E.

$$E = I \times R = (33.3 \times 10^3)(1 \times 10^{-3}) = 33.3 \text{ V}.$$

Example 22. $R = 1.5 \times 10^5 \ \Omega$, $I = 10 \times 10^{-6}$ A; find E.

$$E = I \times R = (1.5 \times 10^5)(10 \times 10^{-6}) = 15 \times 10^{-1} = 1.5 \text{ V}.$$

Example 23. $E = 12$ V, $R = 3.3 \times 10^3 \ \Omega$; find I.

$$I = E/R = 12/(3.3 \times 10^3) = 3.64 \times 10^{-3} = 3.64 \text{ mA}.$$

Example 24. $E = 6$ V, $R = 5.6 \times 10^5 \ \Omega$; find I.

$$I = E/R = 6/(5.6 \times 10^5) = 1.07 \times 10^{-5} = 10.7 \times 10^{-6} = 10.7 \ \mu\text{A}.$$

Example 25. $E = 6$ V, $I = 60 \times 10^{-6}$ A; find R.

$$R = E/I = 6/(60 \times 10^{-6}) = 0.1 \times 10^6 = 100 \text{ k}\Omega.$$

Example 26. $E = 24$ V, $I = 0.012 \times 10^{-3}$ A; find R.

$$R = E/I = 24/(0.012 \times 10^{-3}) = 2000 \times 10^3 = 2 \text{ M}\Omega.$$

3-6 ELECTRICAL POWER

As was mentioned earlier, when current flows through a resistance, power is expended and heat is generated. The heat is generated within the resistance because of friction between the free electrons and also between the free electrons and the atoms in the conductor.

The unit of electrical power is the watt (W), named in honor of James Watt (1736–1819), a Scottish-American inventor. One watt is the work done by an EMF of 1 volt forcing 1 ampere of current through a resistance for 1 second. Power in watts equals voltage times current. The symbol for power is P. Expressed as algebraic formulas, the relationships are

$$P = I \times E \qquad I = \frac{P}{E} \qquad E = \frac{P}{I}.$$

Example 27. A soldering iron is rated at 120 V and 40 W. To determine the value of current, simply substitute values in the proper equation:

$$I = \frac{P}{E} = \frac{40}{120} = 0.333 \text{ A}.$$

Example 28. A heavy-duty soldering iron is rated at 150 W and 120 V. The value of current drawn is

$$I = \frac{P}{E} = \frac{150}{120} = 1.25 \text{ A}.$$

Variations of the basic relationship for electrical power can be derived from the original formula. The following derivations are self-explanatory. Since $I = E/R$, this can be substituted in the expression $P = I \times E$:

$$P = \frac{E}{R} \times E = \frac{E^2}{R}.$$

Then,

$$E^2 = R \times P,$$

so

$$E = \sqrt{RP}.$$

And, again from $P = E^2/R$,

$$R = \frac{E^2}{P}.$$

Also, $P = I \times E$. However, $E = I \times R$, so

$$P = I \times I \times R = I^2R,$$

and from that,

$$I = \sqrt{P/R} \qquad R = P/I^2.$$

Figure 3-4 provides another memory aid to help you remember these power relationships. Use this aid just as you used the one included with Figure 3-3.

As an example of these relationships, the 150-W soldering iron mentioned before has a resistance of 96 Ω. Determine its power rating using voltage and resistance.

Example 29.

$$P = \frac{E^2}{R} = \frac{120^2}{96} = \frac{14{,}400}{96} = 150 \text{ W}.$$

Determine its power rating using current and resistance.

Example 30.

$$P = I^2R = 1.25^2 \times 96 = 150 \text{ W}.$$

If you know the value of two of the following—voltage, current, resistance, and power—you can always determine the other two values.

It should be noted that the generation of electrical power and the consequent dissipation of heat by a resistive load can either be desirable or undesirable. The heat produced by soldering irons or electric stoves is, of course, desirable and, indeed, necessary. However, the heat dissipated by tube-type television receivers, lamps, radios and stereo reproduction units, and so on is totally wasted. In many cases, much engineering effort is devoted to reducing this wasted energy to a minimum. For years, this effort was directed toward such equipment as

Figure 3-4 A memory aid for the power law relationships.

portable or mobile communications equipment. However, the national energy shortage has brought about a rethinking of engineering efforts to increase efficiency of all electrical apparatus. This is evident, for example, in much of the national advertising of electrical appliances and equipment.

3-7 GRAPHICAL RELATIONSHIPS

It is helpful to visualize the relationships between volts, ohms, amperes, and watts. One way of doing this is to plot a graph of three of the quantities, holding one of them constant. As an example, volts, ohms, and amperes will be plotted on a graph to illustrate the relationships. In a circuit similar to Figure 3-3, the voltage will be incrementally changed across a 10-ohm resistance, and the resulting current will be calculated. The voltage will be changed in 2-volt steps, from 2 to 10 volts. Table 3-3 shows the results of these determinations.

The results of the computations shown in Table 3-3 are plotted in Figure 3-5. The graph clearly illustrates the linear relationship between E, I, and R. If the value of resistance stays the same, doubling the voltage results in doubling the current, while halving the voltage results in halving the current. It can also be shown that if the voltage is kept constant, doubling the resistance results in halving the current while halving the resistance doubles the current, thus verifying the inverse relationship between current and resistance.

Finally, Figure 3-6 illustrates how power varies as the voltage is varied, if resistance is held constant. Because power varies as the square of the current (I^2R), this graphical representation is *not* linear. Such a curve is called a *square-law curve,* since it follows the square of one of the functions (I^2R or E^2/R).

TABLE 3-3 DATA FOR THE GRAPH IN FIGURE 3-4

	R	E	I
$I = \frac{E}{R} = \frac{2}{10} = 0.2$ A	10	2	0.2
$I = \frac{E}{R} = \frac{4}{10} = 0.4$ A	10	4	0.4
$I = \frac{E}{R} = \frac{6}{10} = 0.6$ A	10	6	0.6
$I = \frac{E}{R} = \frac{8}{10} = 0.8$ A	10	8	0.8
$I = \frac{E}{R} = \frac{10}{10} = 1.0$ A	10	10	1.0

Basic Circuit Characteristics 55

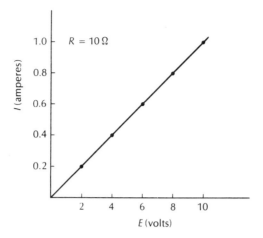

Figure 3-5 Graphical plot of voltage and current with constant resistance.

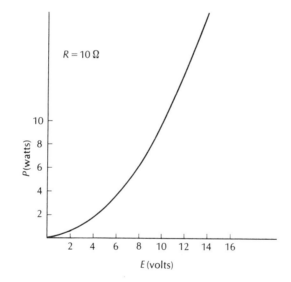

Figure 3-6 Change in power dissipated across a 10-Ω resistor with varying values of voltage.

3-8 MEASURING ELECTRICAL QUANTITIES

In order to troubleshoot electrical equipment or to verify proper performance, it is often necessary to measure voltage, current, or resistance. This is accomplished with a voltmeter, an ammeter, or an ohmmeter. Often these instruments are combined into one, and this is known as a volt-ohm-milliammeter (VOM). Typical VOMs are illustrated in Figure 3-7.

The meter face carries the various scales for the quantities to be measured. The top scale in Figure 3-7a is used for measuring resistance in ohms or kilohms. The meter leads are simply placed across a de-energized resistance with the selector switch on the proper setting, and the pointer (needle) indicates the amount of resistance. The reading is then multiplied by the scale factor indicated by the selector switch. For example, if the needle points to 470 Ω, and the selector switch indicates × 100 (times 100), the resistance being measured is 47 kΩ (47,000 Ω). The ohmmeter section of the VOM provides its own internal power (by means of dry cells); hence, the device being measured must have no other voltage applied.

The voltage and current scales are, of course, used with power applied to the component being measured. The voltage scales are often combined for both direct-current (dc) and alternating-current (ac) scales. In the VOM illustrated, however, the dc and ac scales are separated, with one separate ac scale from 0.0 to 2.5 volts. Ac volts, dc volts, and dc milliamperes are read from the center scale (0.0 to 10; 50, 250). Note that while ac and dc volts and dc milliamperes are all read from the same scale, the selector switch *must* be set to the proper position to provide the correct reading.

The two meter leads (part b of the figure) are color-coded, black for negative and red for positive. If proper polarity is not observed, the needle will not read upscale, and the meter may be damaged. In Chapter 7 this kind of meter, as well as several others, will be discussed in detail.

To use the pocket-size VOM (Figure 3-7b) to measure dc voltage, choose the proper scale setting and move the selector to this position. Place the black lead (marked COM) on the negative part of the circuit and the red lead (marked V-O-M) on the positive part of the circuit. Read the value of dc voltage on the scale corresponding to the switch setting. Note that polarity is not a factor in measuring ac volts.

To measure dc current, the circuit must be broken at a convenient place and the meter must be connected into the circuit so the current flows *through* the meter (observing polarity). The meter itself forms a bridge across the open in the circuit and circuit current is then restored. Read the appropriate scale for the dc current value.

Direct current is the smooth unvarying current delivered by a battery or other such device. Alternating current, on the other hand, is constantly varying in a smooth curve (Figure 3-8). Alternating current has many advantages over direct current. Alternating current is the type delivered to our homes and businesses, so nearly everyone uses ac every day. We shall discuss ac voltage and current quite thoroughly in Chapter 13.

(a)

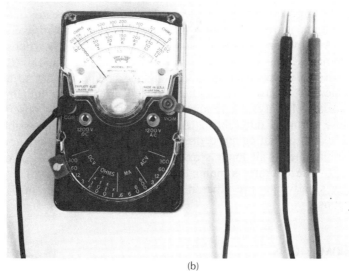

(b)

Figure 3-7 Typical VOMs.

58 Chapter 3

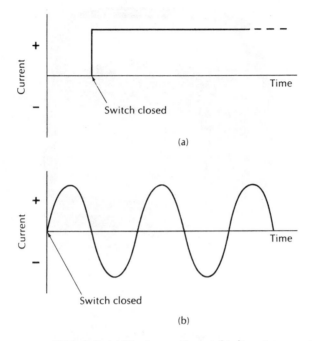

Figure 3-8 (a) Direct current versus (b) alternating current.

As previously noted, there are several ranges for both dc and ac scales. It is important that the scale used be equal to or greater than the value being measured, to avoid "pinning" the needle at an overscale amount. Also, when measuring dc voltages it is necessary to observe proper polarity. Since the meter leads are color coded, this is not difficult. If the wrong polarity is chosen, the needle will try to move downscale below zero, and may be damaged. Finally, the milliampere scales (usually dc only) also require that polarity be carefully observed.

Because so many different meters are available, it is impossible to provide exact operating procedures in this text. However, general rules, applicable to all models and varieties of meters, are given in Chapter 7. The operator's manual for your specific meter, plus your instructor's advice, are the best sources for basic operating instructions.

3-9 APPROXIMATIONS

Because electronic calculators are so inexpensive and so convenient, we tend to rely on them. However, it is also true that they often do not function properly. Batteries need replacing or recharging at frequent intervals; and while calculators are very dependable, they do need occasional service.

In order to be able to solve the kinds of problems you will encounter in working as a technician, you should learn to solve such problems *in your head*. Then, if your calculator is not available for any reason, you will not be completely lost. Performing mental arithmetic may seem at first to be difficult, but many of the problems in this book—as well as in everyday work—can easily be solved this way, with a little practice. Also, unless your calculator is very sophisticated, you cannot solve an entire problem in one calculation. Rather, you solve parts of the problem in intermediate steps, writing down each result as it is obtained, and then properly combining the steps. This is exactly how you solve a problem by approximations, substituting your mind for the calculator.

Nearly all electronic components occur in more-or-less standard sizes, resistors in ohms, kilohms, or megohms, for example. Other components are also used in relatively standard sizes. Hence, if you memorize the relationships between these sizes, you have developed a very real simplification process.

It should be noted that memorizing certain relationships in Table 3-1 (page 44) can ease the arithmetic in solving certain problems. Thus,

$$\frac{\text{volts}}{\text{kilohms}} = \text{milliamperes}$$

$$\frac{\text{kilovolts}}{\text{megohms}} = \text{milliamperes}$$

$$\frac{\text{volts}}{\text{megohms}} = \text{microamperes}$$

$$\text{kilohms} \times \text{milliamperes} = \text{volts}$$
$$\text{megohms} \times \text{microamperes} = \text{volts}$$
$$\text{megohms} \times \text{milliamperes} = \text{kilovolts}$$
$$\text{kilohms} \times \text{microamperes} = \text{millivolts}$$

A few examples will serve to illustrate how this can be done.

Example 31. A voltage of 12 V is impressed across a resistance of 6 kΩ. How much current flows?

Answer: Knowing that $I = E/R$, we know that we must divide 12 by 6000. Whenever units (in this case, 12 V) is divided by values in the thousands (6000), the result is *always* in thousandths of a unit; that is the resulting unit will carry the prefix *milli-*. Hence, 12 divided by 6 is 2, which you can certainly calculate mentally, and since we are solving for current, you know that the unit is amperes. Therefore, the answer is

2 mA. If you memorize the unit relationships given above you can easily solve many similar problems. An alternate way of doing this problem is to use the power-of-ten equivalents.

$$I = \frac{E}{R} = \frac{12}{6 \times 10^3} = \frac{12}{6} \times 10^{-3} = 2 \times 10^{-3} = 2 \text{ mA}.$$

Example 32. A current of 2 mA is flowing in a 12-kΩ resistor. What is the voltage drop across the resistor?

Answer: Since $E = I \times R$, we can say that when *milli-* (1/1000) is multiplied by *kilo-* (1000), the result will be in units. That is,

$$E = I \times R = 2 \text{ mA} \times 12 \text{ k}\Omega = 24 \text{ V}.$$

It is clear that 10^{-3} *(milli-)* and 10^3 *(kilo-)* cancel each other; thus the problem is simply $2 \times 12 = 24$ V. To use the power-of-ten approach:

$$E = I \times R = (2 \times 10^{-3})(12 \times 10^3) = 2 \times 12 \times 10^{-3+3}$$
$$= 24 \text{ V}.$$

Example 33. A voltage of 45 V is impressed across a resistor and 5 μA of current is flowing. What is the value of the resistor?

Answer: Since $R = E/I$, we have $R = E/I = 45/5$ μA. Units divided by microunits results in megaunits. Carrying out the calculation, $45/5 = 9$, which is 9 MΩ. In terms of powers of ten,

$$R = \frac{E}{I} = \frac{45}{5 \times 10^{-6}} = \frac{45}{5} \times 10^6 = 9 \times 10^6 = 9 \text{ M}\Omega.$$

To develop the knack of doing this kind of mental arithmetic, you must simply practice it. Even if multiplication or division must be done on paper (as when 27 mA is divided by 49 μA), the remaining part can be done mentally to verify results. As an aid in learning mental arithmetic, tabulate all possible combinations of metric prefixes. Then practice mental arithmetic on the problems in this book, and you will soon have the relationships memorized.

SUMMARY

- A series circuit is one in which there is only one path in which electrons can move.
- Ohm's law provides the numerical relationship between voltage, current, resistance, and power.
- Multiple and submultiple unit prefixes are used to indicate very large and very small numbers. For example, it is quicker to write—and easier to read—1 pA than 0.000000000001 ampere.

- Powers of ten allow very large or very small numbers to be written conveniently and simplify certain mathematical operations.
- The instrument used to measure electrical quantities in the meter.
- A volt-ohm-milliammeter (VOM) is an instrument that can measure voltage, current, and resistance. The VOM often has several scale factors for each quantity.

QUESTIONS

1. Briefly define a simple electrical circuit.
2. An electrical circuit having only a single path of current flow is known as a _____ circuit.
3. A fuse _____ an electrical circuit.
4. A switch is used to _____ or _____ a circuit.
5. A single electron travels around a circuit very (slowly, fast).
6. Electrons (repel, attract) each other.
7. True or false: Ohm's law was first formulated in the year 1847.
8. Ohm's law states in part that current varies directly with the _____.
9. Ohm's law states in part that current varies inversely with the _____.
10. Ohm's law can be stated as: current = (_____)/resistance.
11. Ohm's law can be stated as: voltage = current × (_____).
12. Ohm's law can be stated as: resistance = voltage/(_____).
13. Briefly explain in your own words the meaning of $I = E/R$.
14. Briefly explain in your own words the meaning of $R = E/I$.
15. Draw the symbol for a battery.
16. Draw the symbol for a resistor.
17. Draw the symbol for a length of wire.
18. One kV equals _____ volts.
19. Ten kΩ equals _____ ohms.
20. One mA equals _____ amperes.
21. A 100-kΩ resistor equals _____ ohms.
22. True or false: 1 µA equals 1/1,000,000 A.
23. The expenditure of electrical power can often be sensed as _____.
24. A resistor that gets hot during operation dissipates more _____ than one that does not.
25. The instrument used to measure volts is called a _____.
26. The instrument used to measure current is called an _____ or _____.
27. The instrument used to measure resistance is called an _____.

PROBLEMS

1. Write the three basic forms of Ohm's law with respect to E, I, and R.
2. Write the three basic relationships for finding power dissipation.
3. Convert 3.3 mA to amperes.
4. Convert 7.8 μA to mA.
5. Convert 12 kΩ to ohms.
6. Convert 120 kΩ to ohms.
7. Convert 25 μW to watts.
8. Convert 750 mW to watts.
9. Convert 25 kV to volts.
10. Convert 175 kV to volts.
11. Convert 350 μs (microseconds) to seconds.
12. Convert 0.25 μs to seconds.
13. A 47-Ω resistor is connected across a 12-V battery. Determine the current value.
14. A 1000-Ω resistor is connected across a 24-V battery. Determine the current value.
15. A 6-V battery is supplying current for a resistance. If the current has a value of 0.015 A, what is the resistor value?
16. A 3-V battery is supplying current for a resistance. If the current has a value of 0.00017 A, what is the resistor value?
17. A resistor is connected across a battery and is drawing 0.036 A. If the resistor has a value of 330 Ω, what is the battery voltage?
18. A resistor is connected across a battery and is drawing 0.00333 A. If the resistor has a value of 1800 Ω, what is the battery voltage?
19. A 100,000-Ω resistor is connected across a 6-V source. What value of current is flowing?
20. A 500,000-Ω resistor is connected across an 18-V source. What value of current is flowing?

CHAPTER 4

SERIES CIRCUITS

In this chapter we investigate the series circuit in nearly every aspect. Two, three, and more resistors will be connected in the series circuit. Voltage, current, resistance, and power relationships will be covered. Such useful configurations as the voltage divider will be dealt with. Additionally, series strings of resistors with two or more power sources will be examined, as will those connected to an equipment ground.

Figure 4-1 illustrates a series circuit taken from a commercially available electronic organ. Shown is the PC board; the general area where this series circuit is to be found is shown in part *a,* and a schematic is given in *b.* Note that there are three resistors in the schematic drawing. The first example to follow uses only two resistors. However, an example is given in Figure 4-4 that is electrically almost identical to the circuit shown in Figure 4-1.

The topics to be covered in this chapter are listed below.

4-1 Series Circuit Relationships
4-2 Total Series Resistance
4-3 Total Circuit Current
4-4 Voltage Drops
4-5 Voltage Drop Polarity
4-6 Analyzing Series Circuits
4-7 Multiple Power Sources
4-8 Reference Voltage
4-9 Ground
4-10 Circuits with Various Unknowns
4-11 Open and Short Circuits

4-1 SERIES CIRCUIT RELATIONSHIPS

A typical series circuit is shown in Figure 4-2*a*. The battery, labeled V_{source}, provides 9 V to the circuit. From now on, we shall use the shorter expression V_s. The two resistors are connected *in series.* Remember that a series circuit is one in which there is but one path for current to take, and the circuit in Figure 4-2*a* can, by inspection, be identified as such. A single electron never has a choice of paths—it

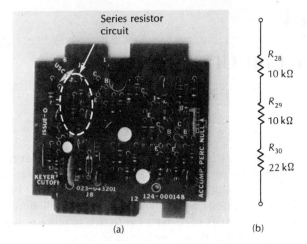

Figure 4-1 (a) This series circuit, of the type discussed in this chapter, was taken from commercial equipment; (b) the schematic diagram of this circuit.

must emerge from the battery, go up through R_2, then R_1, and then return to the positive battery post. Each of the two series-connected resistors has a value of 12 Ω.

4-2 TOTAL SERIES RESISTANCE

The first relationship to become familiar with is the total equivalent resistance offered to the source by these resistors. *The total equivalent resistance offered by two or more resistors in series is the sum of the individual values.* Total resistance is symbolized by R_t. Therefore,

$$R_t = R_1 + R_2 + R_3 + \ldots + R_n,$$

for as many resistors as the series circuit contains. For the circuit in Figure 4-2, $R_t = R_1 + R_2 = 12 + 12 = 24$ Ω.

Insofar as the battery is concerned, it provides the same value of current as it would if a single 24-Ω resistor were used in place of the two 12-Ω resistors. See Figure 4-2b.

4-3 TOTAL CIRCUIT CURRENT

The total circuit current for the circuit in Figure 4-2a, or for any series circuit, is determined by the total applied voltage V_s and the total effective resistance R_t. In the example,

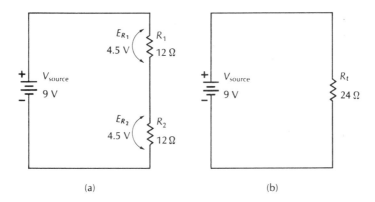

Figure 4-2 (a) A typical series circuit; (b) the equivalent circuit.

$$I_t = \frac{V_s}{R_t} = \frac{9}{24} = 0.375 \text{ A}.$$

For the general case, with any number of resistors,

$$I_t = \frac{V_s}{R_1 + R_2 + R_3 + \ldots + R_n} = \frac{V_s}{R_t},$$

for as many resistors as are in the series circuit.

An important fact regarding series circuits is that, no matter where in the circuit the current is measured, its value is identical with the value measured anywhere else. Figure 4-3 illustrates this point, using five current meters at various points in the circuit.

The reason that the current has the same value no matter where measured in a series circuit is that the number of electrons passing any point in the circuit per unit time must be the same. In a series circuit, it is impossible to have, for example, 1 C/s flowing in one part of the circuit and 2 C/s flowing in another part.

Note in Figure 4-3 how the milliameters are connected in the circuit. As mentioned in Section 3-7, to measure current it is necessary to break the circuit and then insert the meter, observing polarity. The meter completes the circuit and current flows through the meter, indicating the value on the meter face.

4-4 VOLTAGE DROPS

In the circuit of Figure 4-2a, the total applied voltage is 9 V. However, if a voltmeter were placed across R_1, for example, it would indicate 4.5 V. Similarly, the voltage across R_2 would also be 4.5 V. These results lead us to an important definition. A voltage across a battery or other

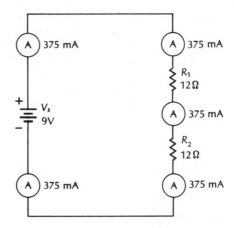

Figure 4-3 Current has the same value anywhere in a series circuit.

generating device is known as a *source* voltage, as previously mentioned. A voltage across a *load*, any device that consumes electrical power, is a *voltage drop*. This name derives from the fact that the voltage across a load is always less than the applied voltage if there are two or more series loads. Thus, as one considers various points around the circuit, the voltage "drops" or decreases. In this book, we shall differentiate between sources and drops by V and E, respectively. Hence, the battery is labeled V_s and the voltage drops across each resistor, E_{R_1} and E_{R_2}.

Another fact can be deduced from the preceding comment. The applied voltage is 9 V, while each of the voltage drops is 4.5 V. Evidently, *the sum of the voltage drops is equal to the applied voltage.* In any series circuit, this rule holds true exactly. For the circuit of the present example, this relationship is expressed as:

$$V_s = E_{R_1} + E_{R_2} = 4.5 + 4.5 = 9.0 \text{ V}.$$

This expression is true no matter how many resistors are in the series configuration.

In the circuit example of Figure 4-2, the voltage drops were given in the text. But if they were unknown they could easily be computed. In analyzing such a circuit the first step is to find the total resistance:

$$R_t = R_1 + R_2 = 12 + 12 = 24 \text{ }\Omega.$$

Next, the total current is determined:

$$I_t = \frac{V_s}{R_1 + R_2} = \frac{9}{24} = 0.375 \text{ A} = 375 \text{ mA}.$$

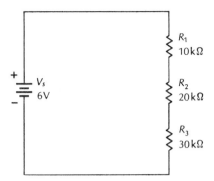

Figure 4-4 A series circuit with three resistors.

Only now can the individual voltage drops be found. Ohm's law is just as valid for this circuit as for any other. Known values are resistance and current, and to find the voltage across each resistor, substitute values as follows.

$$E_{R_1} = I_t \times R_1 = 0.375(12) = 4.5 \text{ V.}$$
$$E_{R_2} = I_t \times R_2 = 0.375(12) = 4.5 \text{ V.}$$

Note that both voltage drops are equal. Inspecting the two equations just above reveals why. Since, in a series circuit, current is the same throughout, if the resistors have the same value, the resulting voltage drops must be the same.

As will shortly be demonstrated, when the resistances are *not* equal in value, the voltage drops will also *not* be equal. Again, this follows from inspection of the voltage drop equation. Another name for the voltage drop is the *IR drop*. This, of course, derives from the equation $E = IR$.

A slightly more complex circuit is shown in Figure 4-4. This circuit includes three resistors, each of different value. The analytical procedure for this circuit is essentially the same as for the earlier circuit. First, determine the total resistance.

$$R_t = R_1 + R_2 + R_3 = 10,000 + 20,000 + 30,000 = 60,000 \text{ }\Omega.$$

Now, find the total current.

$$I_t = \frac{V_s}{R_1 + R_2 + R_3} = \frac{6}{60,000} = 0.0001 \text{ A} = 0.1 \text{ mA.}$$

The individual *IR* drops across each resistor can now be determined:

$$E_{R_1} = I_t \times R_1 = 0.0001(10,000) = 1 \text{ V;}$$
$$E_{R_2} = I_t \times R_2 = 0.0001(20,000) = 2 \text{ V;}$$
$$E_{R_3} = I_t \times R_3 = 0.0001(30,000) = 3 \text{ V.}$$

Finally, verify results:

$$V_s = E_{R_1} + E_{R_2} + E_{R_3} = 1 + 2 + 3 = 6 \text{ V.}$$

Notice a significant fact from the three voltage drops just calculated. *The voltage drop across each resistor is proportional to the resistor ratio.* That is, with a 10-kΩ resistor, which represents $1/6$ of the total resistance, $1/6$ of the total voltage appears across this resistor. Similarly, the 20-kΩ resistor, which represents $2/6$ or $1/3$ of the total resistance develops a drop of $2/6 = 2$ V. Finally, the 30-kΩ resistor represents $3/6$ or $1/2$ of the total resistance, and $1/2 \times 6 = 3$ V is the voltage drop appearing across it.

In a series circuit, the larger the value of one resistor, compared to the others, the larger the voltage drop appearing across this resistor.

4-5 VOLTAGE-DROP POLARITY

When current flows through a resistor, a voltage drop occurs across the resistor. A voltage drop implies a *difference of potential* across the resistor. Since this is true, there must also be a polarity associated with the voltage drop.

Figure 4-5 illustrates the polarity of the voltage drops in a three-resistor series circuit. Three voltmeters are connected to measure the voltage across the individual resistors. Note the polarity signs adjacent to the resistors. These indicate the direction, or polarity, of each voltage drop. As might be anticipated, the end of the resistor into which

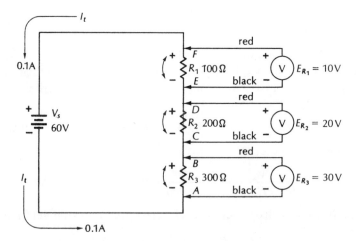

Figure 4-5 A series circuit illustrating voltage-drop polarity.

the electrons enter is the negative end, while electrons emerge from the resistor at its positive end. The same is true for all three resistors.

Note particularly the upper end of R_3 and the lower end of R_2. The same wire carries two different polarity signs! This must not confuse the reader. The plus sign at the top of R_3 is *referenced* to the minus sign *at the lower end of R_3*, not to the one adjacent to it. That is, these signs *must be considered in pairs,* as indicated by the dual-headed arrow. Because the connection between R_3 and R_2 consists of a short piece of wire, which has essentially zero resistance, the upper end of R_3 and the lower end of R_2 are at the same potential relative to any other point in the circuit.

When measuring voltage across the resistors in a series circuit, the reading can be interpreted in one of two ways. Referring to Figure 4-5, note the meter across R_3. It indicates a voltage drop of 30 V. One way of interpreting this reading is to simply state that "the voltage drop across R_3 is 30 volts." Alternatively, one can say, "Point B is 30 volts more positive than point A," or, "Point A is 30 volts more negative than point B."

The first statement specifies only the magnitude of the voltage drop, whereas the second pair of statements specify not only the magnitude, *but also the polarity.* In these statements, one of the two points under measurement *is a reference point.* For example, in the statement, "Point B is 30 volts more positive than point A," point A is the reference used to specify the voltage at the other point in the circuit. There must *always* be two circuit points to be considered when measuring voltage. To say, "Point B is +30 volts," is meaningless without giving a reference point. This is easily proved by touching only one lead of the meter to the circuit point, leaving the other one touching nothing. The meter will give no reading at all. By definition, voltage is a difference of potential, which implicitly states that one point is compared to another. (Reference voltages will be dealt with in detail in Section 4-8.)

The meter connected between points A and B measures the voltage drop across R_3, with A more negative than B. The meter connected between points C and D measures the drop across R_2, with C more negative than D. Finally, the meter connected between points E and F measures the drop across R_1, with point E more negative than F. Note that if point A is more negative than point B, it could also be said that point B is more positive than point A, and so forth.

It should be noted at this point that a circuit such as the one shown in Figure 4-5 is often called a *voltage divider.* This derives from the fact that the applied voltage is *divided by* the series resistors in direct proportion to the resistors' values. Voltage dividers are used extensively in electronic circuits to provide voltages and currents of virtually any value from a single power source.

4-6 ANALYZING SERIES CIRCUITS

As will be discussed in later chapters, there is more than one way of analyzing electrical circuitry. We shall at this point, however, follow the simplest, most straightforward method. In computing unknown values for a circuit such as that shown in Figure 4-5, the first step is to note the known and unknown values. Nearly all values are shown on the drawing, but we shall assume that the only values given are the resistor values and the applied voltage. Hence, the individual voltage drops are unknown, as is total current. If power dissipation is to be determined, this too will be unknown. The following procedural steps will illustrate the most straightforward method of calculating the unknowns for this circuit.

Step 1. Find R_t:
$$R_t = R_1 + R_2 + R_3 = 100 + 200 + 300 = 600 \, \Omega.$$

Step 2. Find I_t:
$$I_t = \frac{V_s}{R_t} = \frac{60}{600} = 0.1 \, \text{A} = 100 \, \text{mA}.$$

Step 3. Find E_{R_1} (Note: It makes no difference which resistor value is computed first.):
$$E_{R_1} = I_t \times R_1 = 0.1(100) = 10 \, \text{V}.$$

Step 4. Find E_{R_2}:
$$E_{R_2} = I_t \times R_2 = 0.1(200) = 20 \, \text{V}.$$

Step 5. Find E_{R_3}:
$$E_{R_3} = I_t \times R_3 = 0.1(300) = 30 \, \text{V}.$$

Step 6. Verify voltage drops:
$$V_s = E_{R_1} + E_{R_2} + E_{R_3} = 10 + 20 + 30 = 60 \, \text{V}.$$

Step 7. Find P_{R_1}:
$$P_{R_1} = I_t \times E_{R_1} = 0.1(10) = 1 \, \text{W}.$$

Step 8. Find P_{R_2}:
$$P_{R_2} = I_t \times E_{R_2} = 0.1(20) = 2 \, \text{W}.$$

Step 9. Find P_{R_3}:
$$P_{R_3} = I_t \times E_{R_3} = 0.1(30) = 3 \, \text{W}.$$

Step 10. Find P_t:
$$P_t = P_{R_1} + P_{R_2} + P_{R_3} = 1 + 2 + 3 = 6 \, \text{W}.$$

4-7 MULTIPLE POWER SOURCES

In many applications of electronic circuits more than one value of voltage is provided, either by separate batteries, or by ac to dc converters for operation with the ac line. In either case, the way this influences the circuits must be investigated.

Two such circuits are shown in Figure 4-6. Figure 4-6a illustrates a circuit in which the two sources are said to be connected in *series-aiding*. This simply means that the total voltage is the *sum* of the battery voltages. Hence, $V_t = V_{s1} + V_{s2} = 3 + 6 = 9$ V, and a 9-V drop will occur across R_1. The total current I_t is a function of V_t and R_1.

$$I_t = \frac{V_t}{R_1} = \frac{V_{s1} + V_{s2}}{R_1} = \frac{9}{100} = 0.09 \text{ A} = 90 \text{ mA}.$$

Figure 4-6b illustrates the *series-opposing* case, where the total applied voltage (V_t) is the *difference* between the two sources:

$$V_t = V_{s2} - V_{s1} = 6 - 3 = 3 \text{ V}.$$

In this instance the total current (I_t) is determined by the effective 3-V source and the 100-Ω resistor.

$$I_t = \frac{V_t}{R_1} = \frac{V_{s2} - V_{s1}}{R_1} = \frac{3}{100} = 0.03 \text{ A}.$$

In series-opposing circuits, the larger (in terms of voltage) of the two sources determines the direction of current flow. In Figure 4-6b, therefore, current flows in the direction dictated by V_{s2} (clockwise). This results in a voltage drop across R_1 as shown. If the two batteries had *equal* voltages, no current would flow and there would be no voltage drop across R_1.

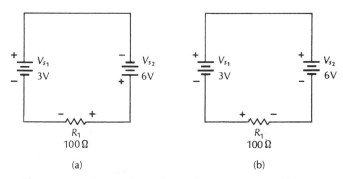

Figure 4-6 (a) Series-aiding voltages; (b) series-opposing voltages.

4-8 REFERENCE VOLTAGE

As mentioned, the polarity of each voltage drop is important and must be kept in mind when measuring the voltage in the circuit or when calculating for it. Also mentioned earlier is the fact that when measuring a voltage between any two points, *one of the two is a reference point*.

For example, in Figure 4-7, meter V_3 is shown measuring the voltage across the voltage divider string. Assuming no voltage drop across the wires, this can be considered to be either the source voltage or the total drop across the loads. Note that the black lead (−) goes to the most negative point in the circuit, while the red lead (+) goes to the most positive. Thus, the meter will read *upscale* a value of 10 V.

This reading can be interpreted in one of two ways: (1) point A is 10 V more positive than point B or (2) point B is 10 V more negative than point A. In the first instance, point B is the reference (zero volts) point, and we say that point A is measured in reference to point B. In the second instance, point A is the reference point, and we say that point B is measured in reference to (or with respect to) point A.

Now, meter V_1 is shown measuring the voltage drop across R_1. Point A is therefore 5 V more positive than point C (C is reference) or conversely, point C is 5 V more negative than point A (A is reference). Similarly, V_2 indicates that point C is 5 V more positive than point B (B is reference) or that point B is 5 V more negative than point C (C is reference).

To determine the intermediate value of voltage at any point in such a voltage divider, first determine the voltage drop across each resistor.

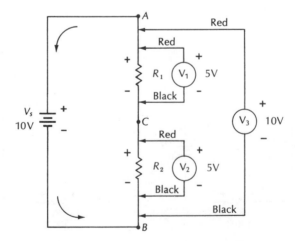

Figure 4-7 Voltage drop and meter polarity.

Then, assign a positive (+) value for each source and a negative (−) value for each voltage drop. For example, in Figure 4-7, assume that point *B* is reference (zero volts). To find the voltages at any intermediate point in the circuit, start at zero volts and progress around the circuit, *adding* source voltages and *subtracting* voltage drops:

$$0 \text{ V} + 10 \text{ V} - 5 \text{ V} - 5 \text{ V}.$$

Hence, point *A* is +10 V relative to *B;* point *C* is +5 V (+10 − 5); point *B* is 0 V (+10 − 5 − 5). To use this simplified method you must proceed around the circuit in a minus-to-plus direction through the sources and a plus-to-minus direction through the loads.

As a further example, assume point *C* is reference. Then, starting at point *C,* the algebraic sum of the source and drops is

$$0 \text{ V} - 5 \text{ V} + 10 \text{ V} - 5 \text{ V}.$$

Hence, point *B* is −5 V relative to reference. Point *A* is +5 V relative to reference.

4-9 GROUND

Another name for reference is *ground,* sometimes referred to as *common*. All three of these terms (reference, ground, common) mean the same thing. All are simply names for the point in the circuit under measurement which is to be considered zero volts. Note, however, that the earth is considered the ideal ground, and equipment is often connected to earth ground by a large conducting pipe driven into damp earth. Inside electrical equipment there is a common *bus,* or wire, that is zero volts. We shall consider this to be ground, even though it may not be at true earth potential.

In order to show the reference point on a schematic diagram, a special symbol is used. Figure 4-8 illustrates the same circuit used previously, but using the ground, or common, symbol. This symbol simply indicates the points in the circuit that are to be considered to be at a level of zero volts. If this symbol is used at more than one point, it also indicates that these points are connected, either with actual wires or by a metal chassis or bus-bar lines on a printed-circuit board. Note that, as indicated in the figure, other symbols exist for ground. You will learn specific differences for these symbols as your studies progress.

Figure 4-9 illustrates another possible circuit configuration using the ground symbol. Here, ground (or zero-volt reference) is placed at an intermediate point that is not one of the power-supply terminals. This may seem at first to greatly complicate the circuit action, but this circuit is no more complex than any other we have dealt with. How-

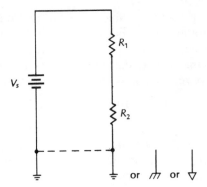

Figure 4-8 A series-circuit schematic including the ground symbol. Note the alternate symbols.

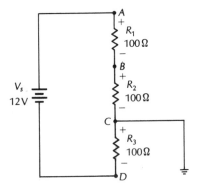

Figure 4-9 A voltage divider with intermediate ground.

ever, as we will find, there are some additional simple considerations to be made.

If the junction between R_2 and R_3 is to be considered zero volts, then all other voltages are measured with respect to this junction. Point C, then, is the reference for all voltage measurements and calculations. Because there are three other points labeled in the circuit, there are three other voltages to be found.

This circuit is analyzed as is any other series circuit, but the results are slightly different. For example, note point D. Knowing the direction of current flow, it can easily be seen that the voltage at D must be more negative than ground (point C). On the other hand, points A and B must have a value more positive than ground, using the same reasoning.

To determine the unknown values, first determine the total current.

$$I_s = \frac{V_s}{R_1 + R_2 + R_3} = \frac{12}{300} = 0.04 = 40 \text{ mA}.$$

The voltage drop across each resistor can now be found.

$$E_{R_1} = I_s \times R_1 = 0.04(100) = 4 \text{ V}.$$

Because each resistor has the same value and the current through each is identical, the remaining voltage drops must also be 4 V. Now, note the polarity signs on the diagram. These will allow the voltages at the various points around the circuit to be determined in reference to ground. Point A is 8 V more positive than ground, while point B is 4 V more positive than ground. Point D, however, is 4 V more negative than ground. Circuits such as these are quite common in electronic equipment.

Using the simplified method to find the voltages at points A, B, and D, we start at point C and move clockwise:

$$0 \text{ V} - 4 \text{ V} + 12 \text{ V} - 4 \text{ V} - 4 \text{ V} = 0 \text{ V}.$$
$$E_A = -4 + 12 = +8 \text{ V}.$$
$$E_B = -4 + 12 - 4 = +4 \text{ V}.$$
$$E_D = 0 - 4 = -4 \text{ V}.$$
$$E_C = -4 + 12 - 4 - 4 = 0 \text{ V}.$$

4-10 CIRCUITS WITH VARIOUS UNKNOWNS

In the circuits used for problems thus far, solving for all required quantities was quite straightforward. Often, however, in a practical instance, solutions are not quite this easy. As an example, see Figure 4-10. At first it seems impossible to solve this circuit for all unknowns. No information is given at all for R_3, and only the resistance value is given for R_1. Also given are the resistance and voltage drop for R_2 and the applied voltage.

To solve for all unknowns in a circuit such as this, find at least one place where an unknown value *can* be determined. At this point, R_1 and R_3 cannot be used to find any values, nor can V_s. However, the voltage across R_2 and the resistance of R_2 can be used to determine the value of current through R_2. Since this is I_t, this will make a good starting point.

$$I_{R_2} = I_t = \frac{E_{R_2}}{R_2} = \frac{4.5}{300} = 0.015 \text{ A}.$$

Since we know the value of R_1, the voltage drop across it can be found now that total current has been calculated.

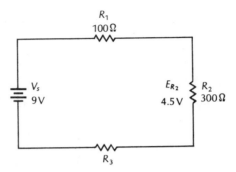

Figure 4-10 Solving for unknowns in this series circuit is slightly more complex; see Section 4-10.

$$E_{R_1} = I_t \times R_1 = 0.015(100) = 1.5 \text{ V}.$$

Because the sum of the individual voltage drops must equal the applied voltage, the voltage across R_3 can now be found.

$$V_s = E_{R_1} + E_{R_2} + E_{R_3} \text{ and } E_{R_3} = V_s - (E_{R_1} + E_{R_2})$$
$$= 9 - (4.5 + 1.5) = 3 \text{ V}.$$

If the current through a resistor and the voltage drop across a resistor are known, its resistance is easily calculated. Thus,

$$R_3 = \frac{E_{R_3}}{I_t} = \frac{3}{0.015} = 200 \text{ }\Omega.$$

Thus, except for power dissipation, all unknown values are found by the use of Ohm's law and a little deductive reasoning.

A similar circuit is given in Figure 4-11. Here, total current is known, as well as E_{R_3}, R_2, and R_1. Unknown are V_s, E_{R_1}, and E_{R_2}. One place in the circuit where there are two known values is at R_3. Its voltage drop and the current through it are known, so its resistance value can be determined.

$$R_3 = \frac{E_{R_3}}{I_t} = \frac{6}{0.02} = 300 \text{ }\Omega.$$

Also known are the values of R_1 and R_2, as well as the current through them. The voltage drop across each is therefore easily found.

$$E_{R_1} = I_t \times R_1 = 0.02(200) = 4 \text{ V}.$$
$$E_{R_2} = I_t \times R_2 = 0.02(500) = 10 \text{ V}.$$

Now V_s can be determined:

$$V_s = E_{R_1} + E_{R_2} + E_{R_3} = 4 + 10 + 6 = 20 \text{ V}.$$

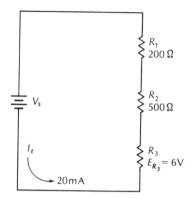

Figure 4-11 Finding the unknowns in this series circuit requires careful analysis.

4-11 OPEN AND SHORT CIRCUITS

In practical series circuits, a short circuit or an open circuit may occur, in which case the circuit will malfunction. Figure 4-12a illustrates a short circuit and Figure 4-12b illustrates an open circuit. Each causes the circuit to malfunction in a different manner.

In effect, the short circuit removes one (or more) of the resistors from the remainder and substitutes a conductor, the resistance of which is essentially zero. As an example, if $R_1 = R_2 = R_3 = 100 \, \Omega$, the total resistance should be 300 Ω. However, when the short circuit develops (perhaps due to twisted bare wires or a drop of solder) the total resistance drops to 200 Ω with the short circuit as shown. This allows a greater drop across the remaining two resistors and more current will flow. More voltage drop and more current result in greater power dissipation, and R_1 and R_3 *may* overheat, depending on their wattage rating and other factors.

An open in a series circuit has just the opposite result. In this instance, essentially no current will flow, assuming V_s is a moderate value and the resistive path between the two open points is dry air or another insulating material. With no current flow anywhere in the circuit, there is no voltage drop across the three resistors. A sensitive voltmeter, however, would indicate the full supply voltage appearing across the open. This is not inconsistent with normal series circuit action, since the greatest voltage will drop across the larger resistance value. The open circuit represents virtually an infinite resistance; hence by comparison the others are of insignificant value.

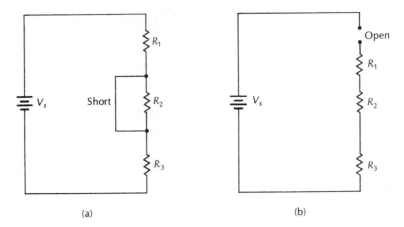

Figure 4-12 (a) A short circuit; (b) an open circuit.

PRACTICE PROBLEMS

The following problems are designed to provide a strengthening of series-circuit problem solving. Solve each problem in detail and check your answer against the answer given.
For the following problems, refer to Figure 4-13a.

1. $R_1 = 470 \, \Omega$, $R_2 = 560 \, \Omega$, $V_s = 20.6$ V; find I_t and E_{R_1}.
2. $R_1 = 1 \times 10^3 \, \Omega$ and $R_2 \times 0.5 \times 10^3 \, \Omega$. If $V_s = 15$ V, find the values of I_t, E_{R_1}, and E_{R_2}.
3. $R_1 = 1 \times 10^5 \, \Omega$ and $R_2 = 1 \times 10^4 \, \Omega$. Total current is 0.1×10^{-4} A. Find V_s.
4. $R_1 = 33$ kΩ and $R_2 = 22$ kΩ. $I_t = 5.5$ mA. Find V_s.

For the following problems, refer to Figure 4-13b.

5. $R_1 = 220 \, \Omega$, $R_2 = 470 \, \Omega$, $R_3 = 330 \, \Omega$, $V_s = 5.1$ V; find I_t.
6. $R_1 = 4700 \, \Omega = R_2 = R_3$. $V_s = 6$ V. Find E_{R_2}.
7. $I_t = 0.015$ A and $V_s = 30$ V. All resistors are of the same value. Find the value of R_1.
8. $I_t = 1 \times 10^{-2}$ A, $R_1 = 100 \, \Omega$, and $R_3 = 300 \, \Omega$. If the drop across $R_2 = 2$ V, find the values of V_s and R_2.
9. $R_1 = 400 \, \Omega$, $R_2 = 600 \, \Omega$, and $R_3 = 500 \, \Omega$. Total current $= 5$ mA and point C is ground. Find V_s and the voltage at B with respect to ground.
10. $R_1 = 1.5 \times 10^3 \, \Omega$, $R_2 = 2.2 \times 10^3 \, \Omega$, and $R_3 = 0.13 \times 10^4 \, \Omega$. If $V_s = 5$ V, find I_t and the voltage at point B with respect to point C.

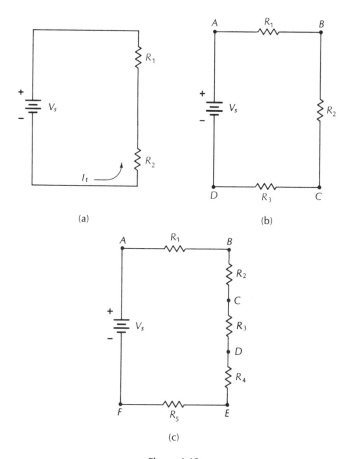

Figure 4-13

For the following problems, refer to Figure 4-13c.

11. All resistors are 470 Ω and V_s is 4.7 V. Find the value of voltage at point B if E is ground.
12. All resistors are 560 Ω and V_s is 8.4 V. A is ground. Find the value of voltage at point D.

Answers: (1) $I_t = 0.02$ A, $E_{R_1} = 9.4$ V; (2) $I_t = 1.0$ mA, $E_{R_1} = 10$ V, $E_{R_2} = 5$ V; (3) $V_s = 1.1$ V; (4) $V_s = 302.5$ V; (5) $I_t = 5$ mA; (6) $E_{R_2} = 2$ V; (7) $R_1 = 667$ Ω; (8) $V_s = 6$ V, $R_2 = 200$ Ω; (9) $V_s = 7.5$ V, $E_B = +3$ V; (10) $I_t = 1 \times 10^{-3}$ A, $E_{BC} = +2.2$ V; (11) $E_B = +2.82$ V; (12) $E_D = -5.04$ V.

SUMMARY

- Total series resistance is the sum of the individual resistance values.
- Total series circuit current equals total voltage divided by total series resistance and has the same value throughout the circuit.
- The sum of the individual voltage drops around a series circuit equals the applied voltage.
- The value of the voltage drop across one resistor in a string of series resistors is proportional to the ratio of its resistance value to the total circuit resistance.
- The polarity of the voltage drop across one resistor in a string of series resistors is negative at the end where electrons enter and positive where they emerge.
- Multiple power sources may be connected in a series-aiding or series-opposing configuration.
- The total voltage when sources are connected in series-aiding is the sum of the individual source voltages.
- The total voltage when sources are connected in series-opposing is the difference in their individual source voltages.
- When relating to a voltage across two points in the circuit, one point is considered the reference; the other point is measured in reference to the first.
- Ground is a common point in a circuit, often the metal chassis or equivalent, that is considered to be at zero potential.
- An open in a series circuit reduces current to zero.
- A short in a series circuit causes excessive current flow.

QUESTIONS

1. Briefly define a series circuit in terms of current flow.
2. Briefly define a series circuit consisting of three resistive loads in terms of voltage sources and drops.
3. Total series resistance is the (sum, difference, product) of all resistance values.
4. The sum of the voltage drops in a series circuit must equal the _____ .
5. The total circuit current value (is different, is the same) wherever measured in a series circuit.
6. A milliameter (may, may not) be inserted in a series circuit without regard to polarity.
7. Briefly define a voltage drop.
8. True or false: A voltage drop is determined solely by the current.

9. The polarity of the voltage drop (is, is not) a function of the direction of current.
10. Briefly define a voltage divider.
11. A voltage divider (is, is not) often used to reduce the supply voltage.
12. Give two other names for ground.
13. The point in a circuit that is called ground is usually at _____ volts.
14. Briefly explain the effect(s) of a short circuit in a series circuit.

PROBLEMS

For the following problems, refer to Figure 4-14.
1. $V_s = 6$ V, $R_1 = 10$ Ω, $R_2 = 10$ Ω; find I_t.
2. $V_s = 6$ V, $R_1 = 20$ Ω, $R_2 = 20$ Ω; find I_t.
3. $V_s = 6$ V, $R_1 = 5$ Ω, $R_2 = 15$ Ω; find I_t.
4. $V_s = 6$ V, $R_1 = 15$ Ω, $R_2 = 25$ Ω; find I_t.
5. $V_s = 60$ V, $R_1 = 100$ Ω, $R_2 = 100$ Ω; find I_t.
6. $V_s = 60$ V, $R_1 = 150$ Ω, $R_2 = 250$ Ω; find I_t.
7. $V_s = 12$ V, $R_1 = 35$ Ω, $R_2 = 85$ Ω; find I_t.
8. $V_s = 12$ V, $R_1 = 45$ Ω, $R_2 = 75$ Ω; find I_t.
9. $V_s = 5$ V, $R_1 = 25$ Ω, $R_2 = 50$ Ω; find I_t.
10. $V_s = 15$ V, $R_1 = 35$ Ω, $R_2 = 10$ Ω; find I_t.
11. $V_s = 24$ V, $R_1 = 50$ kΩ, $R_2 = 22$ kΩ, find I_t.
12. $V_s = 24$ V, $R_1 = 100$ kΩ, $R_2 = 20$ kΩ; find I_t.

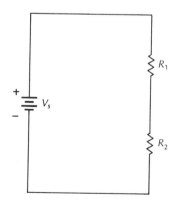

Figure 4-14

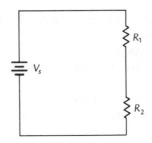

Figure 4-15

13. $V_s = 6$ V, $R_1 = 150$ kΩ, $R_2 = 150$ kΩ; find I_t.
14. $V_s = 6$ V, $R_1 = 33$ kΩ, $R_2 = 20$ kΩ; find I_t.
15. $V_s = 5$ V, $R_1 = 150$ kΩ, $R_2 = 100$ kΩ; find I_t.
16. $V_s = 5$ V, $R_1 = 12$ kΩ, $R_2 = 18$ kΩ; find I_t.

For the following problems, refer to Figure 4-15.

17. $V_s = 4$ V, $R_1 = 10$ Ω, $R_2 = 30$ Ω. Determine the value of: (a) R_t, (b) I_t, (c) E_{R_1}, (d) E_{R_2}, (e) P_{R_1}, (f) P_{R_2}, (g) P_t.
18. $V_s = 10.3$ V, $R_1 = 470$ Ω, $R_2 = 560$ Ω. Determine the value of: (a) R_t, (b) I_t, (c) E_{R_1}, (d) E_{R_2}, (e) P_{R_1}, (f) P_{R_2}, (g) P_t.
19. $V_s = 12$ V, $R_1 = 10$ kΩ, $R_2 = 22$ kΩ. Determine the value of: (a) R_t, (b) I_t, (c) E_{R_1}, (d) E_{R_2}, (e) P_{R_1}, (f) P_{R_2}, (g) P_t.
20. $V_s = 80$ V, $R_1 = 470$ Ω, $R_2 = 330$ Ω. Determine the value of: (a) R_t, (b) I_t, (c) E_{R_1}, (d) E_{R_2}, (e) P_{R_1}, (f) P_{R_2}, (g) P_t.
21. $V_s = 110$ V, $R_1 = 88$ kΩ, $R_2 = 22$ kΩ. Determine the value of: (a) R_t, (b) I_t, (c) E_{R_1}, (d) E_{R_2}, (e) P_{R_1}, (f) P_{R_2}, (g) P_t.
22. $V_s = 12$ V, $R_1 = 18$ kΩ, $R_2 = 22$ kΩ. Determine the value of: (a) R_t, (b) I_t, (c) E_{R_1}, (d) E_{R_2}, (e) P_{R_1}, (f) P_{R_2}, (g) P_t.

For the following problems, refer to Figure 4-16.

23. $V_s = 6$ V, $R_1 = 10$ Ω, $R_2 = 20$ Ω, $R_3 = 30$ Ω. Determine the value of: (a) R_t, (b) I_t, (c) E_{R_1}, (d) E_{R_2}, (e) E_{R_3}, (f) P_{R_1}, (g) P_{R_2}, (h) P_{R_3}, (i) P_t.
24. $V_s = 12$ V, $R_1 = 1$ kΩ, $R_2 = 2$ kΩ, $R_3 = 3$ kΩ. Determine the value of: (a) R_t, (b) I_t, (c) E_{R_1}, (d) E_{R_2}, (e) E_{R_3}, (f) P_{R_1}, (g) P_{R_2}, (h) P_{R_3}, (i) P_t.
25. $V_s = 18$ V, $R_1 = 1500$ Ω, $R_2 = 3000$ Ω, $R_3 = 4500$ Ω. Determine the value of: (a) R_t, (b) I_t, (c) E_{R_1}, (d) E_{R_2}, (e) E_{R_3}, (f) P_{R_1}, (g) P_{R_2}, (h) P_{R_3}, (i) P_t.

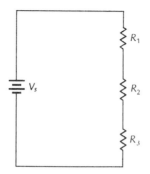

Figure 4-16

26. $V_s = 24$ V, $R_1 = 1500$ Ω, $R_2 = 3000$ Ω, $R_3 = 4500$ Ω. Determine the value of: (a) R_t, (b) I_t, (c) E_{R_1}, (d) E_{R_2}, (e) E_{R_3}, (f) P_{R_1}, (g) P_{R_2}, (h) P_{R_3}, (i) P_t.
27. $V_s = 6$ V, $R_1 = 200$ Ω, $R_2 = 300$ Ω, $R_3 = 400$ Ω. Determine the value of: (a) R_t, (b) I_t, (c) E_{R_1}, (d) E_{R_2}, (e) E_{R_3}, (f) P_{R_1}, (g) P_{R_2}, (h) P_{R_3}, (i) P_t.
28. $V_s = 12$ V, $R_1 = 2000$ Ω, $R_2 = 3000$ Ω, $R_3 = 4000$ Ω. Determine the value of: (a) R_t, (b) I_t, (c) E_{R_1}, (d) E_{R_2}, (e) E_{R_3}, (f) P_{R_1}, (g) P_{R_2}, (h) P_{R_3}, (i) P_t.
29. $V_s = 18$ V, $R_1 = 33$ kΩ, $R_2 = 47$ kΩ, $R_3 = 68$ kΩ. Determine the value of: (a) R_t, (b) I_t, (c) E_{R_1}, (d) E_{R_2}, (e) E_{R_3}, (f) P_{R_1}, (g) P_{R_2}, (h) P_{R_3}, (i) P_t.
30. $V_s = 24$ V, $R_1 = 33$ kΩ, $R_2 = 47$ kΩ, $R_3 = 68$ kΩ. Determine the value of: (a) R_t, (b) I_t, (c) E_{R_1}, (d) E_{R_2}, (e) E_{R_3}, (f) P_{R_1}, (g) P_{R_2}, (h) P_{R_3}, (i) P_t.

For the following problems, refer to Figure 4-17.

31. $V_s = 6$ V, $R_1 = 10$ kΩ, $R_2 = 20$ kΩ, $R_3 = 30$ kΩ. Determine the value of: (a) R_t, (b) I_t, (c) E_{R_1}, (d) E_{R_2}, (e) E_{R_3}, (f) E_A, (g) E_B, (h) E_C.
32. $V_s = 12$ V, $R_1 = 1$ kΩ, $R_2 = 2$ kΩ, $R_3 = 3$ kΩ. Determine the value of: (a) R_t, (b) I_t, (c) E_{R_1}, (d) E_{R_2}, (e) E_{R_3}, (f) E_A, (g) E_B, (h) E_C.
33. $V_s = 12$ V, $R_1 = 4.7$ kΩ, $R_2 = 6.8$ kΩ, $R_3 = 9.1$ kΩ. Determine the value of: (a) R_t, (b) I_t, (c) E_{R_1}, (d) E_{R_2}, (e) E_{R_3}, (f) E_A, (g) E_B, (h) E_C.
34. $V_s = 24$ V, $R_1 = 1.8$ kΩ, $R_2 = 3.3$ kΩ, $R_3 = 2.2$ kΩ. Determine the value of: (a) R_t, (b) I_t, (c) E_{R_1}, (d) E_{R_2}, (e) E_{R_3}, (f) E_A, (g) E_B, (h) E_C.

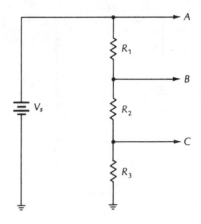

Figure 4-17

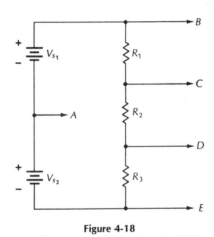

Figure 4-18

For the following problems, refer to Figure 4-18.

35. $V_{s_1} = 6$ V, $V_{s_2} = 12$ V, $R_1 = 10{,}000$ Ω, $R_2 = 12{,}000$ Ω, $R_3 = 14{,}000$ Ω. a. Point A is ground; determine the value of E_C, E_D. b. Point B is ground; determine the value of E_C, E_D. c. Point C is ground; determine the value of E_B, E_D.

36. $V_{s_1} = 3$ V, $V_{s_2} = 6$ V, $R_1 = 1$ kΩ, $R_2 = 1.2$ kΩ, $R_3 = 1.4$ kΩ. a. Point A is ground; determine the value of E_C, E_D. b. Point B is ground; determine the value of E_C, E_D. c. Point C is ground; determine the value of E_B, E_D.

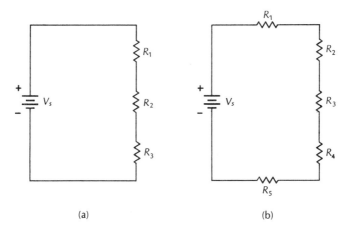

Figure 4-19

For the following problems, refer to Figure 4-19a.

37. $R_2 = 500\ \Omega$, $E_{R_1} = 7.0$ V, $I_{R_3} = 0.01$ A, and $V_s = 18$ V. Find R_1, R_3, E_{R_2}, E_{R_3}, and R_t.
38. $R_1 = 4.7\ \text{k}\Omega$, $E_{R_2} = 16.5$ V, $I_{R_3} = 5$ mA, and $V_s = 51$ V. Find R_2, R_3, E_{R_1}, E_{R_3}, R_t.

For the following problems, refer to Figure 4-19b.

39. $R_1 = 10{,}000\ \Omega$, $E_{R_2} = 2$ V, $I_{R_3} = 0.1$ mA, $R_4 = 40{,}000\ \Omega$; $E_{R_5} = 5$ V, and $V_s = 15$ V. Find E_{R_1}, R_2, E_{R_4}, R_5, R_t, R_3, and E_{R_3}.
40. $E_{R_1} = 2$ V, $R_2 = 20{,}000\ \Omega$, $I_{R_3} = 0.2$ mA, $E_{R_3} = 6.0$ V, $R_4 = 40{,}000\ \Omega$, and $E_{R_5} = 10$ V. Find R_1, E_{R_2}, E_{R_4}, R_5, and V_s.

CHAPTER 5

PARALLEL CIRCUITS

A parallel circuit is illustrated in Figure 5-1. The difference between this and a series circuit is immediately apparent: Each component is individually connected across the source. For this reason, the full supply voltage appears across each resistor, and the value of each resistor determines the current that flows through it. Because parallel circuits are so different from series circuits, different rules apply. We shall therefore investigate the following topics in this chapter:

5-1   Parallel Circuit Voltage
5-2   Parallel Circuit Branch and Total Current
5-3   Total Effective Resistance
5-4   Conductances in Parallel
5-5   Current Dividers
5-6   Power in Parallel Circuits
5-7   Parallel Circuit Problems
5-8   Opens and Shorts in Parallel Circuits

5-1 PARALLEL CIRCUIT VOLTAGE

As is evident from Figure 5-1, the applied voltage appears across each resistor, and therefore the voltage across each component in a parallel circuit is the same. Each of the resistors constitutes a *branch*, so in this particular circuit there are two branch currents, which can be identified as I_{R_1} and I_{R_2}.

Parallel circuits are used in numerous applications. For example, ordinary house wiring uses the parallel connection, because nearly all household appliances require 120 V to operate properly. Also, the electrical circuits of a car are basically of the parallel kind. Many other examples exist, but all have one thing in common—the voltage across all elements is the same.

5-2 PARALLEL CIRCUIT BRANCH AND TOTAL CURRENT

In Figure 5-1c, each resistor draws the amount of current that is determined by the value of the resistor and by the voltage across it. Assuming no voltage drop in the wiring, $E_R = V_s$. For the circuit shown,

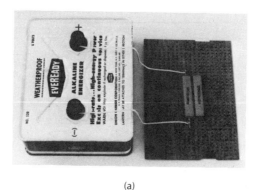

Figure 5-1 (a) A photograph of a parallel connection; (b) one form of schematic for this parallel circuit; (c) the schematic as usually drawn. Electrically, parts b and c are identical.

$$I_{R1} = \frac{E_{R1}}{R_1} = \frac{V_s}{R_1} = \frac{10}{200} = 0.05 \text{ A}.$$

$$I_{R2} = \frac{E_{R2}}{R_2} = \frac{V_s}{R_2} = \frac{10}{200} = 0.05 \text{ A}.$$

Because each resistor in this circuit draws 0.05 A, the source must be providing $0.05 + 0.05 = 0.1$ A. From this we can infer the following rule: *Total circuit current in a parallel circuit is the sum of the branch currents.* In the example given, the two currents are of the same value because the two resistances have the same value. Resistors having unequal values draw values of current that depend upon their specific resistances. In such a case, the total current is still the

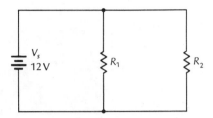

Figure 5-2 A parallel circuit for the examples in text.

sum of the individual values. The following examples will illustrate these points. Refer to Figure 5-2 for these examples.

Example 1. $R_1 = 12\ \Omega, R_2 = 24\ \Omega$; find I_{R_1}, I_{R_2}, I_t.

$$I_{R_1} = \frac{V_s}{R_1} = \frac{12}{12} = 1\text{ A}.$$

$$I_{R_2} = \frac{V_s}{R_2} = \frac{12}{24} = 0.5\text{ A}.$$

$$I_t = I_{R_1} + I_{R_2} = 1 + 0.5 = 1.5\text{ A}.$$

Example 2. $R_1 = 48\ \Omega, R_2 = 96\ \Omega$; find I_{R_1}, I_{R_2}, I_t.

$$I_{R_1} = \frac{V_s}{R_1} = \frac{12}{48} = 0.25\text{ A}.$$

$$I_{R_2} = \frac{V_s}{R_2} = \frac{12}{96} = 0.125\text{ A}.$$

$$I_t = I_{R_1} + I_{R_2} = 0.25 + 0.125 = 0.375\text{ A}.$$

5-3 TOTAL EFFECTIVE RESISTANCE

As was the case with series circuits, a parallel resistance circuit can be replaced with a single equivalent resistance that draws the same amount of current as the original circuit. The value of this equivalent resistance can be determined in a number of ways. The most straightforward of these is outlined below.

Referring to Figure 5-1c again, total equivalent resistance can be determined by first calculating all branch currents. These have previously been found to be 0.5 A each. Hence, total current $I_t = 0.05 + 0.05$ or 0.1 A. Next, substitute voltage and current in the corresponding Ohm's law formula.

$$R_t = \frac{V_s}{I_t} = \frac{10}{0.1} = 100 \text{ } \Omega.$$

Example 3. In Example 1 above, total current was found to be 1.5 A. Therefore,

$$R_t = \frac{V_s}{I_t} = \frac{12}{1.5} = 8 \text{ } \Omega.$$

Example 4. In Example 2, total current was found to be 0.375 A. Therefore,

$$R_t = \frac{V_s}{I_t} = \frac{12}{0.375} = 32 \text{ } \Omega.$$

These examples allow us to formulate a general rule. In Figure 5-1c the two resistors each have a value of 200 Ω, and their equivalent resistance is 100 Ω. This clearly illustrates the following rule: *Two or more resistances of equal value have an equivalent parallel resistance equal to the value of any one resistor divided by the number of resistors.* To apply this rule, *the resistors must have the same value.* In the example,

$$R_t = \frac{R_1}{2} = \frac{200}{2} = 100 \text{ } \Omega.$$

Note specifically that equivalent resistance for resistors of unequal value *cannot* be calculated using this equation. Also, the more resistances added across the source, the greater the value of current and the less the equivalent resistance.

When resistances of differing values are connected in parallel, their combined effective resistance can be determined by the *double reciprocal method*. The total equivalent resistance can be found by the following general relationship:

$$R_t = \frac{1}{(1/R_1) + (1/R_2) + (1/R_3) + \cdots + (1/R_n)}$$

for as many resistances as are connected in parallel. Assume that a 200-Ω resistor and a 300-Ω resistor are connected in parallel. Then

$$R_t = \frac{1}{1/200 + 1/300} = \frac{1}{0.005 + 0.00333} = \frac{1}{0.008333} = 120 \text{ } \Omega.$$

This relationship holds for *any value* resistor and for any number of parallel resistors.

It is interesting and informative to derive this equation. Since total current in a parallel circuit is equal to the sum of the individual branch currents, we can write

$$I_t = I_{R1} + I_{R2} + I_{R3} + \cdots + I_{Rn}.$$

However, $I = E/R$, so E/R can be substituted for I in the original expression.

$$\frac{V_s}{R_t} = \frac{E_{R1}}{R_1} + \frac{E_{R2}}{R_2} + \frac{E_{R3}}{R_3} + \cdots + \frac{E_{Rn}}{R_n}.$$

Because the voltage symbols (V_s, E_{R1}, E_{R2}, and so on) represent the same value, each term is divided by total voltage V_s, which yields

$$\frac{1}{R_t} = \frac{1}{R_1} + \frac{1}{R_2} + \frac{1}{R_3} + \cdots + \frac{1}{R_n}.$$

To clear $1/R_t$, take the reciprocal of all members to give

$$R_t = \frac{1}{1/R_1 + 1/R_2 + 1/R_3 + \cdots + 1/R_n}.$$

Note that $1/(1/n) = n$.

Example 5. Three resistors are connected in parallel. If their values are 100, 200, and 300 Ω respectively, find the equivalent parallel resistance.

$$R_t = \frac{1}{(1/R_1) + (1/R_2) + (1/R_3)} = \frac{1}{1/100 + 1/200 + 1/300}$$

$$= \frac{1}{0.01 + 0.005 + 0.00333} = 54.545 \text{ Ω}.$$

If only two resistors are connected in parallel, the *product-over-the-sum* method can be used. For example, if a pair of resistors, 12 kΩ and 18 kΩ, are connected in parallel, the effective parallel resistance R_t can be determined as follows.

$$R_t = \frac{R_1 \times R_2}{R_1 + R_2} = \frac{(12{,}000)(18{,}000)}{12{,}000 + 18{,}000} = 7200 \text{ Ω}.$$

This equation is often simpler to use than the double reciprocal method. Furthermore, it is possible to use it with three or more resistors by working with only two at a time.

5-4 CONDUCTANCES IN PARALLEL

Conductance *(G)* is the reciprocal of resistance *(R)*. The customary unit of measure of conductance is the *mho;* the SI unit is the siemens (S).

$$G = \frac{1}{R} \text{ and } R = \frac{1}{G}$$

To determine unknown values in parallel circuits, it is often useful to use conductance values. In the preceding example, with 12 kΩ and an 18 kΩ resistors, R_t can be calculated in steps as follows.

$$G_t = G_{R_1} + G_{R_2} = 0.00008333 + 0.00005555 = 0.00013888 \text{ mho},$$

and

$$R_t = \frac{1}{G_t} = \frac{1}{.00013888} = 7200 \; \Omega.$$

In parallel circuits, conductances are added to arrive at a total value because, in the parallel resistance equation using reciprocal values the denominator values *are* conductance values.

$$R_t = \frac{1}{1/R_1 + 1/R_2} = \frac{1}{G_1 + G_2}.$$

An additional example illustrates how conductances are used when four resistors are connected in parallel. Assume that four resistors having values of 1.5 kΩ, 2.5 kΩ, 0.8 kΩ, and 7.5 kΩ are connected in parallel. What is the total conductance?

$$\begin{aligned} G &= \frac{1}{R_1} + \frac{1}{R_2} + \frac{1}{R_3} + \frac{1}{R_4} = \frac{1}{1500} + \frac{1}{2500} + \frac{1}{800} + \frac{1}{7500} \\ &= (0.667 \times 10^{-3}) + (0.4 \times 10^{-3}) + (1.25 \times 10^{-3}) + (0.133 \times 10^{-3}) \\ &= 2.45 \times 10^{-3} \text{ mho (or siemens)}. \end{aligned}$$

$$\begin{aligned} R_t &= \frac{1}{G_1 + G_2 + G_3 + G_4} = \frac{1}{G_t} = 1/2.45 \times 10^{-3} \\ &= 408.2 \; \Omega. \end{aligned}$$

5-5 CURRENT DIVIDERS

If the total current into a parallel branch is known, plus the resistor values, the individual branch currents can be determined by considering the circuit to be a *current divider*. Such a circuit is shown in Figure 5-3. The total current is shown to be 10 mA, and this current will divide according to the inverse ratio of the resistors. In general terms, the smaller resistor will draw the larger current. Hence, R_1 draws a current twice the value of R_2, since its value is half that of R_2. Specifically, for a two-resistor network:

$$I_{R_1} = \frac{R_2}{R_1 + R_2} \times I_t = \frac{400}{600} (10 \text{ mA}) = 6.67 \text{ mA}$$

$$I_{R_2} = \frac{R_1}{R_1 + R_2} \times I_t = \frac{200}{600} (10 \text{ mA}) = 3.33 \text{ mA}$$

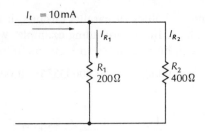

Figure 5-3 A current divider.

Alternatively,

$$I_{R_2} = I_t - I_{R_1} = 10 - 6.67 = 3.33 \text{ mA}.$$

Proving these results,

$$I_t = I_{R_1} + I_{R_2} = 6.67 + 3.33 = 10 \text{ mA}.$$

Once the branch currents are known, the applied voltage can be determined, since its value is the same as the voltage across either resistor.

$$V_s = E_{R_1} = I_{R_1} \times R_1 = 6.67 \times 10^{-3}(200) = 1.33 \text{ V},$$

or

$$V_s = E_{R_2} = I_{R_2} \times R_2 = 3.33 \times 10^{-3}(400) = 1.33 \text{ V}.$$

It is occasionally easier to use conductances to solve for values in a current divider. Still referring to Figure 5-3, conductances can be applied as follows.

$$G_{R_1} = \frac{1}{R_1} = 0.005 \text{ mho (or siemens)};$$

$$G_{R_2} = \frac{1}{R_2} = 0.0025 \text{ mho (or siemens)};$$

$$G_t = G_{R_1} + G_{R_2} = 0.0075 \text{ mho (or siemens)}.$$

$$I_{R_1} = \frac{G_{R_1}}{G_{R_1} + G_{R_2}} \times I_t = \frac{0.005}{0.0075}(10 \text{ mA}) = 6.67 \text{ mA};$$

$$I_{R_2} = \frac{G_{R_2}}{G_{R_1} + G_{R_2}} \times I_t = \frac{0.0025}{0.0075}(10 \text{ mA}) = 3.33 \text{ mA}.$$

5-6 POWER IN PARALLEL CIRCUITS

In parallel circuits, the generation and dissipation of power is computed in the same manner as in series circuits. If two resistors are each consuming 10 W of power, the source must provide 20 W of power

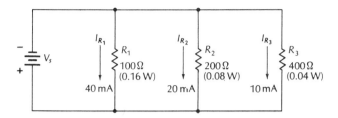

Figure 5-4 A parallel circuit with three branches.

whether the resistors are connected in series or parallel. Such factors as voltage drops, current, and so on, are, of course, greatly influenced by the type of connection, but power is unique among all other circuit characteristics.

Figure 5-4 illustrates the power distribution in a three-resistor parallel network. Current and resistance are known in all the branches.

$$P_{R_1} = (I_{R_1})^2 \times R_1 = 0.04^2 \times 100 = 0.16 \text{ W.}$$
$$P_{R_2} = (I_{R_2})^2 \times R_2 = 0.02^2 \times 200 = 0.08 \text{ W.}$$
$$P_{R_3} = (I_{R_3})^2 \times R_3 = 0.01^2 \times 400 = 0.04 \text{ W.}$$
$$P_t = P_{R_1} + P_{R_2} + P_{R_3} = 0.16 + 0.08 + 0.04 = 0.28 \text{ W.}$$

To determine the supply voltage,

$$E_{R_1} = V_s = I_{R_1} \times R_1 = 0.04 \times 100 = 4 \text{ V} = E_{R_2} = E_{R_3}.$$

5-7 PARALLEL CIRCUIT PROBLEMS

A generalized circuit diagram of a parallel circuit is given in Figure 5-5. By assigning values to various components, the circuit can be used for a number of analytical examples. Note that the answers are rounded off to three significant places.

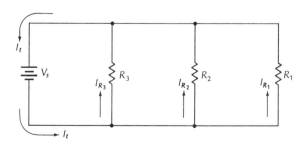

Figure 5-5 A parallel circuit used for the examples in text.

Example 6. $V_s = 10$ V, $R_1 = 500 \,\Omega$, $R_2 = 350 \,\Omega$, $R_3 = 200 \,\Omega$; find I_{R_1}, I_{R_2}, I_{R_3}, I_t.

$$I_{R_1} = \frac{10}{500} = 0.02 \text{ A}.$$

$$I_{R_2} = \frac{10}{350} = 0.0286 \text{ A}.$$

$$I_{R_3} = \frac{10}{200} = 0.05 \text{ A}.$$

$$I_t = 0.02 + 0.0286 + 0.05 = 0.0986 \text{ A} = 98.6 \text{ mA}.$$

Example 7. $I_t = 20$ mA, $R_t = 100 \,\Omega$, $R_1 = R_2 = R_3$; find V_s, I_{R_1}, I_{R_2}, I_{R_3}. Since three resistors, *all of equal value,* are connected in parallel, each must be $3 \times 100 = 300 \,\Omega$.

$$I_{R_1} = I_{R_2} = I_{R_3} = \frac{I_t}{3} = \frac{20 \text{ mA}}{3} = 0.00667 \text{ A} = 6.67 \text{ mA}.$$

$$R_t = 100 \,\Omega; I_t = 20 \text{ mA}; V_s = I_t \times R_t = 0.02 \times 100 = 2.0 \text{ V}.$$

Example 8. $V_s = 10$ V, $R_1 = 100 \,\Omega$, $R_2 = 200 \,\Omega$, $R_3 = 300 \,\Omega$. Using the product-over-the-sum relationship, find R_t.

$$R_{\text{equiv}} = \frac{R_1 \times R_2}{R_1 + R_2} = \frac{100 \times 200}{100 + 200} = 66.7 \,\Omega$$

$$R_t = \frac{R_{\text{equiv}} \times R_2}{R_{\text{equiv}} + R_2} = \frac{66.67 \times 300}{66.67 + 300} = 54.6 \,\Omega.$$

Proof:

$$R_t = \frac{1}{1/R_1 + 1/R_2 + 1/R_3} = \frac{1}{1/100 + 1/200 + 1/300} = 54.6 \,\Omega.$$

PRACTICE PROBLEMS

Work these problems carefully and you will greatly increase your ability to find unknown values in parallel circuits.

For the following problems, refer to Figure 5-6a.

1. Two 30-Ω resistors are connected across a 1.5-V source. Find I_t, R_t, I_{R_1}, and P_t.
2. Two resistors are connected across a 20-V source. $R_1 = 100 \,\Omega$ and $R_2 = 200 \,\Omega$. Find I_t, R_t, I_{R_1}, and P_t.

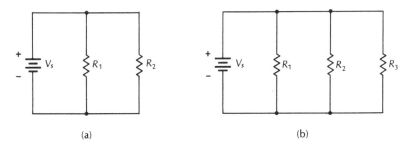

Figure 5-6 Possible open-circuit points in a parallel circuit.

3. A 47-Ω resistor (R_1) and a 68-Ω resistor (R_2) are connected across a 28-V source. Find I_t and R_t.
4. A 3.3 kΩ resistor (R_1) and a 3.9 kΩ resistor (R_2) are connected across a 6-V source. Find R_t and I_t.
5. Two resistors are connected in parallel across a source of emf. Given $I_t = 15$ mA, $I_{R_1} = 10$ mA, and $R_1 = 10{,}000$ Ω, find V_s, I_{R_2}, R_2.

For the following problems, refer to Figure 5-6b.

6. Three resistors are connected in parallel across a 10-V source. Each branch current is 15 mA. Find I_t and R_t.
7. Three resistors are connected in parallel across a 24-V source. R_1 branch current is 15 mA, R_2 branch current is 20 mA, and R_3 branch current is 30 mA. Find the values of I_t and R_t.
8. Three resistors are connected in parallel across a 6-V source. R_1 branch current is 40 mA, R_2 branch current is 60 mA, and R_3 branch current is 80 mA. Find the value of each resistor.
9. A parallel circuit contains three branch resistances of 200, 100, and 50 Ω. The current through the 200-Ω branch is 0.1 A. Find the values of V_s, $I_{R(100)}$, $I_{R(50)}$, I_t.
10. A parallel circuit contains three resistive branches, R_1, R_2, and R_3. The sum of the branch currents is 450 μA and $V_s = 6$ V. Find the value of total equivalent resistance.

Answers: (1) $I_t = 0.1$ A, $R_t = 15$ Ω, $I_{R_1} = 0.05$ A, $P_t = 0.15$ W. (2) $I_t = 0.3$ A, $R_t = 66.6$ Ω, $I_{R_1} = 0.2$ A, $P_t = 6$ W. (3) $I_t = 1$ A (or 1.0075 A), $R_t = 27.8$ Ω. (4) $R_t = 1788$ Ω, $I_t = 3.36$ mA. (5) $I_t = 45$ mA, $R_t = 222$ Ω. (6) $I_t = 65$ mA, $R_t = 369$ Ω. (7) $R_1 = 150$ Ω, $R_2 = 100$ Ω, $R_3 = 75$ Ω. (8) $V_s = 100$ V, $I_{R_2} = 5$ mA, $R_2 = 20{,}000$ Ω. (9) $V_s = 20$ V, $I_R(100) = 0.2$ A, $I_R(50) = 0.4$ A, $I_t = 0.7$ A. (10) $R_{\text{equiv}} = 13{,}333$ Ω.

5-8 OPENS AND SHORTS IN PARALLEL CIRCUITS

Malfunctions in parallel circuits produce circuit responses somewhat different than in series circuits. Figure 5-7 illustrates several possible places where an open-circuit condition may be found. Such opens may occur as a result of a poor solder joint, a broken wire, a loose screw on a terminal strip, a blown fuse, or an open resistor or other component.

The electrical effect of an open circuit at any point is to prevent current flow in that portion of the circuit that would otherwise contain current. For example, an open at the break labeled A would prevent *all* current, since total current must flow here. An open at break D would prevent current only through R_1, and would not affect current flow through R_2 and R_3. However, in this instance, note that *total current* is less than in a perfect circuit by the amount that R_1 would normally draw.

Short circuits occurring in parallel circuits generally cause more severe problems than in series circuits. On the assumption that the wiring has negligible resistance, the effect of *any* of the short circuits illustrated in Figure 5-8 is the same. In a circuit such as this, a short anywhere has the effect of *totally short-circuiting the supply*. At best, the

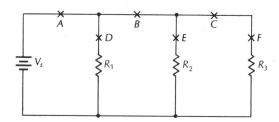

Figure 5-7 Possible short-circuit paths in a parallel circuit.

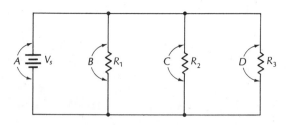

Figure 5-8

voltage across the circuit drops to zero, thus causing all electrical action to stop. At worst, the supply can be overloaded and possibly destroyed, due to excessive current flow caused by the essentially zero resistance of the short. As mentioned, excessively large currents can seriously overheat components, but note that only the wiring and battery are thus affected. The resistors will not overheat, since all current is bypassed, or shunted, around them.

SUMMARY

- The applied voltage appears across all elements of a simple parallel circuit.
- Branch current has a value equal to the applied voltage divided by the branch resistance.
- Total resistance can be found by any of the following:

   (a) $R_t = \dfrac{V_s}{I_t}$.

   (b) $R_t = \dfrac{1}{(1/R_1 + 1/R_2 + 1/R_3 \ldots)}$.

   (c) $R_t = \dfrac{R_1 \times R_2}{R_1 + R_2}$.

- Total branch-current conductance is the sum of the individual parallel conductances.
- Unknown values in a parallel circuit can be found by the current-divider method:

$$I_{R_1} = \dfrac{R_2}{R_1 + R_2} \times I_t.$$

$$I_{R_2} = \dfrac{R_1}{R_1 + R_2} \times I_t.$$

- The total power dissipated in a parallel circuit is the sum of the individual branch power dissipation.
- A short in a parallel circuit places a short across the supply. An open in a branch removes that branch from the circuit action.

QUESTIONS

1. When two resistances of different values are connected in parallel, the current through each (will, will not) be the same.
2. True or false: When two resistances are connected in parallel, the voltage across each will be the same.
3. True or false: When two resistances are connected in parallel, their combined effective resistance is the sum of the value of each.

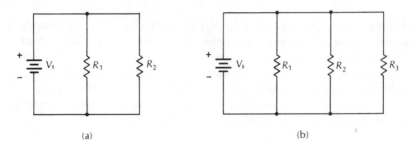

Figure 5-9 Possible open-circuit points in a parallel circuit.

4. When two resistances are connected in parallel, their combined effective resistance is (smaller, larger) than the value of either.
5. True or false: When any two resistances are connected in parallel, each dissipates the same power.
6. When two resistances are connected in parallel, the total power dissipated (does, does not) equal the sum of the individual power dissipation.
7. The total resistance of two parallel resistors (is, is not) equal to $(R_1 + R_2)/(R_1 \times R_2)$.
8. The total resistance of two parallel resistors (is, is not) equal to a value smaller than the smallest value resistor.
9. When total current divides into branch currents, each branch current (is, is not) inversely proportional to branch resistance.
10. Each branch current is (proportional, inversely proportional) to branch resistance.
11. Refer to Figure 5-9b. There is an open circuit in R_3. Briefly describe how the circuit response has changed.
12. Refer to Figure 5-9b. There is a short circuit directly across R_2. Briefly describe how the circuit response has changed.
13. Briefly describe the relationships between the branch and total currents in a parallel circuit.
14. Briefly describe the relationships between the various voltages appearing in a parallel circuit.

PROBLEMS

For the following problems, refer to Figure 5-6a.
1. $V_s = 9\text{ V}, R_1 = 360\text{ }\Omega, R_2 = 360\text{ }\Omega$. Find $R_t, I_{R_1}, I_{R_2}, I_t, P_{R_1}, P_{R_2}, P_t$.
2. $V_s = 9\text{ V}, R_1 = 1800\text{ }\Omega, R_2 = 1800\text{ }\Omega$. Find $R_t, I_{R_1}, I_{R_2} I_t, P_{R_1}, P_{R_2}, P_t$.
3. $V_s = 12\text{ V}, R_1 = 2.4\text{ k}\Omega, R_2 = 3.6\text{ k}\Omega$. Find $R_t, I_1, I_2, I_t, P_{R_1}, P_{R_2}, P_t$.

4. $V_s = 15$ V, $R_1 = 1.5$ kΩ, $R_2 = 3$ kΩ. Find R_t, I_{R_1}, I_{R_2}, I_t, P_{R_1}, P_{R_2}, P_t.

For the following problems, refer to Figure 5-6b.

5. $V_s = 9$ V, $R_1 = 100$ Ω, $R_2 = 200$ Ω, $R_3 = 300$ Ω. Find R_t, I_{R_1}, I_{R_2}, I_{R_3}, I_t.
6. $V_s = 9$ V, $R_1 = 200$ Ω, $R_2 = 300$ Ω, $R_3 = 400$ Ω. Find R_t, I_{R_1}, I_{R_2}, I_{R_3}, I_t.
7. $V_s = 5$ V, $R_1 = 1.5$ kΩ, $R_2 = 2.4$ kΩ, $R_3 = 3.3$ kΩ. Find R_t, I_{R_1}, I_{R_2}, I_{R_3}, I_t, G_t.
8. $V_s = 12$ V, $R_1 = 33$ kΩ, $R_2 = 39$ kΩ, $R_3 = 51$ kΩ. Find R_t, I_{R_1}, I_{R_2}, I_{R_3}, I_t, G_t.
9. $V_s = 150$ V, $R_1 = 330$ kΩ, $R_2 = 470$ kΩ, $R_3 = 680$ kΩ. Find R_t, I_{R_1}, I_{R_2}, I_{R_3}, I_t.
10. $V_s = 15$ V, $R_1 = 1$ megΩ, $R_2 = 1.5$ megΩ, $R_3 = 1.8$ megΩ. Find R_t, I_{R_1}, I_{R_2}, I_{R_3}, I_t.

CHAPTER 6

SERIES-PARALLEL CIRCUITS

In many practical circuit applications the components are not all connected the same way, as has been the case to this point. The *series-parallel* circuit is just such an instance, where some number of parts are connected in series, while others are connected in parallel. Thus, the series-connected components have the same current through them, but different voltages across them; while the parallel-connected components have the same voltage but different values of current. An example of a series-parallel connection is shown below in Figure 6-1a, where the two types of connections are clearly visible.

The topics covered in this chapter include:

6-1 Total Series-Parallel Resistance
6-2 Analytical Procedure
6-3 Series-Connected Parallel Banks
6-4 Parallel-Connected Series Strings
6-5 Series-Parallel Circuit Examples
6-6 Opens and Shorts in Series-Parallel Circuits
6-7 Series-Parallel Ground Connections
6-8 Bridge Circuits

6-1 TOTAL SERIES-PARALLEL RESISTANCE

In the circuit diagram of Figure 6-1a, R_1 and R_2 are connected in series, while R_3 and R_4 are connected in parallel. Furthermore, considered as a whole, *the parallel bank of R_3 and R_4 is in series with R_1 and R_2*. By tracing current flow, the circuit can be understood more easily. Electrons emerge from the negative post of the battery, and the total current flows in the wire up to point A. Here current has a choice of paths and it divides according to the individual value of the resistors. A portion of total current flows through R_3 and the remainder through R_4. At point B the branch currents join and again become the total current, which flows up through R_2 and R_1 in series and returns to the source through the positive terminal. Considering the circuit from the standpoint of direction of current flow, point A can be considered the start of the parallel circuit while point B is the end.

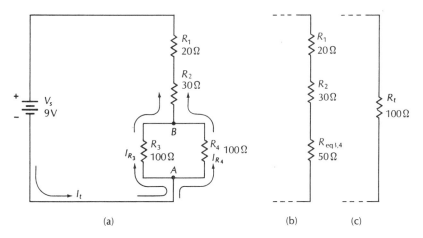

Figure 6-1 (a) A series-parallel circuit schematic diagram; (b) the equivalent series circuit; (c) the equivalent total resistance.

In order to solve for unknown circuit values, one of the first steps to perform is to determine the total effective resistance. One way to do this is to use the *equivalent circuit* method. These basic ideas have been presented before. In the present instance the method simply requires that a value of resistance be found that can be substituted for the parallel bank without changing the total value of current. This, of course, simply means that the equivalent value of resistance be found for R_3 and R_4. When this is substituted in the circuit, the circuit then becomes a *simple series circuit,* as shown in Figure 6-1b. For the values given, $R_{eq\ 3,4}$ is easily determined.

$$R_{eq\ 3,4} = \frac{R_3 \times R_4}{R_3 + R_4} = \frac{100 \times 100}{100 + 100} = 50\ \Omega.$$

The equivalent circuit in Fig. 6-1b can now be used to find the value of R_t, the total circuit resistance, indicated in part c of the figure.

$$R_t = R_1 + R_2 + R_{eq\ 3,4} = 20 + 30 + 50 = 100\ \Omega.$$

Note that the procedures used above constitute nothing new at all, except that we have not previously used both series and parallel circuit concepts together. The method just demonstrated generally holds true for series-parallel circuits: The separate series and parallel processes are simply combined *in a logical sequence* to solve problems in series-parallel combinations. In very complex circuits, the most logical place to start is not always immediately evident, but careful study will reveal it as well as all the steps that follow it. In Section 6-4 several complex examples are given, along with step-by-step solutions.

6-2 ANALYTICAL PROCEDURE

It is not possible to provide a single procedure for all series-parallel circuits. Many series-parallel circuit problems, however, can be solved by using the following general steps.

1. Reduce all parallel combinations to a single equivalent resistance. Label clearly to avoid confusion in later steps.
2. Redraw the circuit using equivalent resistances.
3. Reduce all series combinations to a single equivalent resistance and label clearly.
4. Redraw the circuit using equivalent resistances.
5. Repeat Steps 1 through 4 if necessary until you have computed a single resistance that is the equivalent resistance of the entire circuit.
6. Solve for I_t.
7. Solve for individual voltage drops.
8. Solve for individual branch currents.
9. Solve for individual power dissipation values.
10. Note that in some circuit problems Steps 1 and 3 may have to be reversed.

As might be anticipated, there are many, many kinds of combinations that form series-parallel circuits; Figure 6-1 is but one possibility. The procedure outlined above will work on nearly any series-parallel circuit, but it may not be necessary to perform Step 5 if the circuit is as simple as that of Figure 6-1.

To illustrate the procedure, we shall solve the circuit of Fig. 6-1a for all unknown values. $R_{eq\ 3,4}$ and R_t have previously been found to be 50 and 100 Ω respectively. The number in parentheses following each computation refers to the step in the preceding list.

$$I_t = \frac{V_s}{R_t} = \frac{9}{100} = 0.09 = 90 \text{ mA}. \tag{6}$$

$$E_{R_1} = I_t \times R_1 = 0.09 \times 20 = 1.8 \text{ V}. \tag{7}$$

$$E_{R_2} = I_t \times R_2 = 0.09 \times 30 = 2.7 \text{ V}. \tag{7}$$

$$E_{R_{3,4}} = I_t \times R_{eq\ 3,4} = 0.09 \times 50 = 4.5 \text{ V}. \tag{7}$$

$$I_{R_3} = I_{R_4} = \frac{E_{R_3}}{R_3} = \frac{4.5}{100} = 0.045 \text{ A} = 45 \text{ mA} \quad \text{(or}$$

$$I_{R_3} = I_{R_4} = \frac{I_t}{2} = 0.045 \text{ A} = 45 \text{ mA}). \tag{8}$$

$$P_{R_1} = I^2 \times R_1 = 0.09^2 \times 20 = 0.162 \text{ W}. \tag{9}$$

$$P_{R_2} = I^2 \times R_2 = 0.09^2 \times 30 = 0.243 \text{ W}. \tag{9}$$

$$P_{R_{3,4}} = I^2 \times R_{eq\ 3,4} = 0.09^2 \times 50 = 0.405 \text{ W}. \tag{9}$$

$$P_{R_3} = (I_{R_3})^2 \times R_3 = 0.045^2 \times 100 = 0.2025 \text{ W.} \qquad (9)$$
$$P_{R_4} = (I_{R_4})^2 \times R_4 = 0.045^2 \times 100 = 0.2025 \text{ W,} \quad \text{or} \qquad (9)$$
$$P_{R_3} = P_{R_4} = \frac{P_{R_{3,4}}}{2} = 0.2025 \text{ W.} \qquad (9)$$

6-3 SERIES-CONNECTED PARALLEL BANKS

Fig. 6-2a illustrates four resistors connected in a *series-connected parallel-bank* circuit. R_1 and R_2, R_3 and R_4 constitute parallel-connected banks, but the two banks are connected in series with each other. To solve for unknown values in a circuit such as this, first reduce each parallel branch to a single equivalent resistance, then proceed as in a series circuit.

Example 1. All resistances are 200 Ω.

$$R_{\text{eq } 1,2} = \frac{R_1 \times R_2}{R_1 + R_2} = \frac{(200)(200)}{200 + 200} = 100 \text{ Ω.}$$
$$R_{\text{eq } 3,4} = \frac{R_3 \times R_4}{R_3 + R_4} = \frac{(200)(200)}{200 + 200} = 100 \text{ Ω.}$$

An equivalent circuit (Figure 6-2b) consists of two 100-Ω resistors in series, so $R_t = 200$ Ω. The circuit can now be solved for any required unknown values by conventional means.

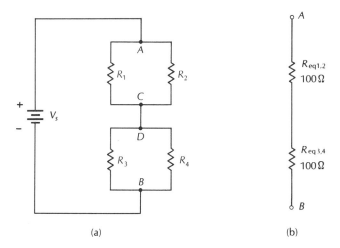

Figure 6-2 Series-connected parallel banks.

6-4 PARALLEL-CONNECTED SERIES STRINGS

Figure 6-3 illustrates a slightly different series-parallel arrangement. Here, *series strings* of two resistors each (R_1 and R_2, R_3 and R_4) are connected in parallel. In this instance, the correct procedure is to reduce the series strings to single equivalent values and then compute the total parallel resistance.

$$R_{eq\ 1,2} = R_1 + R_2 = 100 + 150 = 250\ \Omega.$$
$$R_{eq\ 3,4} = R_3 + R_4 = 200 + 250 = 450\ \Omega.$$
$$R_t = \frac{(R_{eq\ 1,2})(R_{eq\ 3,4})}{R_{eq\ 1,2} + R_{eq\ 3,4}} = \frac{250 \times 450}{250 + 450} = 160.7\ \Omega.$$

Thus, once the series resistances are added together, the problem is simply that of a parallel circuit. At this point, if V_s is given a value, it is a simple matter to determine all other values. For the circuit of Figure 6-3, we shall assign V_s a value of 9 V. First, determine the value of total current:

$$I_t = \frac{V_s}{R_t} = \frac{9}{160.7} = 0.056 = 56\ mA.$$

Now branch currents may be determined, which allows voltage drops to be found.

$$I_{R1,2} = \frac{V_s}{R_1 + R_2} = 0.036 = 36\ mA.$$
$$I_{R3,4} = \frac{V_s}{R_3 + R_4} = 0.020 = 20\ mA.$$
$$E_{R1} = I_{R1,2} \times R_1 = 3.6\ V.$$
$$E_{R2} = I_{R1,2} \times R_2 = 5.4\ V.$$
$$E_{R3} = I_{R3,4} \times R_3 = 4.0\ V.$$
$$E_{R4} = I_{R3,4} \times R_4 = 5.0\ V.$$

Finally, verify the values of voltage drops:

$$V_s = E_{R1} + E_{R2} = 9\ V.$$
$$V_s = E_{R3} + E_{R4} = 9\ V.$$

In this circuit, one application calls for a knowledge of the potential difference between points A and B. This is determined by finding the voltage at each of these points with respect to a common point, say point C. If point C is considered to be at zero volts, or ground, then point A is at a potential of +5.4 V relative to ground. This, of course, is the drop across R_2 as calculated previously. Now, point B is at a potential of +5.0 V with respect to ground. Hence, a voltmeter placed between A and B will read the difference between these two values, with point B the more negative of the two.

$$V_{diff} = E_A - E_B = 5.4 - 5.0 = 0.4\ V.$$

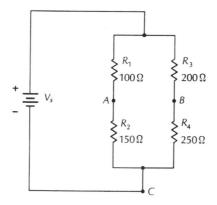

Figure 6-3 Parallel-connected series strings.

A circuit such as this is often called a *bridge* circuit, and this example is an *unbalanced* bridge circuit, since A and B are at different voltages. If they were at the same voltage it would be called a *balanced* bridge. Bridge circuits have wide application in electronic circuitry ranging from ultra-accurate metering circuits to power supplies to communication circuits. We shall cover bridge circuits in greater detail later in this chapter.

6-5 SERIES-PARALLEL CIRCUIT EXAMPLES

In this section somewhat more complex series-parallel circuits will be covered, one of which is given in Figure 6-4a. Although this example contains more components, the previously given rules can still be applied if you note the actual circuit configuration. Generally, the best starting point in analyzing such a circuit is the area farthest away from the source. The steps are given below, followed by the actual computations.

1. Combine R_5 and R_6, series.
2. Combine R_4 and $R_{eq\ 5,6}$, parallel.
3. Combine R_3 and $R_{eq\ 4,5,6}$, series.
4. Combine R_2 and $R_{eq\ 3,4,5,6}$, parallel.
5. Combine R_1 and $R_{eq\ 2,3,4,5,6}$, series.

1. $100 + 100 = 200\ \Omega$.
2. $200/2 = 100\ \Omega$.
3. $100 + 100 = 200\ \Omega$.
4. $200/2 = 100\ \Omega$.
5. $300 + 100 = 400\ \Omega$.

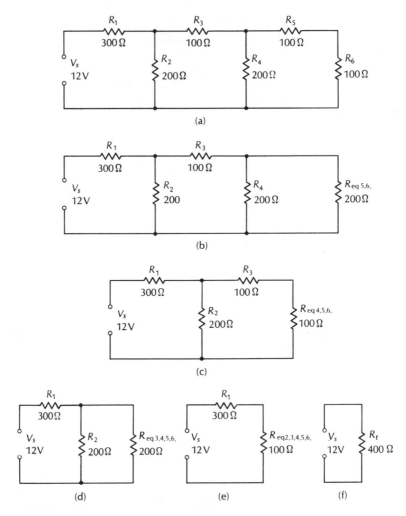

Figure 6-4 (a-f) Successive simplification of a complex series-parallel circuit.

Note in Figure 6-4 how the circuit is simplified step by step. First, R_5 and R_6, series-connected resistors, are combined to equal 200 Ω, part b. This, in parallel with R_4, also 200 Ω, gives an equivalent resistance of 100 Ω, drawing c. In Figure 6-4d, R_3 has been combined with the previous results. Hence, $R_{eq\ 3,4,5,6}$ has an equivalent resistance of 200 Ω. Now, R_2 has a value of 200 Ω, and is in parallel with $R_{eq\ 3,4,5,6}$, for an equivalent resistance of 100 Ω, as shown in part e. Finally, adding the 300-Ω series resistance of R_1 yields a total effective resistance (R_t) of 400 Ω (f).

With total resistance known, the total current can be determined.

$$I_t = \frac{V_s}{R_t} = \frac{12}{400} = 0.03 = 30 \text{ mA}.$$

Once I_t is known, the voltage drops across each resistor can be found. In this circuit, total current flows only through R_1.

$$E_{R_1} = I_t \times R_1 = 0.03 \times 300 = 9 \text{ V}.$$

If 9 V is dropped across R_1, then $12 - 9 = 3$ V must be dropped across R_2, since R_1 and R_2 are in series across V_s. The current through R_2 is

$$I_{R_2} = \frac{V_s - E_{R_1}}{R_2} = \frac{3}{200} = 0.015 = 15 \text{ mA}.$$

Therefore, if 30 mA flows in R_1 and 15 mA flows in R_2, then the current through R_3 is

$$I_{R_3} = I_t - I_{R_2} = 0.03 - 0.015 = 15 \text{ mA (see Fig. 6-4}d).$$

Knowing the current through R_3 allows the voltage drop across it to be found.

$$E_{R_3} = I_{R_3} \times R_3 = 0.015 \times 100 = 1.5 \text{ V}.$$

Across R_4, then, as well as R_5 and R_6, there must be a voltage of

$$E_{R_4} = E_{R_2} - E_{R_3} = 3 - 1.5 = 1.5 \text{ V}.$$

The current through R_4 is

$$I_{R_4} = \frac{E_{R_4}}{R_4} = \frac{1.5}{200} = 0.0075 = 7.5 \text{ mA}.$$

The current through $R_5 + R_6$ is ($E_{R_{5,6}} = E_{R_4}$)

$$I_{R_{5,6}} = \frac{E_{R_{5,6}}}{R_5 + R_6} = \frac{1.5}{200} = 0.0075 \text{ A} = 7.5 \text{ mA}.$$

Finally, the drops across R_5 and R_6 are

$$E_{R_5} = I_{R_{5,6}} \times R_5 = 0.0075 \times 100 = 0.75 \text{ V}.$$
$$E_{R_6} = I_{R_{5,6}} \times R_6 = 0.0075 \times 100 = 0.75 \text{ V}.$$

Note particularly the sequence of the foregoing procedure. While other methods might be applied here, the one given is probably the most straightforward and the simplest.

Another example of a series-parallel circuit is given in Figure 6-5a. It is often beneficial to redraw a complex circuit such as this, in order to see the series-parallel relationships more clearly. Such a simplification is shown in Figure 6-5b, where the series-parallel relationships are a bit easier to see. These two circuits are electrically

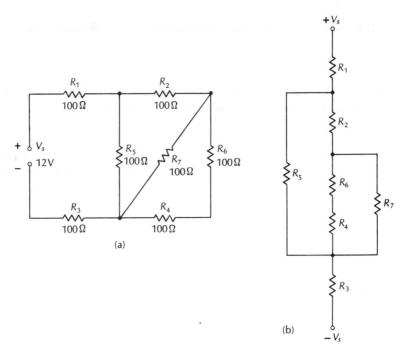

Figure 6-5 Circuit simplification by redrawing.

identical. Again, the circuit will be analyzed step by step, to reinforce your appreciation of these methods.

1. $R_6 + R_4$: $100 + 100 = 200 \, \Omega$.
2. $R_{eq \, 4,6}$ and R_7: $\dfrac{200 \times 100}{200 + 100} = \dfrac{20{,}000}{300} = 66.67 \, \Omega$.
3. $R_2 + R_{eq \, 4,6,7}$: $100 + 66.67 = 166.67 \, \Omega$.
4. R_5 and $R_{eq \, 2,4,6,7}$: $\dfrac{100 \times 166.67}{100 + 166.67} = 62.5 \, \Omega$.
5. $R_1 + R_3 + R_{eq \, 2,4,5,6,7}$: $100 + 100 + 62.5 = 262.5 \, \Omega = R_t$.
6. $I_t = \dfrac{V_t}{R_t} = \dfrac{12}{262.5} = 0.0457 = 45.7 \, \text{mA}$.
7. $E_{R_1} = I_t \times R_1 = 45.7 \, \text{mA} \times 100 = 4.57 \, \text{V}$.
8. $E_{R_3} = I_t \times R_3 = 45.7 \, \text{mA} \times 100 = 4.57 \, \text{V}$.
9. $E_{R_5} = V_s - E_{R_1} - E_{R_3} = 12 - 4.57 - 4.57 = 2.86 \, \text{V}$.
10. $I_{R_5} = \dfrac{E_{R_5}}{R_5} = \dfrac{2.86}{100} = 0.0286 = 28.6 \, \text{mA}$.
11. $I_{R_2} = I_t - I_{R_5} = 45.7 \, \text{mA} - 28.6 = 17.1 \, \text{mA}$.
12. $E_{R_2} = I_{R_2} \times R_2 = 17.1 \, \text{mA} \times 100 = 1.71 \, \text{V}$.

Series-Parallel Circuits

13. $E_{R_7} = E_{R_5} - E_{R_2} = 2.86 - 1.71 = 1.15 \text{ V} = E_{R_{4,6}}$.
14. $I_{R_7} = \dfrac{E_{R_7}}{R_7} = \dfrac{1.15}{100} = 0.0115 = 11.5 \text{ mA}$.
15. $I_{R_{4,6}} = \dfrac{E_{R_{4,6}}}{R_4 + R_6} = \dfrac{1.15}{200} = 0.00575 = 5.75 \text{ mA}$.
16. $E_{R_6} = E_{R_4} = I_{R_{4,6}} \times R_6 = 0.575 \text{ V}$.

Another example of a complex series-parallel circuit is given in Fig. 6-6. Part *a* is the original schematic, while the schematic in *b* is electrically identical but makes the series and parallel relationships more clearly evident. As before, to solve such a circuit for total resistance, it is best to reduce the series-parallel groups individually, one at a time. First, assume that all resistors are 100 Ω in value, to make the following calculations as simple as possible. Then, note that the innermost loop consists of R_3, R_4, R_5, and R_9. The equivalent resistance of these four resistors will be labeled $R_{eq\,1}$.

$$R_{eq\,1} = \dfrac{(R_3 + R_4 + R_5) \times R_9}{R_3 + R_4 + R_5 + R_9} = \dfrac{300 \times 100}{400} = 75 \text{ Ω}.$$

Replacing these four resistors with a single 75-Ω resistor gives the simplified circuit shown in Fig. 6-6*c*.

Next, reduce the four resistances R_2, $R_{eq\,1}$, R_6, and R_{10} to a single equivalent value, $R_{eq\,2}$.

$$R_{eq\,2} = \dfrac{(R_2 + R_{eq\,1} + R_6) \times R_{10}}{R_2 + R_{eq\,1} + R_6 + R_{10}} = \dfrac{27{,}500}{375} = 73.33 \text{ Ω}.$$

Replacing these four resistors with a single 73.33-Ω resistor gives the further simplified circuit shown in Fig. 6-6*d*.

Now simplify the loop consisting of R_{11}, R_1, $R_{eq\,2}$, and R_7.

$$R_{eq\,3} = \dfrac{(R_1 + R_{eq\,2} + R_7) \times R_{11}}{R_1 + R_{eq\,2} + R_7 + R_{11}} = \dfrac{27{,}333}{373.33} = 73.2 \text{ Ω}.$$

This operation yields the simplified circuit shown in Fig. 6-6*e*.

Finally, total resistance is simply the sum of $R_{eq\,3}$ and R_8, as indicated in part *f* of the figure:

$$R_t = R_{eq\,3} + R_8 = 73.2 + 100 = 173.2 \text{ Ω}.$$

If it is necessary to determine voltage and current values for any part of the circuit, the circuit *must be expanded backward* from this point. Assume that V_s is 12 V. Find the value of current in each branch and the voltage drop across each resistor. First, determine total current:

$$I_t = \dfrac{V_s}{R_t} = \dfrac{12}{173.2} = 69.3 \text{ mA}.$$

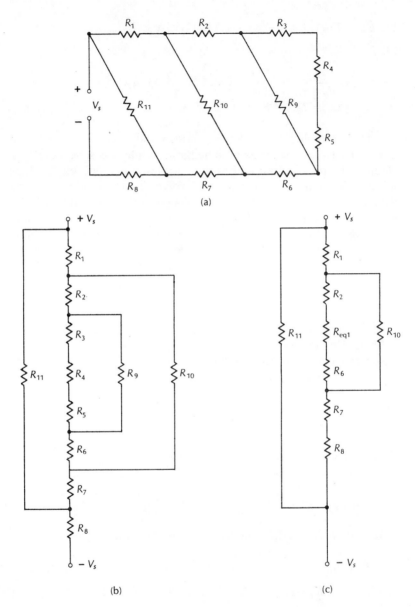

Figure 6-6 (a-f) Successive simplification of a complex circuit.

Applying total current to drawing e, the voltage drop across R_8 can be found.

$$E_{R_8} = I_t \times R_8 = 69.3 \text{ mA} \times 100 = 6.93 \text{ V}.$$

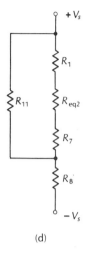
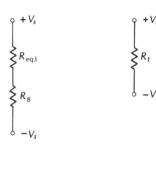

(d)   (e)   (f)

The voltage drop across $R_{eq\,3}$, then, is

$$E_{R_{eq\,3}} = V_s - E_{R8} = 12 - 6.93 = 5.07 \text{ V}.$$

Referring now to drawing d, the branch currents can easily be found. Since 5.07 V appears across the parallel branches, each current has a value determined by the branch resistance.

$$I_{R_{11}} = \frac{E_{R_{11}}}{R_{11}} = \frac{5.07}{100} = 50.7 \text{ mA}.$$

Also,

$$I_{R_1,\,R_{eq\,2},\,R_7} = \frac{E(R_1, R_{eq\,2}, R_7)}{R_1 + R_{eq\,2} + R_7} = \frac{5.07}{273.33} = 18.55 \text{ mA} \cong 18.6 \text{ mA}.$$

The next step is to find the voltage drops across the series string composed of R_1, $R_{eq\,2}$, and R_7 (also in part d).

$$E_{R_1} = I_{R_1} \times R_1 = 18.6 \text{ mA} \times 100 = 1.86 \text{ V}.$$
$$E_{R_{eq\,2}} = I_{R_{eq\,2}} \times R_{eq\,2} = 18.6 \text{ mA} \times 73.33 = 1.36 \text{ V}.$$
$$E_{R_7} = I_{R_7} \times R_7 = 1.86 \text{ mA} \times 100 = 1.86 \text{ V}.$$

Figure 6-6c shows the individual resistances of the $R_{eq\,2}$ branch in more detail, and now the voltages across R_2, $R_{eq\,1}$, R_6, and R_{10} can be found. The voltage across $R_{eq\,2}$ is the same as the voltage across R_{10}, hence $I_{R_{10}}$ can now be found.

$$I_{R_{10}} = \frac{E_{R_{eq\,2}}}{R_{10}} = \frac{1.36}{100} = 13.6 \text{ mA}.$$

Therefore, current in the R_2, $R_{eq\,1}$, and R_6 branch can be found:

$$I_{R2,6} = I_{R_{eq\,2}} - I_{R10} = 18.6 \text{ mA} - 13.6 \text{ mA} = 5 \text{ mA}.$$

Now, the voltage drops in the branch composed of R_2, $R_{eq\,1}$, and R_6 can be found.

$$E_{R2} = I_{R2,6} \times R_2 = 5 \text{ mA} \times 100 = 0.5 \text{ V}.$$
$$E_{R_{eq\,1}} = I_{R2,6} \times R_{eq\,1} = 5 \text{ mA} \times 75 = 0.375 \text{ V}.$$
$$E_{R6} = I_{R2,6} \times R_2 = 5 \text{ mA} \times 100 = 0.5 \text{ V}.$$

Finally, referring to part b, the unknown values still to be found are those concerning R_3, R_4, R_5, and R_9 ($R_{eq\,1}$). The current in this innermost branch is the same as $I_{R_{eq\,1}}$, computed above as $I_{R2,6}$, 5 mA. First, find the current in R_9 as a function of $E_{R_{eq\,1}}$.

$$I_{R9} = \frac{E_{R_{eq\,1}}}{R_9} = \frac{0.375}{100} = 3.75 \text{ mA},$$

and

$$I_{R3,4,5} = I_{R_{eq\,1}} - I_{R9} = 5 \text{ mA} - 3.75 \text{ mA} = 1.25 \text{ mA}.$$

At this point, the voltage drops across R_3, R_4, and R_5 are the only unknowns left.

$$E_{R3} = I_{R3,4,5} \times R_3 = 1.25 \text{ mA} \times 100 = 0.125 \text{ V}.$$

Since R_4 and R_5 are also 100-Ω resistors, each drops 0.125 V.

6-6 OPENS AND SHORTS IN SERIES-PARALLEL CIRCUITS

Opens and shorts in series-parallel circuits react in a variety of ways, depending on individual circuit configurations and the exact location of the short or open. Figure 6-7a shows a series-parallel circuit diagram with both a short circuit and an open. This example illustrates the typical case.

Assuming that only the short circuit exists, drawing b shows that R_3, R_4, and R_5 have effectively been removed from the circuit by the short across R_3. These three resistors will have no current through them simply because the voltage has been reduced to zero by the short. Now, because the resistance of $R_4 + R_5$, in parallel with R_3, has been reduced to zero (or essentially zero), the total resistance of the entire circuit has been greatly reduced. Hence, more total current will flow through R_1 and R_2. This may cause overheating in R_1 and R_2, as well as in the associated wiring and in the source. Overall power dissipation is greatly increased and the short circuit completely upsets the original circuit function.

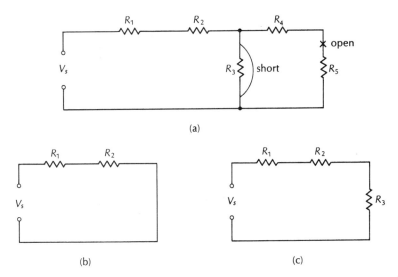

Figure 6-7 (a) Illustrating short-circuit and open-circuit conditions in a series-parallel circuit; (b) considering only the short, R_1 and R_2 constitute the altered circuit; (c) considering only the open, R_1, R_2, and R_3 constitute the altered circuit.

Now, assuming that the open circuit exists, and the short circuit does not, Figure 6-7c shows what happens to the circuit. When this open occurs, no current can flow through R_4 and R_5, thus *reducing* total current flow and increasing total effective resistance. Whether R_1, R_2, and R_3 now have greater or less power dissipation compared to the original circuit depends upon their individual values. However, *total* power dissipation must decrease, since the original dissipation of R_4 and R_5 is nonexistent. Again, the open circuit completely upsets the original circuit function.

6-7 SERIES-PARALLEL GROUND CONNECTIONS

As mentioned earlier, nearly all practical circuits use a common, or ground, return to the supply. If the circuitry is constructed on a metal chassis, the chassis itself usually serves as the common return, as illustrated in Figure 6-8a. Another construction method is illustrated in Fig. 6-8b, which shows a printed-circuit board with the ground *lines* pointed out. In both instances, the common return is used by all circuits on the assembly, thus effecting a considerable savings compared to using individual wires for each circuit.

Electrically, such circuits might appear schematically as shown in Fig. 6-9. Drawing *a* shows a simple series-parallel circuit using a

114 Chapter 6

(a)

Figure 6-8 (a) A metal chassis used as a ground return; (b) a PC-board ground return.

ground return. The solid-line portion of the drawing represents the usual drafting practice, while the dashed lines show the complete return path to V_s. In a circuit such as this, it is important to note that most voltages are measured with respect to ground. That is, in practical circuitry, voltages are given (and measured) with the understanding that ground is one of the points considered. As was described in Chapter 4, such voltages are given as $E_A = +6$ V, $E_B = +3$ V, $V_s = +9$ V, and since we already know that voltage cannot exist at a single point, there must be a reference point. This is, of course, ground. The appropriate meter probe can be temporarily attached to ground and not moved during all measurements, unless it is necessary to measure directly across a resistor that does not return to ground. For example, in Fig. 6-9a, measuring directly across R_4 does not require moving the negative meter lead, while a measurement across R_1, R_2, or R_3 does.

In Figure 6-9b much of the same material applies. All ground points are considered to be directly connected together. The only significant difference (apart from the number of resistors and their values) is that the positive side of the source is grounded. Thus, all voltages are negative with respect to ground.

Ground return

(b)

In Figure 6-9c, however, it can be seen that *neither* terminal of the source is ground, but rather an intermediate point. When measuring a circuit such as this, it is necessary to observe meter polarity even more closely than in the two preceding examples. Note that in one instance the circuit polarity is such that negative is ground, while in the other, positive is ground. To analyze such a circuit, first reduce all series and parallel resistances to a single R_t, so that I_t can be determined. R_3 and R_4 are in series, as are R_5 and R_6, but each group is in parallel with the other. Hence, since each branch has a value of 300 Ω, their equivalent resistance is 300/2 = 150 Ω. This, in series with R_1 and R_2, yields a total resistance of $R_t = 100 + 200 + 150 = 450$ Ω.

Now, find I_t.

$$I_t = \frac{V_s}{R_t} = \frac{4.5}{450} = 0.01 = 10 \text{ mA.}$$

The voltage drop across R_1 and R_2 can now be determined.

$$E_{R_1} = I_t \times R_1 = 0.01 \times 100 = 1.0 \text{ V.}$$
$$E_{R_2} = I_t \times R_2 = 0.01 \times 200 = 2.0 \text{ V.}$$

116 Chapter 6

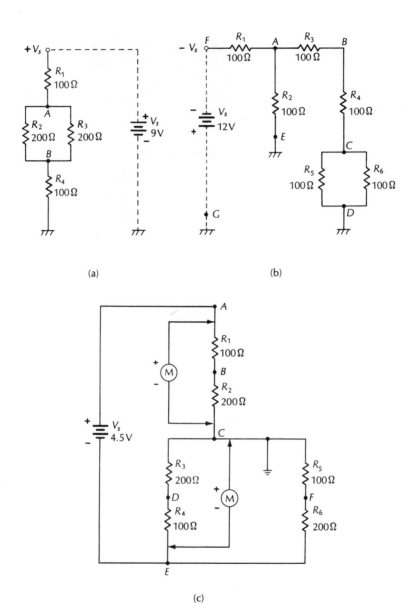

Figure 6-9 Three examples of series-parallel circuits using chassis ground, or common.

Branch currents must be calculated before the remaining voltage drops can be found. The voltage appearing across R_3, R_4 and R_5, R_6 is the value remaining after R_1 and R_2 drops are subtracted from V_s.

$$E_{R_{3,4}} = V_s - (E_{R_1} + E_{R_2}) = 4.5 - (1 + 2) = 1.5 \text{ V.}$$

$$I_{R3,4} = \frac{E_{R3,4}}{R_3 + R_4} = \frac{1.5}{300} = 0.005 = 5 \text{ mA}.$$
$$I_{R5,6} = \frac{E_{R5,6}}{R_5 + R_6} = \frac{1.5}{300} = 0.005 = 5 \text{ mA}.$$

Each of the remaining voltage drops can now be determined.

$$E_{R3} = I_{R3,4} \times R_3 = 0.005 \times 200 = 1.0 \text{ V}.$$
$$E_{R4} = I_{R3,4} \times R_4 = 0.005 \times 100 = 0.5 \text{ V}.$$
$$E_{R5} \doteq I_{R5,6} \times R_5 = 0.005 \times 100 = 0.5 \text{ V}.$$
$$E_{R6} = I_{R5,6} \times R_6 = 0.005 \times 200 = 1.0 \text{ V}.$$

Finally, the voltages at each labeled point can be found.

- Point C is ground, or 0.0 V.
- Point B is more positive than ground by the amount of drop across R_2, or +2.0 V.
- Point A is more positive than ground by the sum of the drops across R_1 and R_2, or +3 V.
- Point D is more negative than ground by the drop across R_3, or -1.0 V.
- Point E is more negative than ground by the drop across either $R_3 + R_4$ or $R_5 + R_6$, or -1.5 V.
- Point F is more negative than ground by the drop across R_5, or -0.5 V.
- Finally, Point D is -1.0 V relative to ground.

All required voltages with respect to ground have been determined.

6-8 BRIDGE CIRCUITS

The Balanced Bridge

As briefly mentioned earlier, a Wheatstone bridge circuit is a useful and common form of series-parallel circuit. This circuit is named after Sir Charles Wheatstone (1802–1875), an English physicist and inventor. The Wheatstone bridge, often called simply the resistance bridge, is illustrated schematically in Fig. 6-10a and b. The distinctive diamond shape (a) is used to emphasize the fact that it is a bridge circuit, but in b it is seen to be a simple parallel-series circuit.

One advantage of this circuit configuration is that a resistor (R_x) can be measured for its resistance value much more accurately than by conventional means. R_x is the unknown value, and R_3 is variable over a given range. In practice, there are several R_3's of different values that can be switched in so a wide range of unknown resistors can be measured. The principle upon which the bridge circuit operates is quite simple. If the two parallel branches (R_2 and R_3, R_1 and R_x) have the

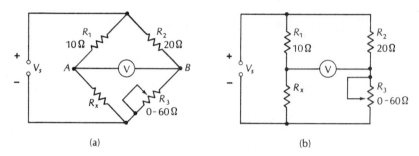

Figure 6-10 The Wheatstone bridge circuit.

same *ratio* of resistances, the voltages at points A and B will be the same. Hence, the unknown value is connected to the bridge, and R_3 is varied until the meter (usually a zero-center movement) reads zero volts. The value of R_x is then read from a calibrated dial on R_3. If the resistors used in the bridge are *very* accurate, and if the voltmeter is *very* sensitive, the indicated resistance value of R_x is similarly very accurate.

An example will serve to illustrate the principles of the resistance bridge. Using the values given in Fig. 6-10, let us see what the value of R_x is if R_3 is adjusted to 30 Ω. Now, when the bridge is *balanced,* the voltage at A will be exactly equal to the voltage at B. Under these conditions, a proportion can be written:

$$R_1 \text{ is to } R_x \text{ as } R_2 \text{ is to } R_3,$$

and

$$R_x \times R_2 = R_1 \times R_3;$$
$$R_x = \frac{R_1 \times R_3}{R_2}.$$

Substituting known values will determine the value of R_x.

$$R_x = \frac{R_1 \times R_3}{R_2} = \frac{10 \times 30}{20} = \frac{300}{20} = 15 \text{ Ω}.$$

If these values exist exactly, there is no potential difference from A to B. Hence, no current flows in the meter, which indicates zero, and the bridge is said to be *balanced.*

PRACTICE PROBLEMS

For the following problems, refer to the Wheatstone bridge of Figure 6-10.

1. R_3 is adjusted to 10 Ω, at which point the meter reads zero. What is the value of R_x?

2. R_3 is adjusted to 50 Ω, at which point the meter reads zero. What is the value of R_x?
3. R_3 is adjusted to 33 Ω, at which point the meter reads zero. What is the value of R_x?
4. R_3 is adjusted to 60 Ω, at which point the meter reads zero. What is the value of R_x?

Answers: (1) 5 Ω, (2) 25 Ω, (3) 16.5 Ω, (4) 30 Ω.

The Unbalanced Bridge

The Wheatstone bridge is often used in the *unbalanced* condition, or with one resistive leg which is variable. In such a case, the network must be analyzed using methods other than the proportion. In Figure 6-11, we shall assign some values to the supply and the resistors to illustrate how the unbalanced bridge circuit is analyzed. Assume that $V_s = 18$ V, $R_1 = 1200$ Ω, $R_2 = 2200$ Ω, $R_3 = 2700$ Ω, and R_4 is 1200 Ω. Determine what the voltmeter will read.

First, find the individual voltages at points A and B. *Note:* the expression $V_s/(R_1 + R_4)$ represents the current in that leg.

$$E_A = V_s - E_{R_1} = 18 - \left(\frac{V_s}{R_1 + R_4} \times R_1\right)$$

$$= 18 - \left(\frac{18}{1200 + 1200} \times 1200\right) = 18 - 9 = 9 \text{ V.}$$

$$E_B = V_s - E_{R_2} = 18 - \left(\frac{V_s}{R_2 + R_3} \times R_3\right)$$

$$= 18 - \left(\frac{18}{2200 + 2700} \times 2700\right) = 18 - 9.92 = 8.08 \text{ V.}$$

The potential difference between A and B, then, is simply the difference between 9 and 8.08 V. $E_{AB} = 9 - 8.08$ V $= 0.92$ V. Thus, the voltmeter will indicate slightly less than one volt.

PRACTICE PROBLEMS

For the following problems, refer to Figure 6-11.

1. The values are $V_s = 24$ V, $R_1 = 10{,}000$ Ω, $R_2 = 33{,}000$ Ω, $R_3 = 66{,}000$ Ω, $R_4 = 18{,}000$ Ω. Determine the voltmeter reading.
2. The values are $V_s = 12$ V, $R_1 = 33{,}000$ Ω, $R_2 = 47{,}000$ Ω, $R_3 = 94{,}000$ Ω, $R_4 = 82{,}000$ Ω. Determine the voltmeter reading.
3. The values are $V_s = 9$ V, $R_1 = 220$ Ω, $R_2 = 330$ Ω, $R_3 = 470$ Ω, $R_4 = 560$ Ω. Determine the voltmeter reading.
4. The values are $V_s = 100$ V, $R_1 = 1.5$ MΩ, $R_2 = 3$ MΩ, $R_3 = 4$ MΩ, $R_4 = 2$ MΩ. Determine the voltmeter reading.

Answers: (1) 0.57 V, (2) 0.56 V, (3) 1.17 V, (4) 0 V.

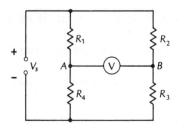

Figure 6-11 The unbalanced bridge circuit.

SUMMARY

- The total equivalent resistance of a series-parallel circuit is the sums of the individual series-connected resistances and the equivalent resistances of the parallel branches.
- The voltage across two parallel banks is the same, and is equal to the value obtained if a single resistor equal in value to the equivalent parallel resistance were used in place of the parallel bank.
- The voltage across series-connected components depends upon their individual resistance values and how these values compare to the total effective resistance.
- An open in a series-parallel circuit will open the entire circuit if it occurs in the series part, but opens only one branch if it occurs in the parallel part.
- A short in a series-parallel circuit effectively removes the shorted element from the circuit, and increases both the current and the resultant voltage drops and power dissipation across the remaining components.
- Ground, or common, is a term used to denote a common return to the power supply. It may consist of the metal chassis or a common strip on a PC board.
- Ground may, or may not, be an actual earth ground.
- A Wheatstone bridge is a sensitive resistance-measuring device.

QUESTIONS

1. Refer to Figure 6-4a, R_1 (is, is not) in parallel with R_2.
2. Refer to Figure 6-4a, R_5 (is, is not) in parallel with R_6.
3. Refer to Figure 6-4a, R_2, R_3, and R_4 (are, are not) all in parallel with each other, considering circuit current.
4. Refer to Figure 6-4a, R_2, R_3, and R_4 (are, are not) all in series with each other, considering circuit current.
5. Refer to Figure 6-4a, R_5 and R_6 (are, are not) in series, while R_4 (is, is not) in parallel with them.

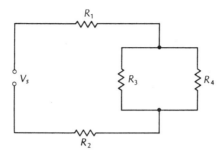

Figure 6-12

6. Refer to Figure 6-4b, R_3 and $R_{eq\ 4,5,6}$ (are, are not) in series with each other, but also they (are, are not) in parallel with R_2.
7. Briefly compare a voltage divider and a current divider.
8. In a balanced bridge circuit the intermediate points (points A and B in Figure 6-10) (are, are not) at the same potential.
9. In a balanced bridge circuit the intermediate points (do, do not) equal zero.
10. In an unbalanced bridge circuit the intermediate points (are, are not) at the same potential.
11. In an unbalanced bridge circuit the potential between the intermediate points (does, does not) equal zero.

PROBLEMS

For the following problems, refer to Figure 6-12.

1. $V_s = 12\ \text{V}, R_1 = 30\ \Omega, R_2 = 60\ \Omega, R_3 = 100\ \Omega, R_4 = 150\ \Omega$. Find $R_t, I_t, I_{R_3}, I_{R_4}, E_{R_1}, E_{R_2}, E_{R_3}, E_{R_4}$.
2. $V_s = 6\ \text{V}, R_1 = 60\ \Omega, R_2 = 120\ \Omega, R_3 = 200\ \Omega, R_4 = 300\ \Omega$. Find $R_t, I_t, I_{R_3}, I_{R_4}, E_{R_1}, E_{R_2}, E_{R_3}, E_{R_4}$.
3. $V_s = 120\ \text{V}, R_1 = 600\ \Omega, R_2 = 300\ \Omega, R_3 = 1\ \text{k}\Omega, R_4 = 1.5\ \text{k}\Omega$. Find $R_t, I_t, I_{R_3}, I_{R_4}, E_{R_1}, E_{R_2}, E_{R_3}, E_{R_4}$.
4. $V_s = 180\ \text{V}, R_1 = 900\ \Omega, R_2 = 450\ \Omega, R_3 = 1500\ \Omega, R_4 = 2.25\ \Omega$. Find $R_t, I_t, I_{R_3}, I_{R_4}, E_{R_1}, E_{R_2}, E_{R_3}, E_{R_4}$.

For the following problems, refer to Figure 6-13.

5. $V_s = 12\ \text{V}$, all resistors are $1.8\ \text{k}\Omega$. Find $R_t, I_t, I_{R_{1,2}}, I_{R_{3,4}}, I_{R_{5,6}}, E_{R_3}$.
6. $V_s = 15\ \text{V}$, all resistors are $2250\ \Omega$. Find $R_t, I_t, I_{R_{1,2}}, I_{R_{3,4}}, I_{R_{5,6}}, E_{R_3}$.
7. $V_s = 5\ \text{V}, R_1 = 100\ \Omega, R_2 = 200\ \Omega, R_3 = 150\ \Omega, R_4 = 225\ \Omega, R_5 = 200\ \Omega, R_6 = 300\ \Omega$. Find R_t, I_t, E_{R_5}.
8. $V_s = 7.5\ \text{V}, R_1 = 150\ \Omega, R_2 = 300\ \Omega, R_3 = 225\ \Omega, R_4 = 337.5\ \Omega, R_5 = 300\ \Omega, R_6 = 450\ \Omega$. Find R_t, I_t, E_{R_4}.

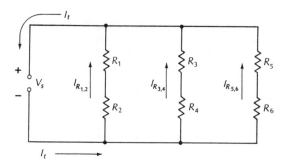

Figure 6-13

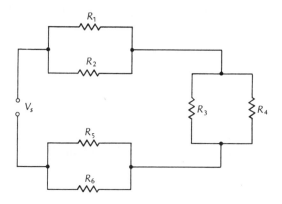

Figure 6-14

9. Briefly describe the fundamental relationships between the various currents in Figure 6-13.
10. Briefly describe the relationships between the various voltages in Figure 6-13.

For the following problems, refer to Figure 6-14.

11. $V_s = 9$ V, all resistors are 3.3 kΩ. Find R_t, I_t, E_{R_3}.
12. $V_s = 12$ V, all resistors are 47 kΩ. Find R_t, I_t, E_{R_3}.
13. $V_s = 12$ V, $R_1 = R_3 = R_5 = 10$ kΩ, $R_2 = R_4 = R_6 = 15$ kΩ. Find R_t, I_t, E_{R_3}.
14. $V_s = 5$ V, $R_1 = R_3 = R_5 = 33$ kΩ, $R_2 = R_4 = R_6 = 47$ kΩ. Find R_t, I_t, E_{R_2}.

For the following problems, refer to Figure 6-15.

15. $V_s = 4.5$ V, $R_1 = R_5 = 1.5$ kΩ, $R_2 = R_3 = R_4$ 1 kΩ, $R_6 = 3$ kΩ. Find $R_t, I_t, I_{R_3}, I_{R_6}, E_{R_6}$.

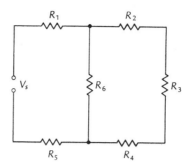

Figure 6-15

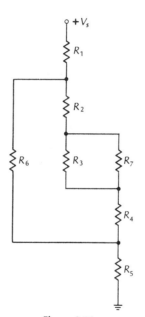

Figure 6-16

16. $V_s = 9$ V, $R_1 = R_5 = 3$ kΩ, $R_2 = R_3 = R_4 = 2$ kΩ, $R_6 = 6$ kΩ. Find R_t, I_t, I_{R3}, I_{R6}, E_{R6}.

17. $V_s = 10$ V, $R_1 = 1500$ Ω, $R_2 = 1800$ Ω, $R_3 = 200$ Ω, $R_4 = 2200$ Ω, $R_5 = 2400$ Ω, $R_6 = 3000$ Ω. Find R_t, I_t, I_{R3}, I_{R6}, E_{R3}.

18. $V_s = 5.9$ V, $R_1 = 15$ kΩ, $R_2 = 18$ kΩ, $R_3 = 20$ kΩ, $R_4 = 22$ kΩ, $R_5 = 24$ kΩ, $R_6 = 30$ kΩ. Find R_t, I_t, I_{R3}, I_{R6}, E_{R3}.

19. Refer to Figure 6-16. $V_s = 30$ V, $R_1 = R_5 = 1$ kΩ, $R_2 = R_4 = 2$ kΩ, $R_3 = R_7 = 3$ kΩ, $R_6 = 5.5$ kΩ. Find R_t, I_t, I_{R6}, I_{R7}, E_{R7}.

20. A Wheatstone bridge similar to that shown in Fig. 6-10 has the following circuit values. $R_1 = 1000\ \Omega$, $R_2 = 2000\ \Omega$, $R_3 = 2700\ \Omega$, $R_4 = 1500\ \Omega$, and $V_s = 6$ V. (a) Is the bridge balanced or unbalanced? (b) What is the voltmeter reading?

21. A Wheatstone bridge similar to that shown in Fig. 6-10 has the following circuit values. $R_1 = 2700\ \Omega$, $R_2 = 3900\ \Omega$, $R_3 = 4700\ \Omega$, $R_4 = 3300\ \Omega$, and $V_s = 18$ V. (a) Is the bridge balanced or unbalanced? (b) What is the voltmeter reading?

22. A Wheatstone bridge similar to that shown in Fig. 6-10 has the following circuit values. $R_1 = 330\ \Omega$, $R_2 = 470\ \Omega$, $R_3 = 560\ \Omega$, and $V_s = 10$ V. (a) What value of R_4 will balance the bridge? (b) Prove that $E_A = E_B$ to four significant figures.

23. A Wheatstone bridge similar to that shown in Fig. 6-10 has the following circuit values. $R_1 = 470\ \Omega$, $R_2 = 330\ \Omega$, $R_3 = 560\ \Omega$, and $V_s = 10$ V. (a) What value of R_4 will balance the bridge? (b) Prove that $E_A = E_B$ to four significant figures.

CHAPTER 7

DIRECT CURRENT METERS

As we have already said, voltage, current, and resistance measurements are usually made with a volt-ohm-milliammeter, or VOM. A typical VOM is shown in Figure 7-1. In this chapter we shall investigate the operation and application of such an instrument. Most *multimeters* will appear superficially different from the one pictured, but nearly all are basically the same, are operated in the same manner, and by the same rules.

Because there are usually three separate sections to a VOM, each will be presented separately. Then, these principles will be combined to form a complete instrument. Note that single-function meters are also used extensively—volts only, amps only, and so on. For example, a voltmeter mounted as an integral part of an overall system may be used to monitor continuously a particular circuit voltage. This is especially true if the value of voltage is critical to proper equipment operation.

After the VOM is thoroughly covered, the digital multimeter (DMM) will be introduced. In many applications the DMM is in widespread use, since it is inherently more accurate, is much safer to use, is virtually damage-proof, and is much easier to read than a VOM. The DMM operates on a completely electronic basis, however, and only the operation of the instrument (not its construction) will be discussed.

In addition to the basic moving-coil meter movement (the D'Arsonval movement) used in nearly all VOMs; and the DMM; we shall also discuss other dc measuring devices and means.

The major subject headings are as follows:

7-1 The Basic Meter Movement
7-2 Meter Scales
7-3 Ammeters
7-4 Voltmeters
7-5 Ohmmeters
7-6 Multimeters
7-7 Digital Multimeters (DMMs)

Figure 7-1 A typical volt-ohm-milliammeter (VOM). Courtesy Triplett Electric Co.

7-1 THE BASIC METER MOVEMENT

Figure 7-2 illustrates the basic D'Arsonval, or moving-coil, meter movement. It is constructed primarily of a strong permanent magnet with a moving coil mounted between the pole pieces. As will be discussed in detail later, when current flows in a wire, a weak magnetic field exists around the wire; the field can be intensified by winding the wire into a coil. Thus, when current is caused to flow in the coil, the coil's magnetic field reacts with that of the permanent magnet, to cause the coil to rotate around a pivot. This action is exactly the same as the attraction or repulsion forces between two magnets. The indicator needle is attached to the coil so that it points to zero on the scale when no current flows in the coil. When current is caused to flow in an amount to cause the needle to read at the top of the scale (full scale), the value of current is known as *full-scale* current (I_{fs}). The movement, or rotation, of the indicating needle is proportional to the value of current. That is, if a current of 50 μA is applied to a 100-μA movement, the needle will indicate a reading exactly halfway between zero and full scale.

The practical meter movement has many mechanical refinements. The moving coil rotates on jeweled bearings to reduce friction. Attached to the pivot is a return spring that returns the needle to zero when no current is applied to the coil. A mechanical zero adjustment is

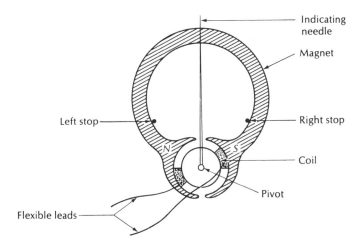

Figure 7-2 The moving-coil (or D'Arsonval) meter movement.

provided to zero the needle directly over the zero mark. At both ends of the scale are mechanical stops, which prevent the indicating needle from traveling too far and being damaged. Each stop has a soft rubber "bumper" attached to absorb the shock when the needle hits it. Two flexible leads connect the coil to the meter circuitry. An alternate method of construction eliminates the jeweled bearings and thus also eliminates much error due to friction. This is known as the *taut-band* meter movement. The coil and indicator assembly are suspended between the pole faces of the magnet by a metal band whose tension is very accurately determined—hence the name *taut*. This tension also eliminates the return spring found on the more conventional movement. Because the taut-band movement eliminates much of the friction found in the moving-coil movements, the coil can be made smaller and lighter, giving the meter a smaller internal resistance for a given sensitivity. Additionally, the taut-band meter is generally more rugged and has equal, if not better, accuracy.

The amount of current required for a full-scale reading for a particular meter movement is determined by the number of turns of wire on the coil, and by the magnetic strength of the magnet. To make a movement that has greater *sensitivity* (smaller I_{fs}) requires more turns of finer wire wound on the coil. Long lengths of very fine wire increase the resistance of the coil as the maximum current is made smaller. For example, a commonly used meter movement has an I_{fs} of 1 mA, and a coil resistance (R_m) of 27 Ω. Compare this movement with one having an I_{fs} of 50 μA and an R_m of approximately 2000 to 3000 Ω.

Moving-coil meter movements are generally available in sensitivities ranging from 10 μA to 1 A. In order to provide for several ranges in one instrument, additional circuitry is added to the basic meter movement. Much of the remainder of this chapter will be concerned with this additional circuitry.

7-2 METER SCALES

A volt-ohm-milliammeter has a multipurpose scale that allows voltage, current, and resistance to be read. Reading such a scale is not difficult, but care must be exercised to make certain the correct scale is used and the correct multiplying factor is applied. Nearly all meter faces and scale factors are arranged so that the multiplying factor is ten, to reduce the chance of error.

A typical meter face with more than one scale is shown in Figure 7-3. The top scale has a range of 0–1.0 mA, while the lower one reads 0 to 5.0 mA. Which scale is used is determined by the switch setting, which usually is labeled according to the full-scale reading. For example, consider the indicator drawn in dashed lines. If the meter's range switch is set to 1.0 mA, then the upper scale is read directly, indicating a current of 0.5 mA. Using the same indicator position, if the range switch is set to 5.0 mA, then the current is 2.5 mA.

Often, the same scales are used for multiples of the printed meter face values. For example, if the meter has a range setting of 50 mA, then the indicated reading would be 25 mA, obtained by multiplying 2.5 by 10. If the range setting is 1000 mA, the reading is 500 mA, using the top scale and multiplying 0.5 by 1000.

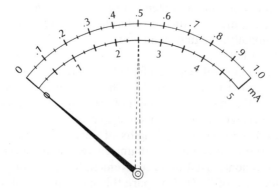

Figure 7-3 A typical meter scale with dual full-scale current values.

If smaller divisions are used on the meter face, remember to take account of the appropriate divisions and to determine the value of the individual divisions for the switch setting in use. This is not unlike reading the minor divisions on an inch ruler, or the divisions on a slide rule.

It should be noted at this point that the most accurate part of the scale is approximately the upper half. Thus, for the most accurate readings, avoid very low readings, near the left end of the printed scale.

7-3 AMMETERS

An *ammeter* is a device used to measure current. Generally, an ammeter measures large values of current (in excess of 1 A). A *milliammeter* is used for values from approximately 1 to perhaps 900 mA, while a *microammeter* is used for values significantly less than 1 mA. All three types are fundamentally the same, differing only in their sensitivity and full-scale current.

The D'Arsonval meter movement is basically a current-measuring device useful only for direct current. However, it is adaptable, and can also be used for dc volts and ac volts, as well as resistance. These additional functions are provided by additional circuitry that allows the meter movement to be much more versatile. Even a practical ammeter (or milliammeter) usually has additional circuitry to provide greater flexibility by providing more than one scale.

The Current Meter Shunt

To increase the full-scale current of a given movement, a circuit arrangement is used that allows some of the current to be *shunted* around the meter itself. When this is done, more total current can be handled and the *effective* full-scale current is increased. Such a device does not increase the I_{fs} of the meter itself, of course, but does increase the total current in the meter leads.

Figure 7-4 illustrates a simple case of a basic 1-mA meter using a shunt to double the measured current. The meter movement has an inherent internal resistance of 27 Ω, and an I_{fs} of 1 mA, as shown in *a*. To double the range, a 27-Ω resistor is added in shunt (or parallel) with the meter, as in *b*. Now, when the total current is 2 mA, each branch current will be 1 mA, since each branch resistance is the same. The total device would now be said to have a full-scale current of 2 mA. Note that the original scale reading of 0 to 1 mA must be doubled. That is, all readings on the original scale are to be multiplied by two. If a switch were to be added, the meter could have two scales, and the

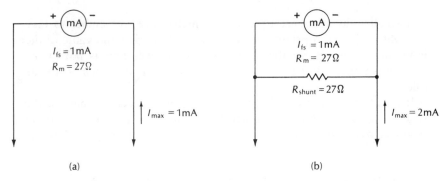

Figure 7-4 A 1-mA meter movement altered to accept 2-mA maximum current.

meter face would be so marked, using two separate scale markings for the two ranges.

To determine the value of a single shunt resistor when the values are not self-evident, Ohm's law can be applied. Figure 7-5a will be used as an example. The meter movement requires 50 μA full-scale and has an internal resistance (R_m) of 3000 Ω. It is required to make the new I_{fs} equal to 500 μA, a tenfold increase.

To determine the value of R_{shunt}, the first unknown to find is the value of voltage, V_{fs}, which, when applied to the movement, just causes full-scale deflection:

$$V_{fs} = I_{fs} \times R_m = (50 \times 10^{-6})(3 \times 10^3) = 0.15 \text{ V}.$$

This simply means that it requires 0.15 V to cause 50 μA of current to flow in the meter movement itself.

Now, because the meter movement and the shunt will be in parallel when R_{shunt} is installed, the same voltage will appear across R_{shunt}. Since no more than 50 μA can flow in the meter, if full scale is to be 500 μA, the shunt must carry 450 μA.

$$R_{shunt} = \frac{V_{fs}}{I_{shunt}} = \frac{0.15}{450 \times 10^{-6}} = 333.33 \text{ }\Omega.$$

Figure 7-5b illustrates a three-scale meter with a switch to select the proper shunt. The value of each can be found in the same way that the single shunt was found above.

In Figure 7-5b, if the switch contacts become dirty, or if they open during switching, the full available current flows through the meter movement, possibly causing damage. To prevent this, the so-called *universal* shunt, Figure 7-5c, is used. Note that if the switch contacts open for *any* reason, no current flows anywhere in the circuit.

With the switch in position 1, the total shunt resistance is 100 Ω, hence the full-scale current is 2 mA, since $R_m = R_{shunt}$. For position 2, (20 mA):

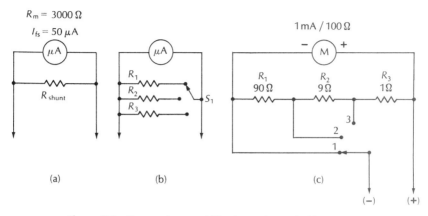

Figure 7-5 Current shunts. (a) The basic shunt; (b) Three-range shunt system; (c) the universal shunt.

$$I_{shunt} = \left(\frac{R_1 + R_m}{R_1 + R_2 + R_3 + R_m}\right)I_t = \left(\frac{190}{200}\right)20 \text{ mA} = 19 \text{ mA}$$

For position 3 (200 mA):

$$I_{shunt} = \left(\frac{R_1 + R_2 + R_m}{R_1 + R_2 + R_3 + R_m}\right)I_t = \left(\frac{199}{200}\right)200 \text{ mA} = 199 \text{ mA}$$

Hence, the shunt bypasses all but 1 mA (I_{fs}) around the movement for any switch position.

Current Measuring Techniques

As has been briefly mentioned, to measure current in a circuit, a circuit connection must be broken and the meter inserted, recompleting the circuit. Also, the dc meter must be inserted so as to produce an upscale reading; that is, polarity must be observed. Note Figure 7-6, where a sequential drawing illustrates these two important rules.

To these two rules must be added a third. *The meter must have a full-scale reading equal to or greater than the circuit current.* If this rule is ignored, meter damage is very likely to occur. In order to apply this rule it is necessary to be able to estimate the total current in the circuit to a reasonable degree of accuracy, and then to use a scale setting equal to or slightly higher than the estimate. If it is not possible to estimate the value of current, then begin the measurement by using the *highest* scale on the meter, and reducing the scale factor a step at a time until a reading is obtained approximately in the upper half of the scale.

For example, using the meter illustrated in Figure 7-1, the available ranges of dc current are 600 mA, 60 mA, 6 mA, 0.6 mA (600 μA). If estimated current is 100 mA, use the 600-mA switch setting. Then read the scale marked 0–60 and multiply by 10. If estimated current is 30 mA, use the 60-mA setting and read the 0–60 scale directly.

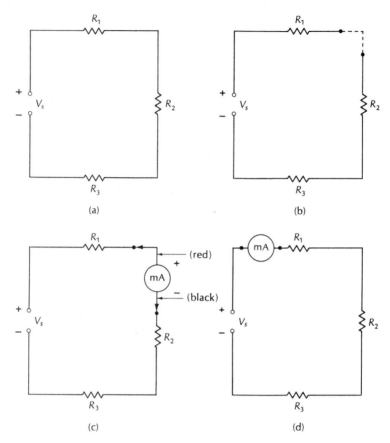

Figure 7-6 Steps in measuring current in a circuit. (a) The circuit in which current is to be measured; (b) break the circuit; (c) insert the ammeter, observing polarity; (d) the same value of current is measured no matter where the measurement is made in a series circuit.

Accuracy of Current Measurements

From the preceding discussion, it can be inferred that *a milliammeter has an appreciable internal resistance*. In some circuits this has a negligible effect. However, in others, the effect can be such as to cause large errors in the reading. To illustrate the possible error, Figure 7-7a shows a circuit with a milliammeter having an internal resistance of 100 Ω. In this circuit, the error will be appreciable if the resistor values are low. For example, assume that each resistor is 100 Ω. Without the meter in the circuit, normal current is 40/400 = 0.1 = 100 mA. However, when

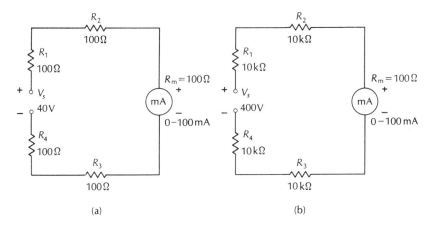

Figure 7-7 Illustrating the importance of a current meter's internal resistance. (a) In a circuit where resistor values are small, the internal resistance of the meter can cause a critical error; (b) with large resistance values the meter's internal resistance has no serious effect. See text for an alternate method of determining current.

the meter is inserted, it will actually read 40/500 = 0.08 = 80 mA, a *20% error*.

Figure 7-7*b* illustrates the same meter in a circuit similar to the preceding one, except that the value of each resistor is 10 kΩ. Now, normal current is 400/40,000 = .01 = 10 mA with the meter out of the circuit. With the meter, the reading will be 400/40,100 = .009975 = 9.975 mA, a much smaller error, about 0.25%. Obviously, current meters must have as low a value of internal resistance as possible.

If there is any doubt as to the accuracy of a current measurement, an alternate method can be used which has several advantages. First of all, the circuit does not have to be broken, which often saves much time. Secondly, the accuracy is determined primarily by the accuracy of the resistors themselves. Using ±5% resistors, the reading itself is usually within that degree of accuracy.

The method is simplicity itself. Simply measure the voltage across any resistor in the circuit, using a sensitive voltmeter. Then apply Ohm's law to find the current. In Figure 7-7*a*, a voltmeter would read 10 V across any resistor. Hence, E_{R_1}/R_1 gives the true value of current:

$$I = \frac{E_{R_1}}{R_1} = \frac{10}{100} = 0.1 = 100 \text{ mA}.$$

Another factor relating to the accuracy of a meter reading is the accuracy of the movement itself, and the repeatability of two successive but identical readings. Because of slight variations in one moving coil compared to another, no two meters will read *exactly* the same.

Most manufacturers exert every effort to reduce this effect to the minimum. Also having a great influence on accuracy are the resistors used as shunts, and so on. Very accurate resistors are used in the meter circuitry to increase overall accuracy as much as possible.

An accuracy of ±5% is typical for many current meters, with ±3% accuracy available. Laboratory instruments with accuracy on the dc scales of ±1% or better can be obtained for a corresponding increase in cost.

7-4 VOLTMETERS

While it is true that the moving-coil movement is fundamentally a current-measuring device, it can be (and is) very effectively used for voltage measurements. To understand this, consider a movement having a full-scale current of 1 mA and an internal resistance of 27 Ω. One might ask the question, What will be the voltage drop E_{fs} across the coil when full-scale current is flowing? This is easily answered.

$$E_{fs} = I_{fs} \times R_m = 0.001 \times 27 = 27 \text{ mV}.$$

The meter movement itself, then, can be considered a millivoltmeter, which will read full scale if placed across a source of 27 mV. One might also ask how the voltage can be increased to make the instrument more practical. Well, since the internal resistance is the limiting factor as far as full-scale current is concerned, it seems logical to simply increase the effective R_m to limit current to, in this case, 1 mA at a higher voltage. While the coil resistance itself cannot be changed, simply placing more resistance in series with the movement will produce the same effect.

A resistor placed in series with the coil to increase the voltage while limiting current to I_{fs} or less is called a *voltage multiplier,* or simply a *multiplier*. A simple example of the use of a voltage multiplier is given in Figure 7-8a. Here, a 1-mA movement is connected in series with a 9973-Ω resistor used as the multiplier. The total effective resistance is 9973 + 27 = 10,000 Ω. The voltage which will cause exactly 1 mA of current to flow through the meter is

$$E_{fs} = I_{fs} \times R_t = 0.001 \times 10,000 = 10 \text{ V}.$$

In this instance, the basic 1-mA movement has been converted to a voltmeter with a full-scale reading of 10 V. The meter scale would, in all probability, be constructed to yield direct voltage measurements, as illustrated in Figure 7-8b.

Any current meter of the moving-coil type can be changed to read voltage by this simple expedient. The desired full-scale voltage is cho-

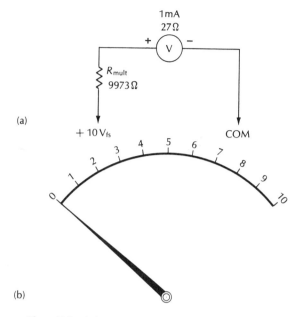

Figure 7-8 A 1-mA movement altered to form a 10-V full-scale meter: (a) the meter circuit; (b) the meter scale.

sen and a value of resistance is added in series with the movement to allow I_{fs} just to flow when the chosen value of voltage V_{fs} is applied. Assume that the meter shown in Figure 7-8 must also have a 100-V scale. The required value of resistance is

$$R_{mult} = \frac{E_{fs}}{I_{fs}} - R_m = \frac{100}{0.001} - 27 = 99{,}973 \ \Omega.$$

The circuit with two ranges might appear as in Figure 7-9, along with the meter face.

Ohms-per-volt Rating

A widely used method of comparing one voltmeter with another is the ohms-per-volt rating. Most VOMs have this information printed on the meter face, thus allowing a judgment to be made in favor of one meter or another.

The ohms-per-volt rating of a particular meter is a number that indicates how much resistance a particular meter movement requires *per volt full scale*. Using the voltmeter in Figure 7-10 as an example, the total resistance in the 1-V full-scale circuit is equivalent to the meter's ohms-per-volt rating. In this instance, this number is 20,000 Ω/V

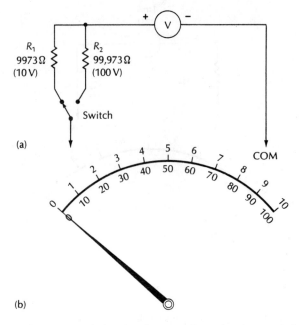

Figure 7-9 A dual-scale voltmeter with a scale-selector switch: (a) the meter circuit; (b) the meter scale.

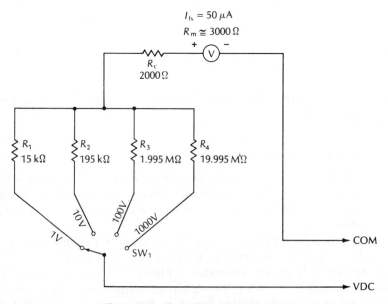

Figure 7-10 The multirange meter circuit.

($R_1 + R_2 + R_3$). The ohms-per-volt rating of any meter with a 1-V full-scale rating can be as easily determined.

An equally simple method requires only a knowledge of the basic full-scale current of the movement. The ohms-per-volt rating is simply the reciprocal of I_{fs}. For the 50-μA movement, this is

$$\Omega/V = \frac{1}{50\ \mu A} = 20{,}000\ \Omega/V.$$

A 1-mA movement is rated at $1/0.001 = 1000\ \Omega/V$. Because ohms-per-volt and I_{fs} are related reciprocally, the greater the Ω/V rating, the more sensitive the meter is and the less current it requires from the circuit being measured.

Multiscale Voltmeter

Figure 7-10 illustrates a circuit for a voltmeter with four ranges, switch selectable by S_1. The ranges are 1, 10, 100, and 1000 V_{dc}. The meter movement has a 50-μA sensitivity and an internal resistance of approximately 3000 Ω. Because it is not practical to manufacture the coil resistance with an accurate internal resistance (coil dimensions and magnetic properties are most important), a separate resistor is often used to calibrate the movement. The 2000-Ω resistor R_c in series with the movement is just such a calibration resistor. If the coil resistance (R_m) happens to be 2950 Ω, then R_c is chosen to be 2050 Ω; the total resistance is exactly 5000 Ω. If the coil resistance is 3100 Ω, R_c is chosen to be 1900 Ω, so again the total is 5000 Ω. R_c, then, can be considered a part of the total meter resistance.

The proper value of each multiplier can now be determined for each scale factor. R_t is first calculated for each value of V_{fs} and the total meter resistance is subtracted from R_t to find the multiplier resistance.

1-V scale:

$$R_t = V_{fs}/I_{fs} = 1.0/(50 \times 10^{-6}) = 20{,}000\ \Omega;$$
$$R_1 = R_t - (R_c + R_m) = 20{,}000 - 5000 = 15{,}000\ \Omega.$$

10-V scale:

$$R_t = V_{fs}/I_{fs} = 10/(50 \times 10^{-6}) = 200{,}000\ \Omega;$$
$$R_2 = 200{,}000 - 5000 = 195{,}000\ \Omega.$$

100-V scale:

$$R_t = V_{fs}/I_{fs} = 100/(50 \times 10^{-6}) = 2\ M\Omega;$$
$$R_3 = 2\ M\Omega - 5\ k\Omega = 1.995\ M\Omega.$$

1000-V scale:

$$R_t = V_{fs}/I_{fs} = 1000/(50 \times 10^{-6}) = 20 \text{ M}\Omega;$$
$$R_4 = 20 \text{ M}\Omega - 5 \text{ k}\Omega = 19.995 \text{ M}\Omega.$$

Voltmeter Loading

As mentioned earlier, a voltage reading is taken by measuring across one of the circuit elements, whether a source or a drop. This places the voltmeter *in parallel* with the device being measured. When measuring a voltage source, the meter does not normally affect the circuit, since the source can normally supply the very small current required by the meter.

However, when measuring across a load, the meter *can* cause a completely erroneous reading. Figure 7-11 illustrates an extreme instance, where this effect is strikingly clear. Figure 7-11a shows a circuit with two equal value resistors. The voltage is to be measured across R_2. The meter to be used is a 1000 Ω/V meter having a 100-V scale. The measurement is being made in Figure 7-11b, where you can see that the *meter reads 40 V*. However, since both resistors are equal, *each should drop 60 V*. The meter has obviously upset the expected reading, and reads 20 V too low, a 33% error. This effect is called *loading*.

To determine why this loading error exists consider Figure 7-11c. Note that when the meter is connected to R_2, the meter itself, in parallel with R_2, has an effective resistance of 50,000 Ω. This is verified as follows. The meter's sensitivity is 1000 Ω/V. Therefore, Ω/V $\times$ V_{fs} = 1000 $\times$ 100 = 100,000 Ω. The meter and its multiplier have a total resistance of 100 kΩ, then. This, in parallel with R_2, yields an effective parallel resistance of 50,000 Ω. Thus, the meter has so drastically changed circuit conditions that it is no longer the same circuit, as long as the meter is connected.

It is therefore evident that *a voltmeter must have as high an internal resistance as possible,* to avoid producing such errors. In order to put this into practice, the meter movement chosen should have the highest degree of sensitivity possible. That is, I_{fs} should be as small as practical.

As a further example, let us assume that another meter is used to make the measurement across R_2 in Figure 7-11. The meter movement chosen has an I_{fs} of 50 μA, a commonly encountered rating. Now, this meter has a sensitivity of $1/50\mu A$ = 20,000 Ω/V, which is significantly better than the original meter. For its 100-V scale, the total resistance $(R_{mult} + R_m)$ must be Ω/V $\times$ V_{fs} = 20,000 $\times$ 100 = 2 MΩ. Hence, on its 100-V scale, this meter will change the circuit values much less. R_2, in parallel with 2 MΩ, will have a parallel effective resistance of

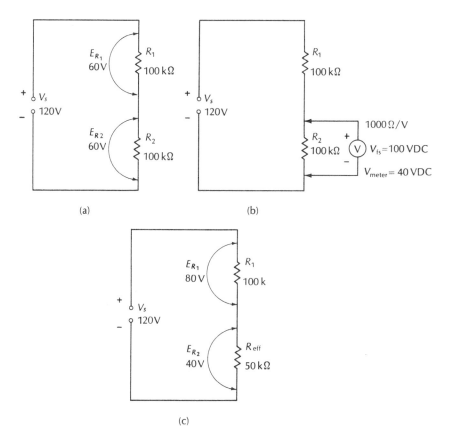

Figure 7-11 Illustrating the effect of poor voltmeter sensitivity: (a) the circuit to be measured; (b) meter loading produces too low a reading; (c) the cause of meter loading.

$$R_{\text{eff}} = \frac{1}{1/R_2 + 1/R_{\text{mult}}} = \frac{1}{1/100 \text{ k}\Omega + 1/2 \text{ M}\Omega} = 95{,}238 \text{ }\Omega.$$

The total circuit resistance is this value in series with R_1.

$$R_t = R_1 + R_{\text{eff}} = 100 \text{ k}\Omega + 95{,}238 \text{ }\Omega = 195{,}238 \text{ }\Omega.$$
$$I_t = \frac{V_s}{R_t} = \frac{120}{195.238} = 0.615 \text{ mA}.$$
$$E_{R_1} = I_t \times R_1 = 61.46 \text{ V}.$$
$$E_{R_2} = I_t \times R_{\text{eff}} = 58.54 \text{ V}.$$

The loading error is much smaller than in the previous example. However, some error does still exist, even using a 20,000 Ω/V meter, which is a relatively sensitive instrument.

When reading voltages across resistors having large values of resistance, it is often of value to use a vacuum-tube voltmeter (VTVM) or a solid-state VOM. These devices are discussed in more detail in Section 7-6; at this point they are mentioned only to point out instruments originally designed to overcome the errors inherent in even a good-quality VOM. The VTVM and the solid-state VOM can be designed to present a much higher resistance to the circuit being measured, typically 10 or 11 MΩ. Laboratory instruments with resistances of 110 MΩ are not uncommon.

When the value of the resistance being measured for voltage drop is relatively small, the loading effect is less pronounced. For example, if the resistors in Figure 7-11a each had a value of 1000 Ω, the error would be much smaller, even using the less sensitive meter. In this instance, the effective resistance is reduced to only 99.9 Ω, a 0.1% error. It is evident that loading is of importance only when the resistor value approaches the value of the internal resistance of the meter.

7-5 OHMMETERS

An *ohmmeter* is a device used to measure the resistance of resistors, wires, and other electrical components. Because a resistor can only limit current, the ohmmeter must contain a source of EMF to provide the meter movement with current. A basic ohmmeter circuit is illustrated in Figure 7-12, where it is clear that this ohmmeter consists of a 3-V battery, the meter movement, and the limiting resistance.

In this simple example, the 2973-Ω series resistor and the 27-Ω R_m comprise the total internal resistance (R_{mt}) of the ohmmeter, which is 3000 Ω. With a 3-V source, the current through the meter will be 0.001 A when the ohmmeter leads are shorted together. Thus, the typical ohmmeter scale indicates the zero-ohm point on the scale at the *right* side of the scale, as shown in Figure 7-13. The scale on nearly all ohmmeters reads from right to left, which is the reverse of voltage and current scales.

To appreciate how the circuit functions, the meter full-scale current I_{fs} flows when the leads are shorted together.

$$I_{fs} = \frac{V_s}{R_{mt}} = \frac{3.0}{3000} = 0.001 = 1 \text{ mA}$$

Now, if an external 3000-Ω resistor to be measured (R_{meas}) is placed between the leads, the current through the meter (I_m) will read exactly half scale:

$$I_m = \frac{V_s}{R_{mt} + R_{meas}} = \frac{3.0}{6000} = 0.0005 = 0.5 \text{ mA}$$

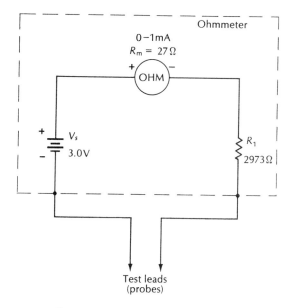

Figure 7-12 A simple ohmmeter circuit.

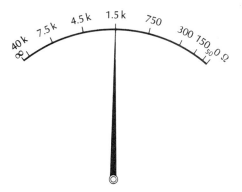

Figure 7-13 The simple ohmmeter scale.

That is, the total resistance is the resistance of the meter (R_{mt}) plus the resistance being measured (R_{meas}), or 6000 Ω. As indicated, only half of I_{fs} will flow, and the needle will therefore indicate only half scale. Other values of resistance placed between the probes will cause a deflection proportional to the resultant current, which is inversely proportional to the resistance being measured.

Inspection of a typical ohmmeter scale reveals that the scale is not linear, as are the voltage and current scales. The resistance scales are

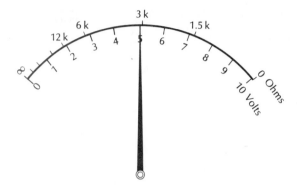

Figure 7-14 Volts and ohms scales compared.

expanded on the right side and compressed on the left side. To assure an accurate reading, care must be exercised in reading the scale, especially on the left, or high ohms, side of the scale. Figure 7-14 illustrates a typical resistance scale compared with a voltage scale. Note the rapid increases in resistance toward the left of the scale. For this reason, the most accurate resistance readings are obtained toward the full-scale side of the meter face (indicator to the right).

Another point to be considered in reading the resistance scale: The infinity sign (∞) at the left end of the scale is relative only. For example, a 500,000-Ω resistor certainly does not have an infinitely high value, but on the lower scales, $R \times 1$ perhaps, so little current flows that no visible movement of the indicator will occur. However, by changing the range switch to a higher value, the 500,000-Ω resistor will certainly cause a suitable deflection.

Ohmmeter Operation

Several meters having typical resistance scales are shown in Figure 7-15. Several refinements are evident, compared to the simple example explained previously. First, there are several scales, $R \times 1$, $R \times 10$, $R \times 100$, $R \times 1$ K, and so on, covering a wide range of resistance values. Second, a *zero-ohms adjustment* is evident on each instrument. This feature is necessary so that, when the leads are shorted together, *exactly* full-scale current can be caused to flow even when the internal battery is weak. Third, the meter in part *c* of the figure has a mirrored scale, which reduces *parallax,* the apparent displacement of the needle when it is viewed from the side. When reading any scale, but especially an unmirrored one, be careful to take all readings from the head-on position.

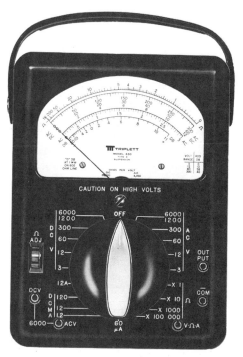

Figure 7-15 Three volt-ohm-milliammeters: (a) a small pocket type; (b) a larger bench type; (c) a bench type with a mirrored scale to reduce parallax. Courtesy Triplett Electric Co.

Operation of the ohmmeter is relatively simple. However, it is important to remember the following rules.

1. Power must be off in the circuit being measured—the ohmmeter provides its own power.
2. On the first measurement, and after changing scales, the meter must be "zeroed" by shorting the leads together and turning the zero adjustment until the indicator needle reads exactly zero ohms full scale.
3. Estimate the correct range to use so that the indication is in the right-hand one-third of the scale. Avoid readings close to ∞ to ensure accurate readings. Remember that ohmmeter readings close to infinity are very inaccurate because the left-hand portion of the scale is so highly compressed.
4. After the measurement is made, multiply the meter reading by the factor indicated by the switch setting.
5. When measuring components not disconnected from the circuit, beware of "sneak" paths that will cause inaccurate readings. If in doubt, disconnect one end of the component. This ensures that the meter current flows *only* through the part being measured.
6. When using an ohmmeter to measure the resistance of resistors, coils, and the like, polarity is normally of no concern. However, as will be discussed later, certain components (such as solid-state diodes and transistors) *are* polarity sensitive. In these instances, it is important to observe polarity by knowing what the polarity of the ohmmeter leads is. Often, it is *not* the same polarity used for dc volts and amperes.

Practical Ohmmeter Circuit

The simple circuit illustrated in Figure 7-12 is not practical, except for a single-range instrument capable of relatively low resistance readings. Furthermore, the 0-to-1-mA movement is not sensitive enough for high-value readings, and there is no provision for battery voltage compensation or for zero-ohms-adjustment.

In Figure 7-16, a circuit is developed step by step to yield a four-range ohmmeter having a zero adjust and capable of readings in excess of 1 MΩ. The only element not illustrated is the switching required to successively switch in additional resistors while leaving in the circuit those already switched in.

In the first drawing the basic meter is a 50-μA, 3000-Ω movement in series with R_1, a 297,000-Ω resistor. The battery provides 15 V. With the meter probes open, there is no complete circuit, hence no current flows. When the probes are shorted together, a current flows in the amount of

Direct Current Meters 145

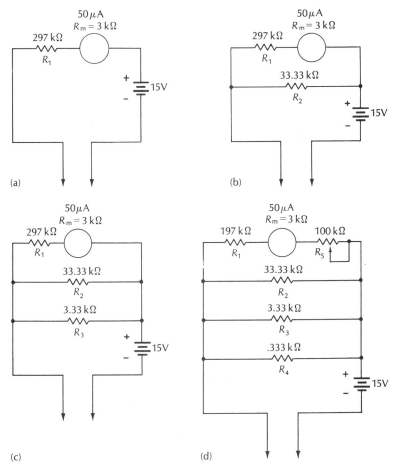

Figure 7-16 The development of a four-range ohmmeter: (a) Midscale is 300 kΩ; (b) midscale is 30 kΩ; (c) midscale is 3 kΩ; (d) midscale is 300 Ω and the zero adjustment has been added.

$$I_{fs} = \frac{V_s}{R_m + R_1} = \frac{15}{297{,}000 + 3000} = 50 \ \mu A.$$

Thus, as long as the values illustrated are exact and the probes are touching, 50 μA flows and the meter indicates full scale, or zero ohms. This, of course, is the value of resistance between the probes when they are shorted together.

Assume now that a 300,000-Ω resistor is placed between the probes. Now, a smaller value of current will flow.

$$I_m = \frac{V_s}{(R_m + R_1) + R_{meas}} = \frac{15}{600{,}000} = 25 \ \mu A.$$

When measuring the 300,000-Ω resistor the meter will indicate exactly half-scale. Values in excess of 1 MΩ can be easily read on this scale.

In the second drawing, a second resistor (R_2) has been added in shunt with the meter and R_1. This completely changes the meter's response to measuring resistance. Total current supplied by the battery when the probes are shorted together is now a function of the total resistance for this circuit.

$$R_t = \frac{1}{[1/(R_m + R_1)] + (1/R_2)} = \frac{1}{1/300 + 1/33.33 \text{ k}\Omega} = 30 \text{ k}\Omega.$$

The branch currents with the probes shorted together can also be determined:

$$I_{R_2} = \frac{15}{33.33 \text{ k}\Omega} = 0.45 \text{ mA};$$

$$I_m = \frac{15}{300 \text{ k}\Omega} = 50 \text{ }\mu\text{A}.$$

On this second scale, then, the shunt has no effect on the meter itself as long as the meter leads are open, because no current flows anywhere in the circuit. If the leads are shorted, $I_m = 50$ μA, or full-scale current.

However, if a 30,000-Ω resistor is placed between the probes, the current through the meter will be much less than 50 μA, since total resistance is now much greater.

$$R_t = R_{\text{int}} + R_{\text{meas}} = 30,000 + 30,000 = 60 \text{ k}\Omega,$$

where R_{int} is the internal meter resistance.

$$I_t = \frac{V_s}{R_t} = \frac{15}{60,000} = 250 \text{ }\mu\text{A}.$$

However, only a portion of this current flows through the meter; the remainder flows through the shunt. The voltage across the shunt resistor R_1 is equal to one-half of the supply voltage, since internal meter resistance equals external resistance.

$$I_m = \frac{7.5}{300,000} = 25 \text{ }\mu\text{A}.$$

The meter will therefore indicate a reading of one-half full scale, which would be calibrated for 30 kΩ. The two scales discussed to this point, then, have mid-scale ranges of 300,000 and 30,000 Ω, and thus differ from each other by a factor of 10.

In Figure 7-16c, a third scale is formed by the addition of another shunt resistor, R_3, leaving R_2 connected in the circuit. As before, total current is a function of total internal resistance and external resistance. Now, when the probes are shorted together, total current is determined by first finding R_{int}.

$$R_{int} = \frac{1}{1/(R_m + R_1) + 1/R_2 + 1/R_3} = \frac{1}{1/300 \text{ k}\Omega + 1/33.33 \text{ k}\Omega + 1/3.33 \text{ k}\Omega}$$
$$= 3 \text{ k}\Omega.$$
$$I_t = \frac{V_s}{R_t} = \frac{15}{3000} = 5 \text{ mA}.$$

As before, the largest part of this total current flows through the shunts, and only a small part through the meter.

$$I_m = \frac{V_s}{R_m + R_1} = \frac{15}{300,000} = 50 \text{ }\mu\text{A}.$$

When measuring an external resistor having resistance of 3000 Ω, meter current can be determined as before. Since $R_{int} = 3000$ Ω, the battery voltage will be divided equally between R_{int} and R_{meas}.

$$I_m = \frac{7.5}{300,000} = 25 \text{ }\mu\text{A},$$

and as before, the meter will read at half scale for the 3000-Ω resistor.

The fourth scale is provided by again adding another shunt, still leaving all other resistors connected in the circuit. This is shown in Figure 7-16d. Also note in the drawing that a further refinement has been added. In the meter leg of the circuit, the 297,000-Ω resistance has been broken up into a 197,000-Ω fixed resistor in series with a 100,000-Ω variable resistor which is the zero-adjust potentiometer. This compensates for aging of the battery and for slight variations in the resistors on the several ranges. To use such a circuit, assume that R_5 is adjusted so that $R_1 + R_5 = 297,000$ Ω. The circuit is analyzed below without further comment.

$$R_{int} = \frac{1}{1/(R_m + R_1 + R_5) + 1/R_2 + 1/R_3 + 1/R_4}$$
$$= \frac{1}{1/300 \text{ k}\Omega + 1/33.33 \text{ k}\Omega + 1/3.33 \text{ k}\Omega + 1/0.333 \text{ k}\Omega} = 300 \text{ }\Omega.$$
$$I_t(\text{leads shorted}) = \frac{15}{300} = 50 \text{ mA}.$$
$$I_m(\text{leads shorted}) = \frac{15}{300,000} = 50 \text{ }\mu\text{A}.$$
$$I_t(R_{meas} = 300 \text{ }\Omega) = \frac{V_s}{R_{meas} + R_{int}} = \frac{15}{600} = 25 \text{ mA}.$$
$$I_m(R_{meas} = 300 \text{ }\Omega) = \frac{7.5}{3000,000} = 25 \text{ }\mu\text{A}, \text{ or } I_m = I_{fs}/2.$$

A final comment on the illustrated ohmmeter circuit is in order. On the two lower ranges there is no need for a current as large as indicated (5 mA and 50 mA). In a practical circuit the 15-V battery would in all probability be switched *out* of the circuit and a 1.5-V cell switched in. This would, of course, change the values of the shunt resistors used, but would conserve current, which in a battery-operated portable meter is important.

7-6 MULTIMETERS

The instrument illustrated in Figure 7-1 is, of course, a multifunction VOM. It is capable of measuring dc volts to 1200 Vdc, dc current to 600 mA, ac volts to 1200 V_{ac}, and ohms to 5 MΩ. Similarly, the instruments pictured in Figure 7-15 are multifunction meters. Figure 7-17 illustrates two other meters, an FET VOM and a digital multimeter (DMM). Each of these multifunction meters will be briefly described and general operating procedures will be provided. Specific operating instructions cannot be given, since a variety of instruments is available, each with somewhat different features and mechanical layout.

VOM

The VOM features portability, compactness, simplicity, and ruggedness. As already said, the VOM measures dc volts, ac volts, dc current, and resistance; and is capable of making such measurements in several ranges. It is usually battery operated, and hence does not depend upon external power.

One feature to be aware of is the fact that on the volts scale, the total internal resistance of the meter ($R_m + R_{mult}$) is different on each switch setting. Thus, the loading effect varies depending upon the full-scale voltage setting. The higher the range setting, the smaller the loading effect.

Also, it should be noted that the range switch on the typical VOM is labeled somewhat differently on the ohms scales than on volts and milliamps. For volts and milliamps, the switch setting indicates the full-scale value; on the resistance scales, a multiplying factor is given that must be applied to the indicated reading.

In general, operating the VOM is similar to operating the single-function meters described previously. All rules given to this point are applicable.

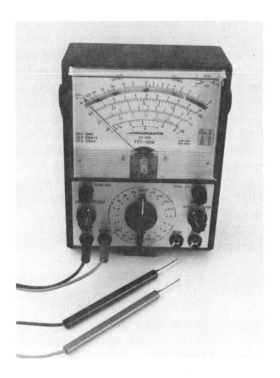

Figure 7-17 (a) An FET-VOM, courtesy of Radio Shack; (b) a 3½ piece-digit digital multimeter (DMM), courtesy of Systron Donner Corp.

Solid State VOM

A solid-state (FET) VOM is a meter that has a very high and constant input resistance on the dc volts ranges. Such a meter is illustrated in Figure 7-17. The typical value of input resistance is 10 MΩ, and this value remains constant on all dc volts ranges. A conventional voltmeter has a different value of input resistance for every range, since the resistance of the multipliers changes with every change of the range switch. The loading effect on the circuit being measured therefore changes with each range change. The FET VOM, however, uses an *amplifier* that consists of one or more transistors that have the effect of greatly increasing the meter's sensitivity. The input circuit to which the probes are connected can therefore be a simple voltage divider, with the range switch arranged to pick off a suitable fraction of the full applied voltage.

The older and nearly obsolete equivalent of the FET VOM is the vacuum-tube voltmeter (VTVM), which uses vacuum tubes instead of the newer and more efficient transistors. Vacuum tubes require a large amount of power to operate their heater, and they cannot be easily arranged for portable use. Due to their increased sensitivity, the VTVM and the FET VOM both have much higher resistance ranges than the conventional VOM. VTVMs and FET VOMs often measure resistance values to 1000 MΩ. Other scales are essentially the same as those of the regular VOM.

7-7 DIGITAL MULTIMETERS

A digital multimeter (DMM), an example of which is shown in Figure 7-17, provides much greater reading accuracy than equivalent analog multimeters. There are several reasons for this, among which are:

1. The electronic circuitry of a DMM is inherently more accurate since in effect it breaks the analog voltage up into thousands of discrete parts, and then measures (or counts) each of them.
2. In a DMM, much of the decision-making is done automatically, since the operator simply selects the range and even the polarity is sensed by the DMM and displayed.
3. There is no chance of reading the wrong scale, or of wrongly interpreting the analog scale reading, or of using the wrong scale multiplier.

Another feature shared by many models is the DMM's capability to measure ac milliamperes, which analog meters seldom have.

TABLE 7-1 MULTIMETER CHARACTERISTICS

Characteristic	VOM	VTVM	DMM
Volts dc	Yes	Yes	Yes
Volts ac	Yes	Yes	Yes
Milliamps dc	Yes	No	Yes
Milliamps ac	No	No	Yes
R_{max}	1 MΩ	1000 MΩ	1000 MΩ
Adjustments	$R = 0$	$R = 0; R = \infty$	None
Power	Battery	AC or battery	AC
Accuracy	2%–3%	5%	0.1%–1.0%
Approximate cost	$10–$100	$50–$200	$200–$2000

Additionally, the DMM usually has higher ranges on the volts and ohms scales, with typical values up to 20 MΩ and 2000 V. Furthermore, with a 3½-digit instrument such as the one illustrated, the accuracy of the indication is greater than can be read from an analog scale. The term *half-digit* refers to the numeral *1* in the most significant position. This *1* cannot be displayed as a *2, 3,* or any other digit. It illuminates only when appropriate. Hence, the maximum number to be displayed on a 3½-digit machine is 1999 with the decimal point displayed wherever appropriate. Other degrees of accuracy are available, with 4½- and 5½-digit DMMs the most popular. However, the greater the number of digits in the display, the greater the cost of the meter.

Operating the DMM is unusually simple. A function is selected (volts dc, volts ac, mA dc, mA ac, or ohms), usually by pressing a pushbutton. Then, an appropriate range push-button is actuated. The two leads are then connected to the circuit to be measured. The leads need not, in most instances, be connected with regard to polarity. The common lead is connected to circuit common (ground) and the volts/ohms or milliamps lead connected to the point to be measured, just as if the instrument were an analog meter. If the voltage is positive, a + sign is displayed, or on some models no sign is displayed when the voltage is positive; but if the voltage is negative with respect to common, a − on the display so indicates. The value is then read directly from the display, which will place the decimal point properly.

To simplify the main points of the preceding discussion, the overall characteristics of these meters are summarized in Table 7-1.

SUMMARY

- The D'Arsonval meter movement is constructed of a powerful magnet with a pivoted coil suspended between the pole pieces. The coil

rotation, and therefore the needle deflection, is proportional to the current flowing in the coil.
- Full-scale current (I_{fs}) is the current required in the movable coil to just cause the indicator to read maximum (full scale).
- The greater the sensitivity, the smaller the current required to deflect the needle to full scale.
- The range of a current meter can be extended (increased) by a *shunt*.
- The range of a voltmeter can be extended by use of a *multiplier*.
- To measure current, the circuit is opened, and the meter itself becomes part of the circuit.
- To measure voltage, the meter probes are simply placed *across* the element to be measured.
- Current meters should have low internal resistance to avoid introducing errors.
- Voltmeters should have high internal resistance to avoid introducing errors.
- The ohms-per-volt rating of a voltmeter is the reciprocal of I_{fs}.
- A vacuum-tube voltmeter or its transistorized equivalent, the DMM, is widely used because of its very high input resistance.
- Typical input resistance of a FET-VOM is 10 MΩ on all ranges.
- An ohmmeter must be used in a *deenergized* circuit; the meter supplies its own power.
- The ohmmeter scale is nonlinear, and zero ohms is full scale (right-hand side).
- The digital multimeter (DMM) is more accurate than the analog type, and thus is preferred where additional cost is not a factor.

QUESTIONS

1. The moving-coil meter movement is constructed so that the indicator (needle) deflection is proportional to _____.
2. The moving-coil meter movement is constructed of a pivoted coil suspended in the field of a _____.
3. The force that moves the indicator upscale results from the _____ _____ surrounding the coil.
4. A meter movement rated at 100 μA full scale will deflect _____% if 33μA is caused to flow in the coil.
5. A meter movement has jeweled bearings to reduce _____.
6. Briefly define I_{fs}.
7. Briefly define R_m.
8. The indicator of a dc voltmeter having a full-scale voltage of 100 V_{dc} is deflected 25%. What is the voltage?
9. The indicator of a dc milliammeter with I_{fs} of 250 mA is deflected 33%. What is the current value?

10. The indicator of an ohmmeter reads 25 Ω and the range switch is set to R × 10K. What is the value of resistance?
11. List three rules for operating a milliammeter.
12. Briefly describe the effect known as voltmeter loading.
13. List six rules for ohmmeter operation.
14. An ammeter or milliammeter must have (high, low) internal resistance.
15. A voltmeter must have (high, low) internal resistance.
16. A voltmeter uses a 25-μA movement. What is its ohms-per-volt rating?
17. A voltmeter uses a 100-μA movement. What is its ohms-per-volt rating?
18. A voltmeter is rated at 2000 ohms per volt. $I_{fs} = $ _____.
19. A voltmeter is rated at 100,000 ohms per volt. $I_{fs} = $ _____.
20. A meter movement having a full-scale current of 50 μA and an R_m of 2000 Ω will require _____ volts for full-scale deflection.
21. Why must a milliammeter be placed in series with the circuit?
22. Why must a voltmeter be placed in parallel with the device to be measured?

PROBLEMS

1. A meter movement has a full-scale current of 100 μA. What is its ohms-per-volt rating?
2. A meter movement has a full-scale current of 10 μA. What is its ohms-per-volt rating?
3. A shunt is to be added to a 50-μA movement having $R_m = 3000$ Ω. The new full-scale current is to be 5 mA (5000 μA). Determine the correct value of shunt resistance.
4. A shunt is to be added to a 10 μA movement in which $R_m = 5000$ Ω. The new full-scale current is to be 100 μA. Determine the value of shunt resistance.
5. A 0.001-A movement has an internal resistance of 50 Ω. What value of voltage will just cause full-scale deflection?
6. A 10-μA movement has an internal resistance of 7000 Ω. What value of voltage will just cause full-scale deflection?
7. The movement in Problem 5 above is to have a multiplier added to cause full-scale deflection with 25 V applied. What should the multiplier value be?
8. The movement in Problem 5 above is to have a multiplier added to cause full-scale deflection with 250 V applied. What is the multiplier value?
9. A 1.0-mA movement having an internal resistance of 27 Ω is to be used for a single-scale ohmmeter similar to the one in Figure 7–12.

The internal supply is a 1.5-V dry cell and the meter will read a mid-scale value of 1500 Ω. What value of internal series resistance will produce this meter action?

10. A 50-μA movement having an internal resistance of 3500 Ω is to be used for a single-scale ohmmeter similar to that in Figure 7–12. The internal supply is a 3.0-V dry cell and the meter will read a mid-scale value of 60,000 Ω. What value of internal series resistance will produce this meter action?

CHAPTER **8**

CONDUCTORS, INSULATORS, AND SEMICONDUCTORS

In this chapter we study in more detail such items as wires, insulators, and a rather unusual group of materials known as semiconductors. Such particulars as wire size and its relationship to the efficient transmission of electrical energy, the fabrication and application of printed-circuit boards, and the relative efficiency of various conductors and insulators are all examined. The chapter concludes with a brief introduction to semiconductor theory.

The major subjects in this chapter are:

8-1 Characteristics of Conductors
8-2 Ion Current
8-3 Insulators
8-4 Basic Semiconductor Principles

8-1 CHARACTERISTICS OF CONDUCTORS

Up to this point in this book, we have considered electrical conductors to be perfect. That is, when a load is connected to a source, we have assumed that the full source voltage will appear across the load. In practical electrical and electronic circuitry this is seldom true. Copper wire (or any other conductor) has some degree of resistance, and it is necessary to know how much resistance a certain length of wire has, so as to reduce the resistance of the wire to tolerable levels. To illustrate the adverse effects of wire resistance, we shall describe and use a circuit that, admittedly, is exaggerated. It nevertheless illustrates the necessity of using the proper kind and size of wire in any given situation.

Figure 8-1 illustrates a circuit using a pilot lamp rated at 7 V and 150 mA. It is necessarily mounted approximately 150 ft from the 7-V source and is connected with number 30 gauge wire. Now, as will shortly be shown, 30-gauge wire has a resistance of 103.2 Ω/1000 ft, or 0.1032 Ω/foot. In the circuit shown, the total wire length is 300 ft (150 ft each leg) for a total wire resistance of $300 \times 0.1032 = 30.96 \cong 31$ Ω. The lamp resistance is $E/I = 7/0.15 = 46.7$ Ω. Total circuit resistance

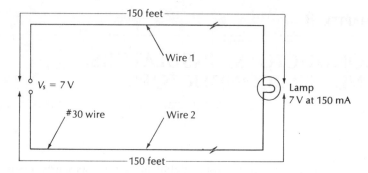

Figure 8-1 Transporting electrical current.

should therefore be 31 + 46.7 = 77.7 ohms. With 7.0 V applied, current will be

$$I_t = \frac{V_s}{R_t} = \frac{7}{77.7} = 0.09 = 90 \text{ mA},$$

or slightly more than half the required amount. Therefore, the voltage supplied to the lamp is the source voltage *less any voltage dropped in the wires* themselves.

$$E_{\text{wire 1}} = I_t \times R_{\text{wire 1}} = 90 \text{ mA} \times (150 \text{ ft} \times 0.1032 \text{ }\Omega/\text{ft}) = 1.39 \text{ V};$$
$$E_{\text{wire 2}} = I_t \times R_{\text{wire 2}} = 90 \text{ mA} \times (150 \text{ ft} \times 0.1032 \text{ }\Omega/\text{ft}) = 1.39 \text{ V};$$
$$E_{\text{lamp}} = V_s - E_{\text{wire 1}} - E_{\text{wire 2}} = 7 - 1.39 - 1.39 = 4.22 \text{ V}.$$

Note the relatively large voltage drops across the wires. As stated earlier, these drops reduce the energy delivered to the lamp to the extent that the lamp will emit much less light than it should.

To illustrate the result of a better choice of wire, assume that the same circuit is to be rebuilt using 20-gauge wire. The Ω/ft resistance of 20-gauge wire is 0.01015 Ω, so total wire resistance is

$$R_{\text{wire}} = 300 \times 0.01015 = 3.045 \text{ }\Omega;$$
$$R_t = R_{\text{lamp}} + R_{\text{wire}} = 46.7 + 3.045 = 49.75 \text{ }\Omega;$$
$$I_t = \frac{V_s}{R_t} = \frac{7}{49.75} = 0.140 = 140 \text{ mA}.$$

Since the total current is nearly the rated value for the lamp, only a small voltage drop is occurring across the wires.

$$E_{\text{wire}} = I_t \times R_{\text{wire}} = 140 \times (300 \times 0.01015) = 0.426 \text{ V}.$$
$$E_{\text{lamp}} = V_s - E_{\text{wire}} = 7 - 0.426 \cong 6.6 \text{ V}.$$

An even larger size wire would reduce losses even further.

From the foregoing discussion it can be seen that the wire gauge number has a direct bearing upon the electrical losses in a given circuit.

Generally, the smaller the gauge number, the larger the wire diameter and the smaller the losses. Table 8-1 illustrates a portion of the American Wire Gauge (AWG) table of standard annealed bare copper wire. The gauge number of a particular wire specifies the size of round wire according to its diameter and cross-sectional area. Note that, for example, 20-gauge wire can also be written as no. 20 wire or as #20 wire (# means number).

Wire Sizes

The size of standard round copper wire can be specified by any of several units, as shown in Table 8-1. The actual diameter of the wire in inches is one, the actual diameter in mils is another, while the circular-mil area is a third. A fourth unit, widely used in the power transmission field, is the kcmil (thousand circular mils).

The most widely used unit, other than the gauge number, is the circular-mil area. A *mil* is one-thousandth of an inch (0.001 in.). One *circular mil* is the cross-sectional area of a round wire having a diameter of one mil. The circular-mil area of a wire with a diameter of 10 mils is

TABLE 8-1 AWG WIRE TABLE FOR ROUND COPPER WIRE AT 70°F

Gauge	Diameter (in.)	Area (Cir Mil)	Ω/1000 ft	Ω/ft	Feet/Ohm
8	0.1285	16,510	0.6282	0.0006282	1592
9	0.1144	13,090	0.7921	0.0007921	1262
10	0.1019	10,380	0.9989	0.0009989	1001
11	0.09074	8,234	1.260	0.001260	794
12	0.08081	6,530	1.588	0.001588	629.6
13	0.07196	5,178	2.003	0.002003	499.3
14	0.06408	4,107	2.525	0.002525	396.0
15	0.05707	3,257	3.184	0.003184	314.0
16	0.05082	2,583	4.016	0.004016	249.0
17	0.04526	2,048	5.064	0.005064	197.5
18	0.04030	1,624	6.385	0.006385	156.5
19	0.03589	1,288	8.051	0.008051	124.2
20	0.03196	1,022	10.15	0.01015	98.5
21	0.02846	810.1	12.80	0.01280	78.11
22	0.02535	642.4	16.14	0.01614	61.95
23	0.02257	509.5	20.36	0.02036	49.13
24	0.02010	404.0	25.67	0.02567	38.96
25	0.01790	320.4	32.37	0.03237	30.90
26	0.01594	254.1	40.81	0.04081	24.50
27	0.01420	201.5	51.47	0.05147	19.43
28	0.01264	159.8	64.90	0.06490	15.41
29	0.01126	126.7	81.83	0.08183	12.22
30	0.01003	100.5	103.2	0.1032	9.691

100 circular mils. *The circular mil area of a wire is equal to the square of the diameter in mils.* Because the circular mil is a unit of area obtained by squaring the diameter, the circular-mil area increases as the square of the diameter. Thus, doubling the diameter increases the circular-mil area by 4; increasing the diameter 5 times increases the circular-mil area by 25. Because the areas of two circles are always proportional to the squares of their diameters, it is evident that measuring round wire in circular mils is equivalent to measuring true cross-sectional area.

When choosing a wire size for a relatively long distance, the losses in the wire should be no greater than 5 percent for reasonably efficient operation of the load. In some critical applications, losses are held to 1 percent or less. Usually, the primary consideration is the minimum voltage required for the load. If there are several loads at the end of a long wire, the problem becomes quite complex—factors to be considered include the probability of loads being energized simultaneously, the minimum voltage for the most critical load, and so on. In instances such as these, *worst-case* conditions are often used, where it is assumed that all loads operate at the same time, and that all load resistances have the same minimum value equal to the lowest.

Wires and Cables

As previously mentioned, most wire used in electrical or electronic applications is made of copper. However, gold, aluminum, and silver are all used to a lesser degree. Copper is used so widely because it is both plentiful and relatively inexpensive, as opposed to both gold and silver. Aluminum wire is used primarily in applications where its light weight is an advantage, as in long-distance transmission lines used to transport electrical energy across country. Gold and silver wire are understandably used only where absolutely necessary. One application of gold wires drawn very fine ($\approx$ 1 mil diameter) is the connection of a transistor or integrated circuit to the pins on its metal or plastic package. Obviously, only a tiny bit of wire is used per device, so the cost is not prohibitive.

Single wires, either bare or insulated, come in a variety of configurations. In one form such wire is called *hook-up wire* since it is widely used for point-to-point connections on low-voltage electronic equipment. It is available in many sizes and with several kinds of insulation, ranging from plastic to waxed cloth. Hook-up wire can be obtained as either solid or stranded wire, as illustrated in Figure 8-2a. Stranded wire is used where its greater flexibility is an advantage. Solid wire is more likely to break when flexed too often, producing an open circuit. If the open in the wire is concealed by the insulation, it may be very difficult to locate.

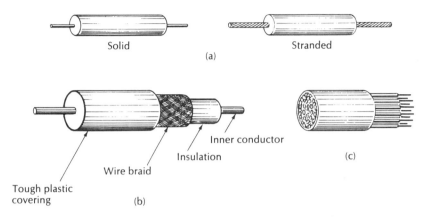

Figure 8-2 (a) Hook-up wire, (b) coaxial cable, and (c) multiconductor cable.

Figure 8-3 A printed-circuit (PC) board.

When two or more conductors are bound together with a common outer covering, the assembly is called a *cable*. Those illustrated in Figure 8-2 are coaxial cable and multiconductor cable.

Printed-Circuit Boards

In modern electronics one is more likely to encounter wiring in the form of printed-circuit (PC) boards, an example of which is shown in Figure 8-3. Because this type of wiring can be produced by automated mass-production techniques, it has several advantages over point-to-point wiring with conventional hook-up wire.

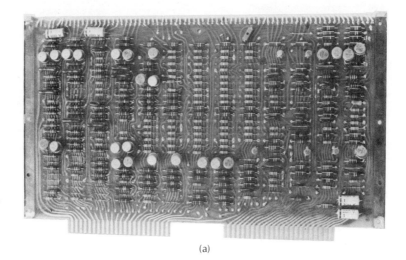
(a)

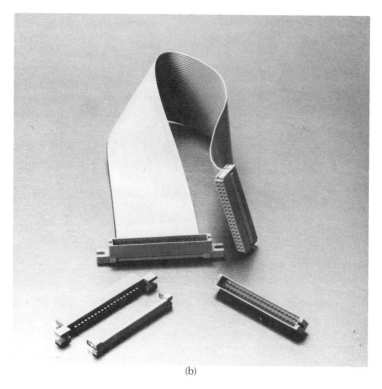
(b)

Figure 8-4 (a) A loaded PC board; (b) and (c) flexible cables used to interconnect PC boards. Courtesy of 3M® Corporation.

(c)

PC boards are produced rapidly and with great precision by a combination of techniques. The technique, in general terms, is as follows. The board, generally a good quality glass epoxy (or other strong insulating medium), is first coated with a thin sheet of solid copper which is firmly bonded to the surface. A photosensitive coating is then applied to the copper and a negative that carries the desired pattern is placed over it. The board is then exposed to a strong light rich in ultraviolet (sunlight works very well for small quantities) and the photosensitive coating is hardened where the light strikes it.

The exposed board is then immersed in an acid bath to etch away the unwanted areas, leaving the desired pattern of copper traces. After rinsing, the board is drilled as required to accept various components, and is then ready to be *loaded* with the desired components. After soldering, either by hand in small quantities, or by *wave soldering*, which solders the entire board at once, the PC board is ready for use. Figure 8-4a shows a completely soldered PC board. Several kits are on the market that provide all materials necessary to produce experimental PC boards. Using such a kit is an excellent way to gain a full understanding of the basic principles involved.

The interconnecting of multiple PC boards is often accomplished by *flat* cables, such as are shown in Figure 8-4*b* and *c*. These cables are called *Scotchflex*® by one manufacturer, and are flexible, allowing great freedom of movement when servicing the equipment. Furthermore, the end of the cable can easily be terminated with the proper plug or jack. Hence, up to 50 connections can be done at once, speeding up manufacturing times and thus saving costs.

Resistance of Wire

As stated earlier, the total resistance of a length of wire depends upon its length, its diameter, the material from which it is made, and to a lesser extent the temperature of the wire. Table 8-1, the AWG wire table, gives the resistance of standard round copper wire at room temperature for gauge numbers 8 through 30. Tables are available for gauge numbers 0000 (4/0) through 56. Gauge 0000 wire is nearly ½ in. in diameter and is seldom used except in large ac power installations. Gauge 56 wire is approximately ½ *mil* in diameter and is so fine and fragile that it, too, is used only rarely. The numbers included in Table 8-1 are those in common use.

As has been mentioned, the longer a given size of wire, the greater the resistance. However, the greater the diameter of a given length of wire, the lower the resistance. For the wire sizes given in the table, any length of a given size of wire has a certain resistance. Using the previous example, the resistance of 300 ft of 30-gauge wire is determined as follows. Reading across the table for 30-gauge wire, resistance value is 0.1032 Ω/ft. This value times the total number of feet, gives the wire resistance:

$$R_{wire} = \Omega/\text{ft} \times \text{feet} = 0.1032 \times 300 = 30.96 \ \Omega.$$

Specific Resistance

Specific resistance, sometimes known as *resistivity,* is the amount of resistance offered by a material having a known standard volume at a reference temperature. The specific resistance ρ (the Greek letter rho) is a measure of the resistance in ohms of material having a length of one foot and a cross-sectional area of one circular mil, usually specified at 20°C (68°F). Table 8-2 lists the specific resistance of several materials of importance in ascending degree of resistance.

TABLE 8-2 SPECIFIC RESISTANCE OF CONDUCTING MATERIALS

Material	ρ Approximate Specific Resistance (Ω/cm ft) at 20°C (68°F)
Silver	9.8
Copper	10.4
Gold	14.7
Aluminum	17.0
Tungsten	33.2
Nickel	50.0
Iron	58.0
Steel	96.0
Nichrome	660.0
Carbon	20,000 (approx.)

Note that silver has the lowest resistance of all substances listed, with copper the next lowest. This, of course, is why copper is used so widely as a conductor. To use the table to determine the resistance of a length of wire requires only that the circular-mil area of the wire in question be known. The resistance can then be calculated as follows.

$$R = \rho \frac{l}{A},$$

where R = resistance in ohms,
l = length of wire in feet,
A = circular-mil area,
ρ = the specific resistance in Ω/cir-mil ft of the material at 20°C.

For example, what is the total resistance of 1000 ft of 18-gauge round copper wire? From Table 8-2 the specific resistance (ρ) value for copper is 10.4 and from Table 8-1 the circular-mil area for 18-gauge copper wire is found to be 1624 circular mils. Substituting these values in the formula yields

$$R = \rho \frac{l}{A} = (10.4) \frac{1000}{1624} = 6.4 \, \Omega.$$

For comparison, we shall determine the resistance of aluminum wire of the same size and length as above.

$$R = \rho \frac{l}{A} = (17) \frac{1000}{1624} = 10.47 \, \Omega.$$

Note that the same circular-mil area is used as for No. 18 copper wire, since the cross-sectional area is independent of the *kind* of material.

Temperature Coefficient of Resistance

All metals, and indeed all conductors, react to temperature changes. In electrical circuits, the way temperature affects wiring is of some importance. The effects of temperature change are designated by the temperature coefficient of the material. This factor is represented by α (the Greek letter alpha), which indicates how much the resistance changes per degree centigrade of temperature change. A positive α means that the resistance increases as the temperature increases, while a negative α means that resistance decreases as the temperature increases. A value of zero means that resistance remains the same.

Table 8-3 lists the α for several materials commonly used. All pure metals have positive α values, while materials called *semiconductors* (carbon, germanium, silicon) have negative values of α. Certain metalic alloys can be formulated so that $\alpha = 0$. Such alloys are extensively used for precision wire-wound resistors that exhibit constant resistance with changes in temperature. One such alloy is *manganin,* a carefully controlled blend of manganese, nickel, and copper.

To determine the value of resistance at some temperature other than 20°C, which is considered room temperature, the following relationship may be applied:

$$R_{\text{new } T} = R_{20°C} + R_{20°C}(\alpha \Delta T).$$

This indicates that the resistance at the new temperature is equal to the resistance at 20°C plus the resistance at 20°C, times α, times the change in temperature (Δ means *change in* and is read "delta").

TABLE 8-3 TEMPERATURE COEFFICIENT OF COMMON MATERIALS

Material	Temperature Coefficient (α)
Aluminum	0.0039
Carbon	−0.0005
Copper	0.00393
Gold	0.0039
Nichrome	0.00016
Nickel	0.006
Steel	0.003
Silver	0.0038
Tungsten	0.0045

As an example, assume a length of copper wire having a resistance of 6.4 Ω at 20°C is to be used continuously in an environment of 100°C. What is the resistance of the wire at this elevated temperature? From Table 8-3 $\alpha = 0.00393$ for copper, and $\Delta T = 100 - 20 = 80$. Therefore,

$$R_{\text{new } T} = R_{20°C} + R_{20°C}(\alpha \Delta T) = 6.4 + 6.4(0.00393 \times 80)$$
$$= 8.41 \text{ Ω}.$$

At 100°C the resistance of the wire has increased significantly with an 80°C increase in temperature. A related factor that must be considered is hot versus cold resistance. A tungsten light bulb can be used as an example of how resistance changes at elevated temperatures. The resistance of a 100-W bulb when nonenergized is approximately 10 Ω. However, Ohm's law indicates that, when operating, the bulb's resistance is much higher.

$$P = IE \text{ and } I = \frac{P}{E} = \frac{100}{120} = 0.833 \text{ A},$$
$$R = \frac{E}{I} = \frac{120}{0.833} = 144 \text{ Ω}.$$

If the temperature of the hot filament is desired, the temperature-coefficient formula given above can be transposed. From Table 8-3, $\alpha_{\text{tungsten}} = 0.0045$, so

$$T = \frac{R_{\text{new } T} - R_{20°C}}{R_{20 C} \times \alpha} + R_{20°C}$$
$$= \frac{144 - 10}{10 \times 0.0045} + 10 = \frac{134}{0.045} + 10 = 2987 \cong 3000°C.$$

8-2 ION CURRENT

When we talk about electrical current we are usually referring to electron current flowing in a wire. By definition, however, the concerted movement of *any charged particles* in a conducting medium is current. Thus, current can consist of negative electrons, positive ions, or negative ions. Recall that an ion is an atom or molecule having an excess negative or positive charge. An ion exists whenever electrons are added to or taken away from an atom or molecule.

Ion current can exist in both liquids and gases. Liquids such as water in which acids, alkalies, or salts have been dissolved generally can be made to pass ion current. A study of ion current in the electrolyte of a battery is made in Chapter 10. Nearly everyone has seen a neon sign at one time or another. Such a sign is an excellent example of ionic current in a gas.

The flow of ions constituting movement of charged atoms or molecules is only possible where the atoms or molecules can move about more or less freely. Hence, ion current cannot exist in a solid, where the atoms or molecules are firmly locked in place. In most solids, only electron current can exist. (An exception is semiconductor "hole" current; see Section 8-4.)

Conduction in Gases

Under normal conditions a gas is an insulator. However, if a strong enough potential field exists, the gas will ionize and will start passing current, and in the path of conduction the resistance becomes low. Frequently, the gas emits light (for example, neon or argon) when current passes through it.

The amount of voltage required to ionize a gas depends upon the kind of gas and the gas pressure. For example, a neon glow lamp requires a voltage of approximately 70 V. Such a lamp is first highly evacuated, then the neon gas is introduced under low pressure, which results in the low operating voltage. These lamps, for example an NE-2, are often used as panel lights or other such indicators. Although 70 V or more is required to ionize the neon, after conduction starts the voltage across the lamp will drop to about 62 to 67 V, where it will remain over a wide variation of current through the gas. For this reason, gas tubes (such as mercury vapor tubes) are often used to provide a relatively constant voltage source. When used for this purpose, the light emitted is generally superfluous.

Figure 8-5 illustrates the mechanism of current conduction in a gas. When a sufficiently high voltage is applied to the terminals, through a suitable current-limiting resistor, the valence electrons are dislodged from some atoms, which are attracted toward the positive terminal. Each electron that is dislodged leaves behind a positive ion, and these are attracted toward the negative terminal. However, because of the much greater mass of the ion, its movement is slower than that of the electron. It does, nevertheless, contribute to current flow within the lamp.

A *corona* discharge will occur into the air if a very high voltage is applied to a sharply needle-pointed electrode. The breakdown voltage for dry air at standard temperature and pressure is approximately 20 kV per inch. Thus, to prevent corona discharge in high-voltage circuits, all surfaces are smoothly rounded to eliminate sharp points.

Conduction in Liquids

While the mechanism of conduction in a liquid is somewhat different from that of conduction in a gas, the end result is still the same. Positive

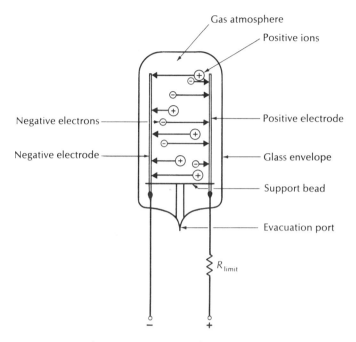

Figure 8-5 Current conduction in a gas.

ions are attracted to the negative electrode and negative ions or electrons are attracted to the positive electrode. In Chapter 10 this kind of conduction is dealt with in some detail in connection with electrochemical cell systems. At this point, we are concerned only with conduction in a liquid medium.

Pure water does not conduct current well at all. However, adding some common table salt causes the solution to become highly conductive. When the salt goes into solution, the sodium and chlorine atoms dissociate and in the process become ionized. The sodium atom loses an electron to the chlorine atom; hence the sodium atom attains a positive charge while the chlorine atom attains a negative charge. If a voltage is impressed upon two electrodes in the solution, current will flow. The negative ions move toward the positive electrode, and the positive ions move toward the negative electrode. This action is depicted in Fig. 8-6a, where the direction of the ion movement is shown. As the positive ion touches the negative plate, it accepts an electron from the plate, thus contributing to continuous current flow. Also, the negative ion gives up an electron at the positive plate. As long as an EMF is applied, current will flow—conventional electron flow exists in the external circuit, and ionic current flow in the solution. Neutral atoms of sodium and chlorine are shown in Figure 8-6b.

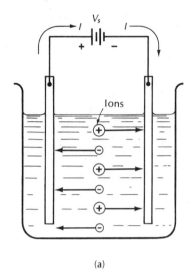

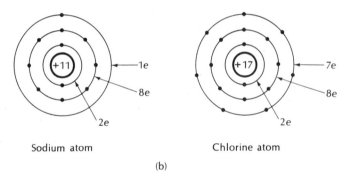

Figure 8-6 (a) The mechanism of conduction in a salt solution; (b) sodium and chlorine atoms in the neutral state.

8-3 INSULATORS

Materials that do not conduct electrical current at all well are called *insulators,* or *dielectrics.* Insulators are used, of course, to keep electrical conductors apart. Insulators can take a variety of shapes, ranging from the huge porcelain or ceramic insulators used with power transmission lines or in a power substation, to the plastic- or rubber-covered rip (or zip) cord used for household appliances.

Insulators possess very high resistance, caused by the fact that there are no free electrons in their molecular structure. However, at high enough voltages some of the insulating material begins to ionize, thus liberating current carriers (electrons). An arc discharge can then occur,

TABLE 8-4 DIELECTRIC STRENGTH OF INSULATORS

Material	Dielectric Strength (V/in.)
Dry air	20,000
Porcelain	100,000
Paraffin wax	250,000
Bakelite	375,000
Hard rubber	450,000
Mica	800,000
Shellac	900,000
Glass	1,000,000
Paper	1,250,000

either directly through the solid material or across its surface. This can severely damage the insulator, possibly causing it to crack. Alternatively, the path taken by the arc can carbonize, or burn, some of the material, leaving a partially conducting path for future breakdowns.

Different materials break down at various voltages. The minimum amount of voltage required to cause an arc discharge is called the *dielectric strength* of the material. Table 8-4 lists the dielectric strength of several common insulating materials. All values are approximate, since there are many variables, including impurities in the material, moisture content of the material and the surrounding air, and the shape of the insulator. Also, as might be inferred from the table, the greater the thickness of the insulator, the higher the voltage it can withstand. One other variable has some effect on dielectric strength: the temperature of the insulator. Very high temperatures can cause an insulator to begin to conduct, since the vibratory motion of the particles increases with increasing temperatures.

The resistance of an insulator cannot conveniently be measured with conventional meters, since the resistance is beyond their capability. However, an instrument is available to perform this measurement if necessary. A *megger* is a high-voltage low-current meter that is, in effect, a very high range ohmmeter. With the megger, it is actually possible to measure insulator resistances in the range of 10,000 to 100,000 MΩ.

8-4 BASIC SEMICONDUCTOR PRINCIPLES

Semiconductors are in such widespread use in today's technology that a brief introduction to the subject is necessary at this point. Semiconductors are broadly classified as either passive or active devices. A *passive* device is similar in circuit action to a resistor, a conductor, and the like. We shall restrict our attention at this time to a study of passive semiconductors such as the various kinds of *diodes*. Active devices require more time and space than are allowed in beginning studies.

The subject of semiconductors rightfully begins with an investigation into the atomic structure of these devices. Only by these means can the devices' operation be explained. Once the structure of a semiconductor diode is understood, it can be applied to a number of circuits that are then more easily understood. From a very basic standpoint, a diode is a unidirectional device; that is, it readily passes current in one direction, but if the polarity of the current is reversed, essentially *no* current flows through it. In this respect, it is radically different from any device that you have studied thus far. Let us first, then, investigate the atomic structure of these remarkable devices so as to be able to understand what it is that makes them work.

Atomic Structure

As you already know, a conductor or an insulator has its own special characteristics that depend upon the electron configuration of its constituent atoms. A material is a good conductor if, in its outer shell of electrons, the individual electrons are loosely bound to the nucleus. When this is true, the electrons can easily be moved to form an electric current. On the other hand, a material with tightly bound electrons that cannot be easily moved is an insulator. As you might expect, semiconductors fall somewhere in between conductors and insulators.

Two elements used most commonly to manufacture semiconductors are germanium (Ge) and silicon (Si). These are known as *tetravalent* elements. That is, each atom has a total of four electrons in its outer orbit. Figure 8-7 illustrates a simplified method of depicting silicon and germanium atoms. Only the valence shell is shown separately; the balance of the atom, including the inner shells, is shown as the nucleus. Because we are only interested in the valence, or outer, shell of electrons, this simplification makes the subsequent drawings far easier to understand.

Valence Electrons

The valence electrons in a given substance determine the chemical characteristics of the substance. Also, they have a large bearing upon whether the substance is a conductor or an insulator, or neither. Recall from your earlier studies that the reason copper is a good conductor is simply that it has only one electron in its valence shell. On the other hand, any material having a full outer ring of electrons makes a good insulator. It has been determined that materials with eight electrons in the valence shell tend to hold these electrons in a tight bond.

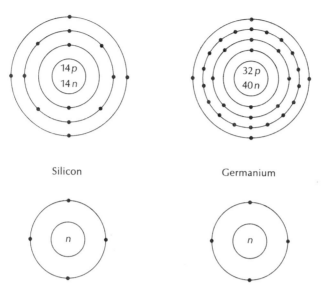

Figure 8-7 Conventional and simplified representations of silicon and germanium atoms. Note that the simplified representations show only the valence electrons.

Generally speaking, the fewer the valence electrons the better the conductor, while the closer there are to eight valence electrons the poorer the conductor. *The conductivity of the general class of semiconductors can be set to any desired value by the practical application of the foregoing statements.*

To understand how a piece of silicon, for example, can be made to have virtually any desired value of resistance, we must look at a few atoms of the material to see what effect the atomic structure has on the conductivity.

Crystals

When comparing solids, liquids, and gases, we find that the primary difference between each of these is the *distance between the atoms*. When the natural state of a material causes the atoms to be separated by more than the diameter of an atom, the material is a gas. When the atoms are at distances from each other about equal to the diameter of an atom, the material is a liquid. However, when the atoms are very close to each other, the attractive forces are large, and the material is a solid. Going one step further, when the atoms are so close together that they actually overlap, the solid material is said to be in its *crystalline* form. In general, a crystal is stronger and harder than the same material in its noncrystalline form.

Materials having four valence electrons crystallize quite readily in most cases. One exception to this is carbon, which becomes crystalline only with the application of very high temperatures and pressures. Carbon that has crystallized is, as you may already know, a diamond. A diamond is excellent evidence that crystals are strong and hard (although they may be quite brittle).

With high temperature and pressure applied, the carbon atoms are tightly squeezed together into the most compact form possible. This form is called a *lattice structure*. Because of the attractive force between atoms, once in this compact lattice form, they stay in place indefinitely, and the end result is the diamond. In its normal state, carbon is a good conductor of electrical current, but as a crystal, carbon is an excellent insulator. To understand why this drastic change in conductivity occurs, the atomic structure of semiconductor materials must be investigated.

Semiconductor Materials

Such devices as diodes, transistors, and integrated circuits (ICs) are formulated from germanium and silicon. One or the other of these elements is used (they are never mixed) along with minute quantities of other selected materials that alter the electrical characteristics of the main body of Ge or Si. During initial processing, ingots of very pure germanium or silicon are "grown" from a molten mass, using a *seed* crystal. A seed crystal is simply a small piece of Ge or Si that has been already formed in its crystalline state.

One such method of growing a crystal ingot uses the seed crystal in contact with the surface of molten germanium or silicon. The seed is then slowly withdrawn from the molten surface, and surface-tension forces cause the molten material to adhere to the face of the seed. As the seed is slowly lifted, the tiny quantity of molten material that is drawn above the surface solidifies and becomes an extension of the seed itself. In so doing, the newly solidified material *assumes the same atomic configuration as the seed*. Because the seed is crystalline, so is the new material that attaches itself to it. After several hours, the end result of this process is an ingot of pure crystalline Ge or Si that is typically six to eight inches in length and perhaps two inches in diameter.

Pure, or *intrinsic,* semiconductors in the crystalline form have an atomic configuration not unlike that of a diamond. That is, the atoms are compressed to the point that they occupy as little space as possible. Under these conditions, the valence electron shells actually overlap.

These electrons are said to be *shared* with adjacent atoms. Note Figure 8-8, where the overlapping of valence electrons is clearly shown. Also, using any of the atoms near the center of the drawing, note that each atom *appears* to have eight electrons in its valence shell, even though both Ge and Si actually have only four valence electrons. Because the atoms seem to have eight valence electrons, the electrical conductivity of the material is now like that of an *insulator*. There are essentially no free electrons, since each valence electron is locked in a *covalent bond*, which is simply the name given to the force that holds the electrons in apparently complete valence shells. Thus, pure crystalline semiconductors are excellent insulators.

N-Doping

To be useful, pure semiconductor material must be altered slightly during manufacture. Without going into detail on how this is done, we

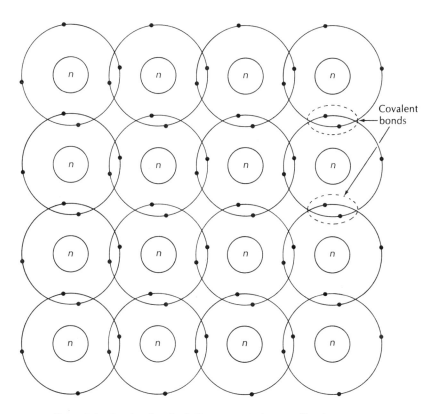

Figure 8-8 Covalent bonds of silicon atoms in the crystalline form.

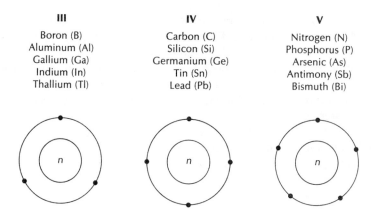

Figure 8-9 Trivalent, tetravalent, and pentavalent elements.

can nevertheless easily illustrate the end result of adding tiny amounts of other materials to the otherwise pure Ge or Si during crystal growth.

Figure 8-9 illustrates a small portion of the periodic table of the elements. Each column is headed with the valence number of the elements in that column. Note that carbon, germanium, and silicon are in the valence-of-four (tetravalent) column. The other columns contain elements with valences of three (trivalent) and five (pentavalent). These are the materials used to alter the electrical qualities of pure Ge or Si.

If, during the crystal-growing phase of manufacture, a *very* small amount of arsenic (As) is added to the molten material homogeneously, the growing crystal will contain a few arsenic atoms. Now, arsenic has a valence of *five* electrons, not four, and this drastically changes the electrical characteristics of the resulting crystal.

The arsenic atom fits itself into the crystal structure very nicely, as illustrated in Figure 8-10. Four of the five valence electrons will become tightly bound in their respective covalent bonds. However, *this leaves one electron that is bound to nothing*. It, then, becomes a *free electron* and is quite capable of contributing to any current flow if a potential is applied across the material.

The conductivity of the Ge or Si, then, can be accurately controlled during manufacture by adjusting the amount of *impurity* (in this instance, arsenic) introduced during crystal growth. The process of introducing the foreign material is known as *doping* and the material itself is called a *dopant*. The resulting doped material is known as *n*-type semiconductor, because the current carrier released is an electron, which carries a negative charge. By itself, *n*-type semiconductor is of little use. It has a conductivity about halfway between a good conductor and a good insulator, the actual value depending upon the number of impurity atoms introduced.

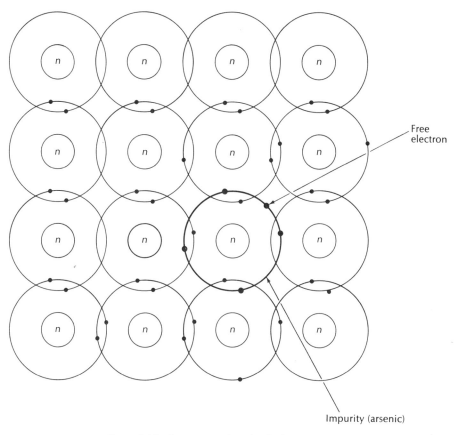

Figure 8-10 A representation of n-doping.

P-Doping

If, instead of doping the germanium or silicon with pentavalent material, we use a trivalent material such as boron (B), a different effect occurs. Figure 8-11 illustrates the generation of a *hole*. When a trivalent atom is introduced into the crystalline structure, the three valence electrons form covalent bonds with their adjoining neighbors. However, where there would normally be a fourth covalent bond, *there cannot be,* since boron only has three valence electrons. This produces a so-called "hole," which is simply an incomplete covalent bond. The hole carries a positive electrical charge, but has no mass. That is, it does not exist as a thing, but it does exist as a *unit positive charge* (that is, the hole has the same *positive* value that the negative electron carries). The hole alters the electrical conductivity of the crystal of which it is a part; *it contributes to current flow through the material.*

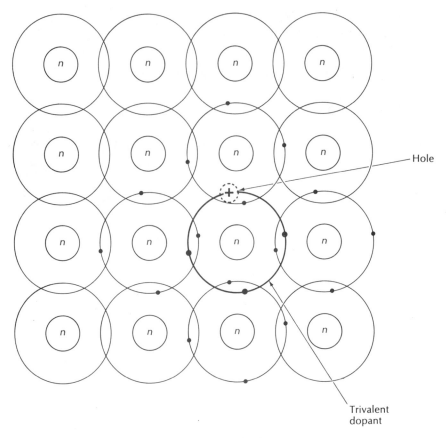

Figure 8-11 A representation of p-doping.

Because the hole carries unit positive charge, a material so doped is known as p-type semiconductor. Again, the p-type material is not of great utility by itself.

Conduction Mechanics

We now have produced two separate and distinct types of Ge and Si—n-type and p-type. The next question to answer is simply this: How does current flow in each type of material? By placing an imaginary battery across each we can describe the mechanism of conduction quite easily.

N-type Conduction

The conduction of electrical current through the n-type semiconductor is nearly the same as conduction through copper wire. The available

Conductors, Insulators, and Semiconductors   177

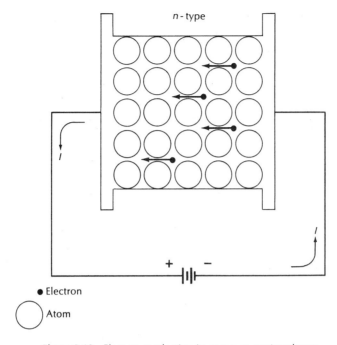

Figure 8-12 Electron conduction in an *n*-type semiconductor.

electrons, liberated in a number equal to the number of atoms of dopant, simply drift around the circuit, both in the semiconductor and in the wire. The direction of flow is, of course, determined by the source. See Figure 8-12.

About the only difference between conduction in wires and in *n*-type or *p*-type semiconductors is that the wire contributes free electrons in a number about equal to the total number of atoms. In the semiconductors the number of dopant, or impurity, atoms determines the number of current carriers. Hence, the resistance of the semiconductor is much higher than that of the wire, but not nearly as high as that of an insulator.

P-type Conduction

Because the hole in *p*-type semiconductors *is mobile,* although to a lesser extent than an electron in a conductor, it can contribute toward current flow. However, the mechanism of this conduction is somewhat different. As electrons are injected into *p*-type material, they literally jump from hole to hole in their progress through the material. However, it is also true that the holes themselves move in the material *in the direction opposite to the electrons.* Now, a very important rule relating to *p*-type material is that *the holes cannot exist outside of the semiconduc-*

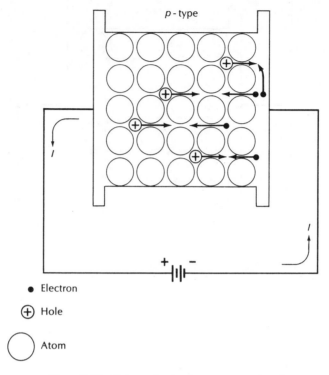

Figure 8-13 Hole conduction in a p-type semiconductor.

tor. Hence, the disposition of the holes that might be expected to emerge at the negative side (see Figure 8-13) must be explained.

As a hole reaches the limit of the semiconductor, an electron from the connecting piece of material enters, joins with the hole, and destroys its very existence, since at this point there is one unit positive charge and one unit negative charge with the net result of zero. The hole is said to be *annihilated*. However, note that at the opposite end of the p-type semiconductor an electron is emerging at the very same instant. As the electron leaves the semiconductor, *it leaves behind a newly created hole!* This must be true, since the number of holes within the volume of the semiconductor is determined by the number of impurity atoms introduced during manufacture, which can never change. Hence, as one hole is annihilated at the negative side, a new one is created on the positive side. Simultaneously, then, electrons move toward positive while the holes move toward negative.

An excellent analogy to hole movement exists in a game which consists of interlocking blocks upon which a letter or number is imprinted. Usually, there are 15 blocks and 16 spaces, as in Figure 8-14. Only one

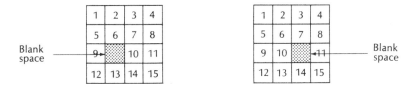

Figure 8-14 A method of illustrating hole movement.

block can be moved at a time, and the object of the game is to rearrange the blocks to spell a word or sentence or to rearrange the order of the numerals. However, if the blocks are considered electrons and the single space is thought of as a hole, then you can appreciate that to move the hole from, say, left to right, the right-hand adjacent block (electron) must be moved toward the left. This is almost exactly analogous to hole movement in a semiconductor. When block 10 is moved to the left, the space appears to move to the right.

The Junction Diode

In order to construct a useful device, the two types of semiconductors must somehow be joined together. They cannot be simply placed together, since the crystal structure must be continuous from the n-side through to the p-side. There are several methods used to create pn junctions, only one of which will be described at this time.

If, during the crystal-pulling phase of manufacture described earlier, the molten mass has added to it at intervals small amounts of trivalent and pentavalent dopants, the resulting crystal will consist of a layered ingot as in Figure 8-15a, each layer alternately doped n, then p, then n, then p. When the crystal is properly cut into dice, illustrated in part b of the figure, each cube, or die, represents a piece of semiconductor that is n on one end and p on the other. The junction between the two types is called, appropriately enough, a pn junction. The device itself, when leads are attached to it and it is encapsulated in plastic or glass, is known as a *junction diode*. This is the device, mentioned earlier, that will readily pass current in one direction, but not the other.

Diode Action

In order that the unidirectional characteristic of a diode can be explained the rearrangement of charges at the junction that occurs at the moment of junction formation must be investigated. The boundary, or junction, between the n end and the p end is the basis for explaining nearly all semiconductor phenomena.

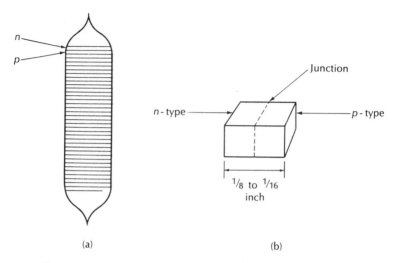

Figure 8-15 One method of producing *pn* junctions: (a) The ingot contains alternate layers of *p*-type and *n*-type dopants; (b) the ingot is cut into individual dice.

Junction Characteristics. Several things occur at the instant the junction is formed that give the diode its distinctive electrical qualities. During the following description, keep the fact in mind that the semiconductor material is going from the molten to the solid state; the temperature is therefore very high, and this provides greater vibratory energy to the particles, assuring greater mobility.

At the instant the junction is formed, the free electrons in the *n* side are attracted toward the positive holes in the *p* side. By the same token, the holes in the *p* side are attracted toward the *n*-side electrons. The two types of current carriers meet in the vicinity of the junction and each annihilates the other. After all possible collisions have occured, there is a space (perhaps 0.0005 in.) on either side of the junction consisting of a region in which *there are no current carriers,* either *n* or *p*. This is called the *barrier* region. Because the material is now a solid, this condition continues throughout the life of the material. That is, this migration of carriers occurs only once, and the result is as depicted in Figure 8-16. The barrier region, which typically has a width of 0.001 in., is now devoid of current carriers. It can therefore be thought of as an area with high insulating qualities; that is, no current can flow if a very small voltage is impressed across the device, even though the remaining volume of the device does contain electrons or holes.

A diode with no voltage (*bias*) applied, Figure 8-17*a*, accomplishes nothing. . . . However, the condition existing within the diode, with

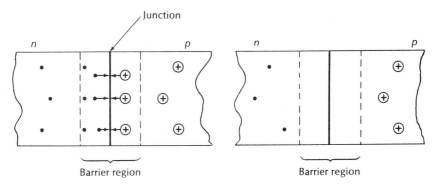

- Electrons
- ⊕ Holes

Figure 8-16 The *pn* junction (a) during formation and (b) after formation.

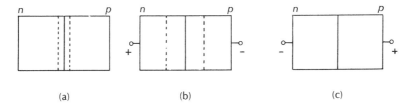

Figure 8-17 (a) A diode with no voltage applied; (b) reverse bias applied; (c) forward bias applied.

no voltage applied, is important. As illustrated, the *n*-material is separated from the *p*-material by a region (barrier, or depletion, region) in which there are no current carriers. This barrier region is, in effect, a *potential* barrier, and as we shall see, it directly influences the device's conduction characteristics. By applying a voltage, first of one polarity and then the other, we can better understand how this device functions. The first polarity to be applied is illustrated in Figure 8-17*b*, where negative voltage is applied to the *p* side, and positive voltage to the *n* side. This is called *reverse bias*, or *reverse voltage*.

Note that the applied voltage has *widened* the barrier region. The negative voltage tends to attract the holes, which moves them away from the junction. Similarly, the positive voltage attracts the electrons, with the same result. There is now a greater volume of material which can be called an insulator. Obviously, no current can flow in the diode or the external circuit. With reverse bias applied, then, *no current flows*.

When the battery or other source is reversed, as shown in Figure 8-17c, the barrier region is *compressed*, and if the voltage exceeds about ½ V, *it disappears altogether*. Holes and electrons are forced toward each other and *will combine at the junction*. Now, current *is* flowing, limited in value only by some external resistance. The small voltage required to fully compress the barrier region represents the energy required to overcome the barrier potential. In devices made from germanium, the barrier potential is one- to five-tenths of a volt, while silicon devices exhibit a barrier potential of seven-tenths of a volt to, perhaps, one volt. The mechanism of conduction is now similar to that explained earlier for the *p*-type material. When all voltage is removed from the diode, the barrier assumes its original dimensions.

When it is conducting, the diode has a very low resistance, typically 1 to 10 Ω. Hence, when forward biased, an external resistance must be in the series circuit to prevent excessive current, since the diode is now nearly a short circuit.

Diode Characteristics

A pictorial drawing and the schematic symbol of a diode are shown in Figure 8-18. We shall use this symbol in the following discussion.

Assuming a complete path, as in the circuit in Figure 8-19, current will flow if it is directed toward the cathode, or *n*-end, of the diode. If current is directed toward the anode, or *p*-end, of the diode, no current can flow. In Figure 8-19a, the source causes the diode to be forward biased, and current flows as indicated. In part *b* of the figure, however, the source is reversed, and no current flows, except that labeled leakage current; this will be explained shortly.

The voltages that exist across the resistor and diode are of interest, so that circuit action can be described in more detail. For the forward-

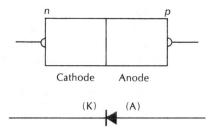

Figure 8-18 (a) A pictorial view of a diode; (b) the diode's schematic symbol. Note the letters in parentheses: (A) = anode and (K) = cathode; these letters are added to the symbol when necessary.

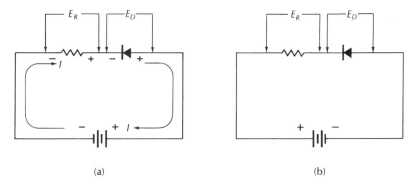

Figure 8-19 (a) A diode with forward bias applied; (b) with reverse bias applied.

biased condition, current is flowing through both the resistor and the diode. If the battery voltage is, for example, 6 V, then E_R might well be 5.4 V, while E_D is 0.6 V. Note that there is some drop across the diode, but that it is small relative to the drop across the resistor.

Suppose now that the battery voltage is increased to 30 V. Then E_R might be expected to be 29.3 V and E_R would be 0.7 V, a very slight increase. *The voltage across a conducting diode is relatively constant.* However, if the source were to be increased by a larger amount, say to 200 V, then the voltage drop across the diode might increase to 0.9 or 1.0 V. Nevertheless, in practical circuit applications the greatest part of the total voltage always appears across the resistor.

In the reverse-biased instance, since no current flows, there is no drop across the resistor and, because the diode is essentially an open circuit, the full value of V_S appears across it.

In order to illustrate most effectively the conductive characteristics of a diode, a curve is used. Figure 8-20 illustrates a diode conductance curve with typical values indicated. Both forward and reverse directions are indicated. Forward voltage is incremented in $1/10$-V parts, while reverse voltage is incremented in 10-V parts. This simply keeps the drawing to a reasonable physical size.

To begin to describe the curve, note first that, starting at the point of origin (all values equal to zero), if 0.1 V or less is applied to the germanium diode, *no current flows,* even though this is forward bias. This region of operation is explained by realizing that at very small voltages, the barrier region is compressed but still exists. Only when the applied voltage is large enough to force the electrons and holes to meet, thus eliminating the barrier, can current flow. For the silicon case, conduction begins at about 0.5 V. From this point on, as applied voltage is increased (or series resistance decreased) the voltage drop

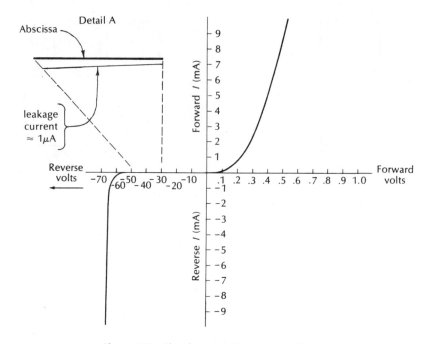

Figure 8-20 The characteristic curve of a diode.

across the diode increases very slowly to a maximum that depends upon the type of diode, but is typically 0.5 V for Ge and 1.0 V for Si.

In the reverse direction, no current flows *until some critical value* of applied voltage is reached. Then, current *will* flow in the reverse direction. Normally, high values of reverse current are to be avoided, since many diodes are instantly destroyed if this is allowed. For the diode illustrated, no more than about 50 V of reverse voltage should be applied, to avoid destruction. The maximum reverse voltage that can safely be applied to a diode is called the *peak-inverse-voltage* (PIV). Values of PIV to several hundred volts are commonly available.

Up to now, we have been saying that with small values of reverse bias applied, no current flows. This is not strictly speaking true. While it is not evident on the graph of Figure 8-21, a tiny current *does* flow when the diode is reverse biased, as shown in detail A. Its value is too small under normal conditions to show on the graph, but it can easily be measured with a microammeter. This current is called *reverse current,* or more commonly, *leakage current.* Its value is on the order of microamps for Ge devices, but only in the range of 10 to 100 nanoamps for Si devices. This is usually negligible for most circuits, and it can frequently be ignored. Leakage current increases very slowly as the reverse voltage is increased until the PIV is approached; it then increases more rapidly until a large current is flowing. For most devices, operation in this region means total destruction because of increased power dissipation.

Leakage current exists in an amount proportional to the *temperature* of the device. At room temperature there are a few free electrons on the p-side of the junction and a few free holes on the n-side. These carriers are liberated in direct proportion to the temperature. At absolute zero (0° K) there are no current carriers in the barrier region. As the temperature is increased to room temperature and beyond, these carriers increase in number in direct proportion to the temperature. Leakage current, then, is very small at low temperatures, moderate at or near room temperature, and very large at high temperatures. The temperature liberated carriers that exist within the confines of the barrier region are called *minority* carriers. Because of the existence of leakage current, the application of semiconductors of all kinds requires that special attention be given to the circuitry. Very elaborate circuits have been devised to minimize the adverse temperature effects in semiconductor circuits. In a diode circuit, operation at very high temperatures can mean that the diode will conduct in *both* directions of applied voltage, hence is simply a conductor under these conditions.

Diodes are used in a wide variety of applications, generally in conjunction with alternating current. In Chapter 18 we investigate such circuits in some detail. For the moment, we shall simply show how the circuit and the characteristic curve of a diode are correlated by giving a few examples.

Example 1. Assume in Figure 8-19a, that $R = 1500 \, \Omega$ and $V_s = 6$ V, and that the curve in Figure 8-20 represents the diode. Find the value of total current and the drops across the resistor and the diode.

First, find the approximate current flowing in the circuit disregarding the diode:

$$I = \frac{V_s}{R} = \frac{6}{1500} = 4 \text{ mA}.$$

Now, read from the curve that at 4 mA of forward current, E_D is about 0.37 V. Use this value to determine a more nearly correct current through R:

$$I_R = \frac{V_s - E_D}{R} = \frac{6 - 0.37}{1500} = 3.75 \text{ mA}.$$

Determine actual E_D; from the curve, $E_D = 0.35$ V. Finally, determine actual I_R:

$$I_R = \frac{V_s - E_D}{R} = \frac{6 - 0.35}{1500} = 03.77 \text{ mA} \cong 3.8 \text{ mA}.$$

Note that one must approximate the unknown values before finding accurate ones. In the following example, this is not necessary.

Example 2. Using the given diode and curve, devise a circuit that allows 5 mA to flow through the resistor and diode, given that $V_s = 10$ V.

From the curve, $E_D = 0.4$ V at 5 mA. Using this value, calculate the voltage drop across the resistor:

$$E_R = V_s - E_D = 10.0 - 0.4 = 9.6 \text{ V}.$$

Now the value of the required resistor can be found.

$$R = I = \frac{9.6}{0.005} = 1920 \text{ }\Omega.$$

SUMMARY

- Copper or any other type of wire has some resistance, and over a long span of wire a voltage drop is distributed along its length.
- Commercial conductors are measured by either American Wire Gauge (AWG) sizes or circular mil area.
- An assembly of two or more wires bound together but insulated from each other is called a cable.
- Printed-circuit boards are widely used to replace standard copper wiring because they are readily mass produced.
- Commonly used metals used as conductors are silver, copper, gold, and aluminum in ascending order of resistance.
- The resistance of a given length of wire is $R = \rho(l/A)$.
- The temperature coefficient of resistance (α) indicates whether the resistance of a material increases or decreases with an increase in temperature.
- Pure metals have a positive α; semiconductors have a negative α.
- Ion current can exist in liquids and gases.
- Insulators have very high resistance because they contain no free electrons to carry current.
- Semiconductors are made from a carefully controlled combination of tri-, tetra-, and pentavalent elements.
- Valence electrons determine the special semiconductor characteristics.

Conductors, Insulators, and Semiconductors 187

- Pure crystalline Ge and Si make very poor conductors, because their valence electrons are locked in covalent bonds.
- *N*-doping frees some electrons to act as current carriers.
- *P*-doping releases some holes to act as current carriers.
- A hole carries unit positive charge but has no mass.
- Conduction in *n*-type material is by conventional electron flow.
- Conduction in *p*-type material is by hole flow in the direction opposite to that of electron flow.
- Holes cannot exist outside of the semiconductor.
- In a *pn* junction diode, current will flow if electrons are directed *into* the cathode, but not in the opposite direction.
- The characteristic curve of a diode describes how the diode acts with both forward and reverse bias.

QUESTIONS

1. In general, metals (are, are not) good conductors.
2. The most efficient electrical conductor (is, is not) gold.
3. Copper, widely used in electrical applications, (is, is not) the most efficient conductor.
4. The most efficient conductor is _____ .
5. (True or false) One mil is 0.001 in.
6. The circular-mil area of a wire (is, is not) the diameter squared.
7. A wire having a diameter of 20.1 mils has a cross-sectional area of _____ circular mils.
8. Gauge number 22 round copper wire is (larger, smaller) than gauge number 20 wire.
9. Gauge number 18 round copper wire is (larger, smaller) than gauge number 26.
10. Of the several metals listed in Table 8-2, which has the lowest resistance?
11. The circular-mil area of a wire (does, does not) vary as the diameter.
12. The circular-mil area of a wire (does, does not) vary as the square of the diameter.
13. The temperature coefficient of a conductor (is, is not) a measure of how hot it gets.
14. The temperature coefficient of a conductor (is, is not) a measure of how its resistance varies with temperature.
15. The temperature coefficient of a material (may, may not) be positive or negative.
16. The specific resistance of copper (is, is not) the same as that of silver.

17. The specific resistance of a conductor (is, is not) a measure of the amount of resistance per unit volume.
18. Pure metals have (positive, negative) temperature coefficients.

Indicate whether each of the following is true or false.

19. Semiconductors are made from tri-, tetra-, and pentavalent elements.
20. The inner shells of electrons determine the special semiconductor characteristics.
21. Pure crystalline Ge or Si is a very poor conductor.
22. When pure Ge or Si is crystallized, the valence atoms are locked in covalent bonds.
23. N-doping releases some atoms to act as current carriers.
24. P-doping releases some holes to act as current carriers.
25. A hole carries unit positive charge but has no mass.
26. Conduction in n-type material is by conventional electron flow.
27. Conduction in p-type material is by hole flow in the direction opposite to that of electrons.
28. Holes can exist in a copper conductor.
29. In a pn junction diode, current flows if electrons are directed into the anode, but not in the opposite direction.
30. The characteristic curve of a diode normally describes how the diode acts with both forward and reverse bias.

PROBLEMS

For the following six problems, refer to Figure 8-1.

1. The distance from V_s to the lamp is 500 ft, lamp rating is 15 V at 0.5 A, and the wire is #20. Find the correct V_s to provide 15 V across the lamp.
2. The distance from V_s to the lamp is 500 ft, lamp rating is 6 V at 0.15 A, and the wire is #26. Find the correct V_s to provide 6 V across the lamp.
3. The distance from V_s to the lamp is 100 ft, lamp rating is 6 V at 1 A, and the wire is #20. What value of V_s will provide 6 V across the lamp?
4. The distance from V_s to the lamp is 250 ft, lamp rating is 12 V at ½ A, and the wire is #20. Find the correct V_s to have 12 Volts across the lamp.
5. V_s is 12 V and the lamp rating is 9 V at 0.5 A. Using #30 wire, what is the maximum distance from the source at which the lamp will still operate properly?

6. V_s is 16 V and the lamp rating is 12 V at 0.67 A. Using #30 wire, determine the maximum distance from the source at which the lamp will operate properly.
7. A 100-ft length of #20 copper wire has how much resistance?
8. A 500-ft length of #30 copper wire has how much resistance?
9. A 300-ft length of #30 aluminum wire has how much resistance?
10. A 225-ft length of #30 tungsten wire has how much resistance?
11. A length of copper wire has a resistance of 15 Ω at 20°C. What is its resistance at 100°C?
12. A length of copper wire has a resistance of 100 Ω at 20°C. What is its resistance at 80°C?
13. A length of aluminum wire has a resistance of 12 Ω at 20°C. What is its resistance at 50°C?
14. A length of tungsten wire has a resistance of 65 Ω at 20°C. What is its resistance at 300°C?
15. A conducting diode is connected in series with a 500-Ω resistor. The source is 12 V and $E_D = 0.5$ V. Find total current and E_R.
16. A conducting diode is connected in series with a 1200-Ω resistor. The source is 18 V and $E_D = 0.7$ V. Find total current and E_R.
17. A conducting diode is connected in series with a resistor. Total current is 1 mA, $E_D = 0.6$ V, and $E_R = 5.4$ V. Find R.
18. A conducting diode is connected in series with a resistor. Total current is 10 mA, $E_D = 0.8$ V, and $E_R = 11.2$ V. Find R.

CHAPTER 9

RESISTORS

Up to now we have assumed that all resistors are of one kind, and that their only characteristic is that of providing resistance for a circuit. Actually, there are many, many kinds of resistors, each suitable for a particular application. Resistors are made from a variety of materials in a variety of sizes. Some are fixed in value, some are variable and, for example, are used for volume or tone controls in radio receivers and audio equipment. Some are capable of dissipating tens, or even hundreds, of watts, while others can safely handle only a few milliwatts. All of these characteristics, plus several others, will be investigated in this chapter.

The major topics are:

9-1 Types of Resistors
9-2 Color Coding
9-3 Design Principles
9-4 Nondiscrete Resistors
9-5 Testing Procedures
9-6 Potentiometers

9-1 TYPES OF RESISTORS

The resistors mentioned in earlier chapters have not been described fully as to size, shape, composition, or other physical or electrical attributes. It has been assumed that the only property of resistors is that of resistance. However, these components are available in a wide variety of kinds, sizes, and materials. Our first purpose, then, is to become familiar with a variety of resistors, especially those most commonly encountered.

Probably the most widely used resistor is the *carbon-composition,* or simply, *carbon* resistor, illustrated in Figure 9-1. Three sizes are shown in part *a* of the figure, and two smaller sizes (¼ W and ⅛ W) commonly exist. As already mentioned, the size of a resistor determines the amount of power it can safely dissipate. Thus the smallest shown in the figure will dissipate no more than ½ W, while the largest can handle up to 2 W. Carbon resistors are seldom found with higher power ratings.

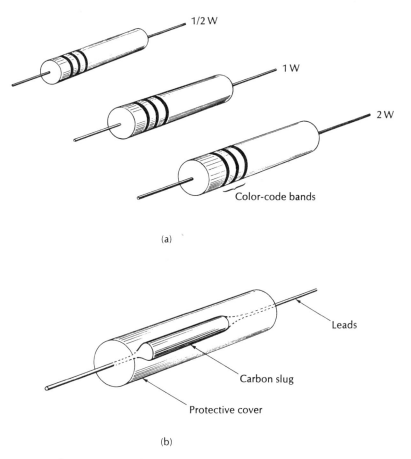

Figure 9-1 (a) Carbon-composition resistors; (b) construction of a carbon-composition resistor.

Part *b* of the figure illustrates construction of the carbon-composition resistor.

The circular bands at one end of the resistor are the color-code bands that indicate resistance value. Section 9-2 describes the color code in detail.

Variable resistors are used in applications where the resistance must be changed in value during operation. One such application is the volume control on a radio receiver. Such resistors are known as *potentiometers* (pots) or rheostats. Generally, a rheostat is used to control a large amount of power and is therefore capable of handling, perhaps, 50 or 100 W. A potentiometer, on the other hand, is often rated at 2 W or less. Figure 9-2 illustrates several variable resistors, with Figure 9-2*a*

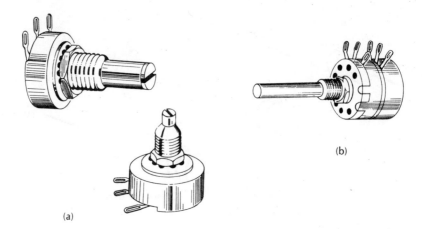

Figure 9-2 (a) Small low-power potentiometers; (b) dual in-line potentiometers.

showing one that might be used as the volume control on a radio set. Figure 9-2b illustrates a *ganged* potentiometer, where a single shaft rotates two otherwise separate potentiometers. Such a control varies the resistance of two separate circuits simultaneously. If each resistor element is connected to a separate shaft, the outer one encompassing the inner one, then the two ganged potentiometers can be varied independently. We shall show examples of the use of each connection and discuss potentiometers more fully in Section 9-6.

Wire-wound Resistors

As the name implies, *wire-wound* resistors use wire for the resistive element. Such wire is made from alloys, such as nickel and chromium (Nichrome); or copper and nickel (Advance); or nickel, copper, and manganese (manganin). The resistance wire is usually wound on a ceramic form, with a slight space between turns. The wire is bare, and must not touch, turn to turn, or the current path will be shorted. Then, terminals are attached and the entire resistor is coated with a ceramic paste and fired to harden the ceramic. In single-value resistors, the wire is completely coated to prevent shorts. However, it is often required to have a *movable tap*, so intermediate values of resistance can be set. This is illustrated in Figure 9-3, along with the fixed type. The tap can only be adjusted when the equipment is deenergized. Also shown is the schematic representation of each type.

Wire-wound resistors find widest use in high-power applications. Wattage ratings to several hundred watts or more are readily available.

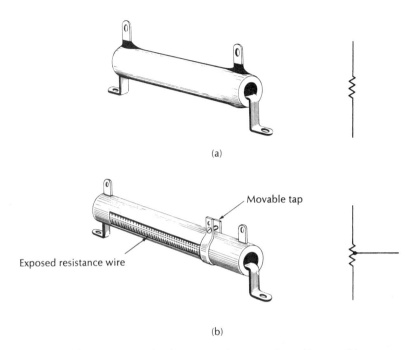

Figure 9-3 (a) A fixed wire-wound power resistor; (b) a movable-tap power resistor.

The smallest practical value available is 3 W since carbon resistors are available to 2 W.

The continuously variable power potentiometer, Figure 9-4, is used in applications where the tap must be easily adjusted over the complete range of resistance while the equipment is operating.

Wire-wound resistors are found in another application, that of highly accurate precision resistors, either fixed or variable. These are used in precision laboratory instruments. Current shunts and voltage multipliers frequently consist of fixed-value precision wire-wound resistors made of *insulated* wire (frequently enameled) wound on a plastic *bobbin*. Such resistors are measured on a very accurate resistance bridge to attain the precision required.

Miscellaneous Resistors

The resistors presented thus far are the ones most generally encountered. Many other types exist, however; three fixed-value types are deposited-carbon, metal film, and metal glaze. All of these are formed by vacuum-depositing the resistance material onto a ceramic form,

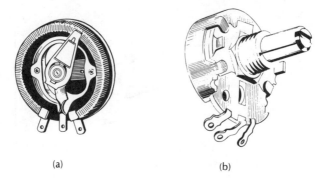

Figure 9-4 (a) A power rheostat; (b) a wire-wound variable resistor.

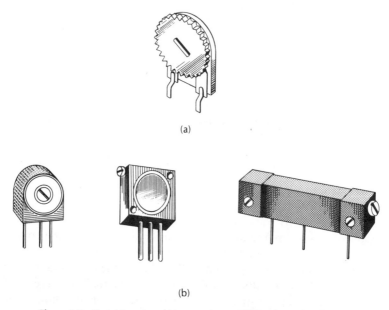

Figure 9-5 Variable resistors for mounting on PC boards: (a) thumb-wheel type; (b) screwdriver-adjustable types.

then cutting away a spiral groove to form a coil of metal or carbon. Such resistors can be obtained in tolerances of 0.1%, 0.25%, 0.5%, or 1%, and so are useful in precision equipment.

So many variable types are available that it is impossible to mention more than a few in this space. A *trimming potentiometer*, Figure 9-5, is a small variable resistor that often has leads designed to fit directly on a PC board, and thus can only be adjusted when access to the board is

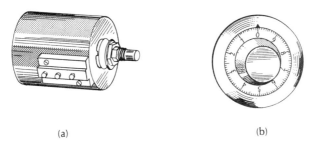

Figure 9-6 (a) A ten-turn potentiometer; (b) a ten-turn knob.

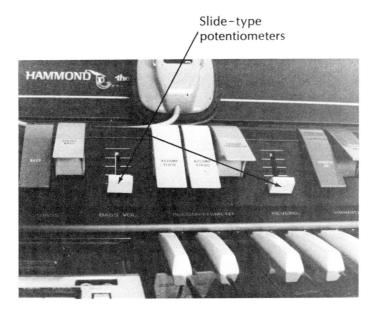

Figure 9-7 Slide-mount volume controls.

provided. Ten-turn potentiometers, Figure 9-6, are frequently used where very accurate settings must be made with excellent repeatability. Slide-actuated potentiometers, Figure 9-7, are in wide use in equipment ranging from radio-and TV-studio control consoles to electronic organs. Slide-actuated potentiometers do not operate by rotation, but instead are operated by moving a lever in a straight line. This type of construction is inherently less noisy and in some situations more convenient to operate.

9-2 COLOR CODING

Carbon resistors are almost always identified by a color coding consisting of brightly colored bands around one end of the resistor. This is useful, since many resistors are too small to have the value printed directly on them. The color bands tell the user what the resistance *value* is, and what the *tolerance* is. The value is the ohms value, while the tolerance tells how much the actual value can differ from the coded value.

The color code used for resistors uses different colors to indicate the digits of the number value, and the number of zeros following the digits, as well as the tolerance. The body of the resistor is usually a light tan color, to avoid confusion with the color bands. Table 9-1 illustrates the color code. The first column represents the most significant digit of the resistor value, and the second column represents the next most significant digit. The third column is the multiplier, and indicates the *number of zeros* to be affixed to the two digits. Figure 9-8 shows several examples of the first three bands and the resultant resistor values. Resistor *a* is orange, orange, brown, reading from the end of the resistor toward the center. Therefore, according to Table 9-1, the first digit is 3, and the second is 3, thus the two most significant digits are 33. To this must be added one zero (brown), giving a resistor value of 330 Ω. The second example *(b)* illustrates a 470-Ω resistor, the third *(c)* is 51,000 Ω, and resistor *(d)* is a 680-Ω resistor.

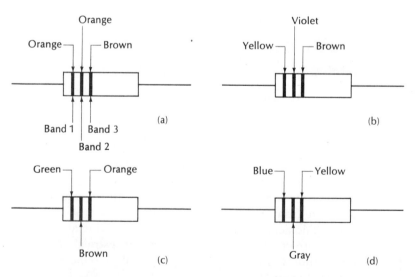

Figure 9-8 Examples of the resistor color code: (a) 330Ω; (b) 470Ω; (c) 51kΩ; (d) 680kΩ.

TABLE 9-1 RESISTOR COLOR CODE

Color	1st Digit (1st Band)	2nd Digit (2nd Band)	Multiplier (3rd Band)	Tolerance (4th Band)
Black	0	0	—	—
Brown	1	1	10 (10^1)	—
Red	2	2	100 (10^2)	—
Orange	3	3	1,000 (10^3)	—
Yellow	4	4	10,000 (10^4)	—
Green	5	5	100,000 (10^5)	—
Blue	6	6	1,000,000 (10^6)	—
Violet	7	7	10,000,000 (10^7)	—
Gray	8	8	100,000,000 (10^8)	—
White	9	9	—	—
Gold	—	—	0.1 (10^{-1})	±5%
Silver	—	—	0.01 (10^{-2})	±10%
No color	—	—	—	±20%

Color	First Band	Second Band	Third Band	Fourth Band	
Black	0	0	—		
Brown	1	1	0		
Red	2	2	00		
Orange	3	3	000		
Yellow	4	4	0000		
Green	5	5	00000		
Blue	6	6	000000		
Violet	7	7	—		
Gray	8	8	—		
White	9	9			
			X0.1	Gold	±5%
			X0.01	Silver	±10%
				No Band	±20%

As an aid in memorizing the color code, it is of help to recognize that the colors have ascending values in the same order as they appear in the color spectrum.

If the resistor value is less than 10 Ω, the gold or silver multipliers are used in the third band. For example, if a resistor has the color bands red, violet, gold, the value is 27 × 0.1 = 2.7 Ω. If the color bands are red, violet, silver, the value is 27 × 0.01 = 0.27 Ω. Do not confuse the gold and silver colors in the *third* band and the same colors in the *fourth* band. Fourth-band colors represent *tolerance only*.

Resistor Tolerance

The fourth color band indicates whether a resistor's tolerance is ±20%, ±10%, or ±5%. It is not economical to manufacture resistors with exact values, and the greater the allowable variance, the less expensive the resistor. The tolerance color band uses gold (±5%), silver (±10%), and no band (±20%) to indicate this parameter. Precision resistors of other types usually have the resistance and tolerance printed on the resistor body.

A ±20% resistor with a color code indicating 100,000 Ω may have a true value anywhere between 80,000 and 120,000 Ω. The same resistor with ±10% tolerance may have a true value of 90,000 to 110,000 Ω, while a ±5% resistor with the same color code has an actual value of between 95,000 and 105,000 Ω.

Where greater precision is required, carbon-composition resistors are not used. Construction techniques such as *deposited carbon* or *metal film*, both of which are described above, are used to provide tolerances from ±1% to ±0.10%. Resistors with such close tolerances are used only where necessary, since the better the tolerance, the higher the cost. General design practices for normal circuitry allow the use of ±5% or ±10% resistors with no ill effects on circuit performance.

Temperature Coefficient

Carbon-composition resistors react to changes in temperature, as do most substances. However, pure carbon has a *negative* temperature coefficient, as opposed to most metals.

Whether the resistance of a carbon resistor will increase or decrease with an increase in temperature is, unfortunately, a complex function. The resistive element in a carbon-composition resistor is composed of both carbon *and* binder. Low-value resistors have a small amount of binder relative to the carbon, while in high-value resistors the reverse is true. Thus the actual response of a carbon resistor depends upon its value and the type of binder used, along with the temperature coefficient of the binder. Generally, large-value resistors have a positive temperature coefficient, while low-value resistors have a negative temperature coefficient.

9-3 DESIGN PRINCIPLES

Choosing a resistor for a particular application requires that several items be considered. Most important, of course, is the proper resistance value, usually determined by Ohm's law. Next, the actual power dissipated must be calculated, and a reasonable safety factor applied. Even though a resistor has, for example, a 1.0-W rating, its useful life will be shortened considerably if it is run continuously at full rated value. Third, the type of resistor must be chosen, usually in favor of the lowest-cost unit that will give satisfactory service. It would be pointless to use an expensive wire-wound resistor when a simple carbon one would do the job. Finally, the tolerance is determined, consistent with the required circuit performance.

These points must always be considered when choosing a resistor. In certain applications, however, it is necessary to consider still other characteristics. The maximum voltage that can be safely applied is sometimes of importance if large voltage drops will appear across the resistor. This is of significance in circuits developing voltages in the range of 1000 to 50,000 V, such as power supplies for TV sets. Manufacturers provide specific data relating to maximum voltage for the specific resistor in question; if more than 200 V appears across the resistor, manufacturer's data should be consulted to determine if the voltage rating is being exceeded.

Other factors, such as temperature coefficient, voltage coefficient, generated noise, and frequency response are sometimes of importance, but are beyond the scope of this book.

To illustrate the application of some of the foregoing points, a simple example of choosing a resistor is given. The example in Figure 9-9 illustrates a circuit with a missing resistor. What value resistor should be used for R_1? Total circuit current is known to be 1 mA, and the voltage across R_1 must fall between 76 V and 88 V. First, the total resistance value must be determined.

$$R_t = \frac{V_s}{I_t} = \frac{150}{0.001} = 150,000 \ \Omega.$$

Now, the value of R_1 can be found.

$$R_1 = R_t - R_2 = 150,000 - 68,000 = 82,000 \ \Omega.$$

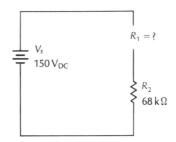

Figure 9-9 Simple design example.

Actual power dissipation is

$$P_{R1} = (I_t)^2 \times R_1 = 0.001^2 \times 82{,}000 = 0.082 \text{ W}.$$

Apply a safety factor of at least 100%:

$$P = P_{R1} \times 2 = 0.082 \times 2 = 0.164 \text{ W}.$$

Choose the next larger size, ¼ W (0.25 W), or any larger size that is convenient, if it will physically fit. In this instance we shall assume that, since R_2 is a carbon resistor, R_1 will also be carbon.

The specs call for R_1 to fall between 76 and 88 kΩ, so a tolerance must be selected. Will a $\pm 20\%$ ¼-W resistor be satisfactory? The value of R_1 is 82,000 Ω, and 20% of this value is 16,400 Ω. So

$$R_1 - R_1(0.20) = 82{,}000 - 16{,}400 = 65{,}000 \text{ }\Omega,$$
$$R_1 + R_1(0.20) = 82{,}000 + 16{,}400 = 98{,}400 \text{ }\Omega,$$

so the $\pm 20\%$ resistor is inadequate.

What about a resistor with a tolerance of $\pm 10\%$? The calculation is the same:

$$R_1 - R_1(0.10) = 82{,}000 - 8200 = 73{,}800 \text{ }\Omega,$$
$$R_1 + R_1(0.10) = 82{,}000 + 8200 = 90{,}200 \text{ }\Omega,$$

which is still inadequate. In order to conform to the specifications, a ¼-W resistor with a tolerance of $\pm 5\%$ must be chosen.

9-4 NONDISCRETE RESISTORS

In modern electronic circuitry there are new and very sophisticated methods of forming resistors. While it is beyond the scope of this book to discuss integrated circuits (ICs), it is at least interesting to note that a typical IC may contain dozens, if not hundreds, of resistors. Since a typical IC chip may be 1/10 to 1/20 of an inch on each side with a thickness of perhaps 10 mils (.01 in.), the process of producing these tiny marvels is worth mentioning.

While completely ignoring the other components on the IC chip, such as transistors and diodes, a brief investigation into these resistors will be made. Fundamentally, the resistors are formed by first defining the geometry of the resistor, and then drawing it magnified about five hundred times. Figure 9-10 illustrates a typical configuration. At the same time, the other components are drawn to the same scale. Then the image is reduced photographically five hundred times (to its actual size). The basic component for ICs is specially processed silicon and the next step is to photographically sensitize a thin, round slice of this material. This light-sensitive slice of silicon is then exposed to the

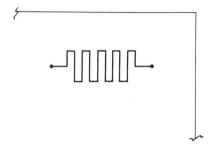

Figure 9-10 Configuration of an IC resistor.

reduced image and placed in an acid bath to etch away the unwanted coating.

The slice of silicon is then subjected to very high temperatures in a furnace with a controlled atmosphere. This high-temperature atmosphere contains traces of elements used to form the various parts of the circuit. The atmospheric element, perhaps arsenic, bombards the surface of the silicon only in the area desired for the particular component by a process called *masking*. A layer of silicon dioxide is formed over the entire surface, then etched in the desired shape for the particular component to be formed. The etched area reveals the surface to be treated, and all other areas are masked by the silicon dioxide. The high temperature causes the atoms in the atmosphere to attain a high velocity, and when they strike the etched area of the surface they actually penetrate and become a part of the silicon. The resistance of this area is controlled by the geometry of the area, the amount of time it is exposed to the atmosphere in the furnace, and the amount of the element in the atmosphere.

The resistors formed by these methods have all the properties of conventional resistors, even though they cannot be seen by the naked eye. Some degree of maximum power dissipation, perhaps only a few milliwatts, must be observed. The correct value of resistance is, of course, a prime requirement, and associated with this is some degree of tolerance. Resistors of this kind have typical tolerances of ±25%, which at first may seem too large to be useful. However, by clever circuit design this apparent disadvantage becomes an advantage.

9-5 TESTING PROCEDURES

Failure Modes

Carbon resistors fail in relatively few ways. If some other component fails, a resistor may overheat due to excessive current flow. In severe cases, the body of the resistor becomes so hot that the color bands burn and turn brown. In even more severe cases the resistor actually breaks in two. In these cases the symptom is readily visible, although simply replacing the resistor without locating the fault may only cause a repeat of the trouble.

Often, however, a carbon resistor simply opens up with no visible symptom. It is necessary then to measure the resistor with an ohmmeter to determine whether or not its actual value is equivalent to its color code. If it is mounted in a piece of equipment, it is usually best to disconnect one end of the resistor, since parallel paths through other resistors usually exist. With one end free, the resistor value can be measured accurately. (Also, *do not touch the probes with both hands* if measuring high-value resistances, since the body has a resistance of perhaps 20,000 to 100,000 Ω.) If the value is beyond the tolerance marked on the resistor, replace it with a new one. If its value is within tolerance, simply reconnect it to its original point in the circuit. Remember, when measuring resistance in a circuit, *power must be off*.

Another possible failure mode is that of an *intermittent* failure, which can be very difficult to locate. One approach is to disconnect one end of the suspected resistor, then connect the probes across it with alligator clips so the resistance measurement can be made with no hands. Then, alternately heat and cool the resistor while watching the meter. Heater-blower devices, such as a hand-held hair dryer, are excellent for heating; pressurized cans containing Freon are excellent for cooling. Any gross variation in resistance is cause for replacement. Performing the same operation on a known good resistor will allow a comparison.

Carbon resistors have a maximum allowable voltage rating, and if this value is exceeded, the life of the resistor is reduced. These are:

1. $\frac{1}{8}$ W = 150 V.
2. $\frac{1}{4}$ W = 250 V.
3. $\frac{1}{2}$ W = 350 V.
4. 1 W = 500 V.
5. 2 W = 500 V.

When replacing a resistor in a circuit, check to be sure that the voltage drop across it is less than the value specified above for the resistor size in question.

Variable carbon resistors often become *noisy*. If the control is used in audio equipment, turning the control results in a scratchy noise from the loudspeaker. If the control is used in the picture-producing part of a TV set, the effect is often excessive "snow" on the screen, either continuously or when the control is rotated. Whatever the symptom, the effect is called noise. Usually, noise indicates a worn element, which calls for replacement. However, if the potentiometer is not sealed, the same effect can be caused by dust entering over a period of time. Special preparations are available to clean and lubricate the control, usually without removing it from the chassis.

Testing Techniques

As mentioned earlier, a resistor is normally checked by measuring its resistance with an ohmmeter. For resistances with tolerances of $\pm 5\%$, $\pm 10\%$, or $\pm 20\%$, this is the proper procedure, and the method and cautions to be observed have been previously given.

However, if a resistor with a tolerance of $\pm 1\%$ or better is checked with an ohmmeter with no better than 5% accuracy, it is impossible to know whether the resistor is good or bad. In such a case, a *resistance bridge* must be used. Specific operating procedures accompany each instrument, and should be followed exactly.

In summary, then, a resistor may be open (*infinite* ohms); shorted (*zero* ohms), although this is not likely; or its value may have changed and hence be beyond normal tolerance. Note that, on rare occasion, a resistor is wrongly color-coded during manufacture. The resistor is perfectly good in all probability but must not be used in commercial equipment where service personnel could be misled by the erroneous color code.

9-6 POTENTIOMETERS

Applications

When it is necessary to provide a continuously variable resistance, a potentiometer (pot) is used. To apply Ohm's law to such a circuit, Figure 9-11 is offered. Here, the pot is used to provide a manually adjustible dc voltage between zero and some maximum value. As noted on the drawing, E_{min} is zero volts in reference to ground and E_{max} is some positive value. E_{max} is determined simply by Ohm's law, treating R_2 as though it were a fixed 14-kΩ resistor.

$$I_t = \frac{V_s}{R_1 + R_2} = \frac{12}{24,000} = 0.0005 \text{ A} = 0.5 \text{ mA}$$

$$E_{R_2} = I_t \times R_2 = 0.5 \times 14,000 = 7 \text{ V}.$$

When the movable arm (called the *wiper arm*) is fully clockwise, as indicated by the small arrow pointing to cw in Figure 9-11a, the output voltage at E_r is +7 V. When the arm is fully counterclockwise, E_r = 0.0 V. When the potentiometer shaft is rotated between these limits, the output voltage will be continuously variable between 0 and +7 V, assuming that R_2 is a carbon-element pot. If the pot is a wire-wound resistor, then the output consists of small *steps* produced as the wiper

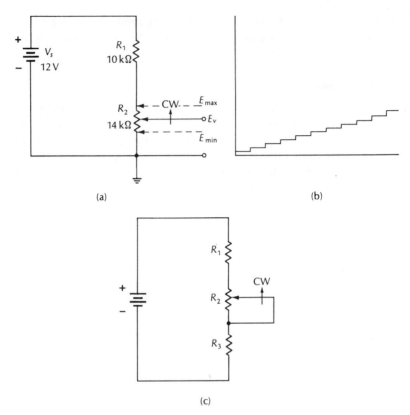

Figure 9-11 Potentiometer application. (a) A circuit incorporating a potentiometer; (b) the output of a variable wire-wound potentiometer; (c) a different potentiometer connection.

arm contacts new turns of wire, but of course only while the arm is being rotated. The output, therefore, would appear as indicated in Figure 9-11b for cw rotation.

Figure 9-11c illustrates an alternate connection for a potentiometer. Here, the application does not call for a *variable voltage divider* as shown in Figure 9-11a, but instead a simple variable resistance. As R_2 is turned fully cw, it has no resistance value, so total circuit resistance is $R_1 + R_3$. When R_2 is turned fully ccw, its full value is evident, and $R_T = R_1 + R_2 + R_3$. At intermediate values, total resistance is $R_1 + R_3$ *plus* whatever percentage of R_2 is evident in the circuit. This connection is useful, for example, in varying current in the circuit between some maximum and minimum limits.

In the case of carbon potentiometers, many of which are rated for a power dissipation of 2 W, remember that they are capable of this value of dissipation *only* when adjusted for full resistance. If connected as

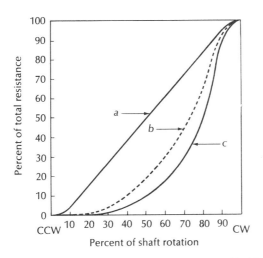

Figure 9-12 Potentiometer tapers: (a) linear (b) modified logarithmic 20 percent curve; (c) modified logarithmic 10 percent curve.

shown in Figure 9–11c, the power dissipation is proportional to the shaft rotation. If this is a 2-W pot and the shaft is halfway between minimum and maximum, maximum dissipation is 1 W.

Taper

Carbon-composition variable resistors are available in a variety of tapers. *Taper* is the degree of rotation versus change in resistance. A *linear* taper is one in which the resistance increases linearly as the shaft is rotated. Such a potentiometer is useful for electrical controls, but is not useful for volume or tone controls in audio equipment. The human ear responds to sound loudness in a logarithmic fashion, so volume controls must have a logarithmic resistance-change characteristic. Three of the most widely used tapers are illustrated in Figure 9-12. Curve *a* is, of course, the linear taper. Curve *b* is the *modified logarithmic 20% curve,* while *c* is the *modified logarithmic 10% curve.* The value at 50% rotation determines the percentage rating. As the control is rotated, the percentage change in resistance is quite different in each instance. In the linear case, the resistance at 50% rotation of the shaft is 50% of the total value; at 80% rotation the resistance is 80% of the total, and so on. The modified logarithmic 10% curve, however, reaches only $R = 10\%$ at 50% rotation, 20% at 60% rotation, 38% at 70% rotation, and 63% at 80% rotation. Notwithstanding the extreme nonlinearity of this curve, if it is used as a volume control, the *increase in audio intensity will sound to your ear as a linear increase.*

SUMMARY

- Carbon-composition resistors are the most widely used of fixed-type resistors.
- Carbon resistors are commonly available in five sizes: ⅛, ¼, ½, 1, and 2 W.
- Potentiometers are used as volume, tone, brightness, and contrast controls, to name but a few applications.
- Resistors rated at more than 2 W are usually wire-wound.
- Wire-wound resistors can be fixed, tapped, or continuously variable and are available at ratings ranging from three to several hundred W.
- A carbon resistor's life will be reduced if voltages much in excess of 200 V are applied to it continuously.
- Common tolerances for fixed carbon resistors are ±20%, ±10%, and ±5%.
- Tolerances better than ±5% are available in other types of resistors, such as deposited-carbon or metal-film resistors.
- The temperature coefficient of pure carbon is negative; but the temperature coefficient of a carbon-composition resistor may be positive or negative, depending on the type of binder and the proportion of binder to carbon.

QUESTIONS

1. Normally, carbon-composition resistors (are, are not) suitable for power dissipation of 10 W.
2. True or false: The size of a carbon resistor determines its ohmic value.
3. True or false: The power rating of a carbon resistor is a direct function of its size.
4. A typical value for a carbon resistor (is, is not) 5000 Ω, 10 W.
5. Two parallel 1000-Ω resistors have a total resistance (less, greater) than 1000 Ω.
6. Two parallel 1000-Ω resistors have a total power dissipation (less, greater) than that of either one.
7. Two series 1000-Ω resistors have a total power dissipation (less, greater) than that of either one.
8. Wire-wound resistors generally have (less, more) power capability than carbon resistors.
9. Wire-wound resistors generally are physically (smaller, larger) than carbon resistors.
10. A 50,000-Ω carbon resistor (would, would not) be a proper choice if current through it is 100 mA.

PROBLEMS

1. Determine the color code of a 33-Ω resistor.
2. Determine the color code of a 47,000-Ω resistor.
3. Determine the value of a resistor whose color code is red, yellow, red.
4. Determine the color code of a 1.8 MΩ resistor.
5. Determine the color code of a 110 kΩ, ±10% resistor.
6. Determine the value of a resistor whose color code is yellow, orange, yellow.
7. Find the limiting values of a 2.7 kΩ, ±10% resistor.
8. Find the limiting values of a 33 kΩ, ±5% resistor.
9. A resistor of unknown value is to be replaced in a circuit similar to that of Figure 9-5. $V_s = 12$V, $R_2 = 20,000$ Ω, $I_{R1} = 25$ μA. Find the value of R_1.
10. A resistor of unknown value is to be replaced in a circuit similar to that of Figure 9-5. $V_s = 90$ V, $R_2 = 1800$ Ω, $I_{R1} = 36$ mA. Find the value of R_1.
11. Refer to Problem 9. The power rating of the replacement resistor should be _____ , using standard carbon-composition sizes.
12. Refer to Problem 10. The power rating of the replacement resistor should be _____ , using standard carbon-composition sizes.
13. A resistor of unknown value is to be replaced in a circuit similar to that of Figure 9-5. $V_s = 1010$ V, $R_2 = 10,000$ Ω, $I_{R1} = 0.1$ A. (a) Find the value of R_1. (b) Determine the power rating of R_1. (c) Determine the least costly kind of resistor.
14. A resistor of unknown value is to be replaced in a circuit similar to that of Figure 9-5. $V_s = 12$ V, $R_2 = 10$ Ω, $I_{R1} = 1$ A. (a) Find the value of R_1. (b) Determine the power rating of R_1. (c) Determine the least costly kind of resistor.

CHAPTER 10

CELLS AND BATTERIES

A *cell* is a single unit of an electrochemical system for generating voltage and current at the expense of chemical energy. A *battery* consists of two or more cells connected either in series or parallel. In this chapter we shall discuss several different electrochemical systems, among which are carbon-zinc, lead-acid, alkaline-manganese, and nickel-cadmium. The fundamentals of a simple system will be explained, and the electrical characteristics of all systems in common use will be compared.

The major topics in this chapter are:

10-1 Basic Electrochemical Action
10-2 Battery and Cell Classification
10-3 Carbon-Zinc Characteristics
10-4 Alkaline-Manganese Characteristics
10-5 Mercury Cell Characteristics
10-6 Lead-Acid Characteristics
10-7 Nickel-Cadmium Characteristics
10-8 Other Systems
10-9 Internal Resistance of Sources

10-1 BASIC ELECTROCHEMICAL ACTION

A very simple electrochemical cell can be easily constructed from readily available materials—copper, zinc, sulphuric acid, and water. While the description that follows is accurate as far as it goes, it has been simplified.

First, it is important to note that when certain metals are immersed in dilute acid, a chemical action occurs that includes the dissolving of the metal surface. In the process of literally eating away some of the metal, a disruption of the surface atomic configuration occurs. In this process, there is a redistribution of electrons, in such a manner that the metal strip actually attains a voltage with respect to the acid solution. Such action is known as *separation of charges*.

The electromotive series of metals is shown in Table 10-1. Note that some metals become negative with respect to the acid solution, while

TABLE 10-1 ELECTROMOTIVE SERIES OF METALS

Material	EMF
Potassium	−2.92
Sodium	−2.71
Magnesium	−1.55
Zinc	−0.76
Tin	−0.13
Lead	−0.12
Hydrogen	0.0
Bismuth	+0.2
Copper	+0.34
Mercury	+0.80
Silver	+0.80

others become positive. Note also that hydrogen (EMF = 0) is used as the reference. These materials, all of which develop a potential with respect to the acid, are used in the manufacture of commercial cells and batteries.

Figure 10-1 illustrates an elementary wet cell. It consists of an inert container with a dilute solution of sulphuric acid (H_2SO_4) called the *electrolyte,* into which is immersed a strip of copper and a strip of zinc. As soon as the strip of zinc is immersed, the acid begins to dissolve it, and each zinc atom that is dissolved *leaves behind two of its electrons.* The zinc strip therefore accumulates electrons until it becomes −0.76 V with respect to the solution. The zinc particles that are dissolved into the electrolyte solution are positive ions, each having lost two electrons. When the zinc strip attains its −0.76-V potential, the dissolving action is altered: For every atom dissolved away from the surface, another atom already in solution is attracted back to the surface. This occurs because the zinc ions in solution carry a positive charge, and the zinc metal has accumulated a negative charge.

Now, when the copper strip is immersed in the solution, a similar action occurs, but with the opposite electrical effect. The copper plate also partially dissolves, releasing copper atoms into the solution. However, in this instance, the atom is quickly attracted back to the copper strip, leaving two electrons in the solution. This action continues until the copper strip attains a potential of +0.34 V with respect to the solution. When the copper strip reaches its potential of +0.34 V, further dissolving of the metal is prevented as before. This condition of balance is known as *equilibrium.*

Once equilibrium is established, a potential difference of 1.1 V exists between the two metal strips. This is the *difference* between the

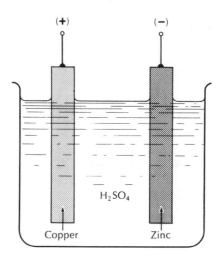

Figure 10-1 The elementary wet cell.

two electromotive series numbers for the two metals being used: $(+0.34) - (-0.76) = 1.10$ V. If the metals are absolutely pure, no further action occurs as long as there is no load attached to the metal strips.

Now, when a resistor or other load is connected between the copper and zinc strips, properly called *plates,* there is a conductive path external to the cell, and current will flow. As the excess electrons in the zinc begin to flow through the load toward the copper plate, the potential of the zinc plate drops slightly. As soon as this occurs, the zinc again begins to dissolve, supplying more electrons to the external circuit.

At the copper plate a similar action occurs. The electrons arriving through the load neutralize the positive charges and the potential drops slightly. Copper is again dissolved at the surface, releasing more electrons to the solution, and keeping the copper plate at its positive potential.

The cell action just described occurs at a rate sufficient to maintain current flow. As the load resistance decreases, more current will flow and chemical action is increased. Conversely, as the load resistance increases, chemical action decreases.

While current is flowing in the external circuit, it also flows, of course, in the acid solution. The positive zinc ions that go into solution displace H_2 ions originally associated with SO_4 particles. The H_2 ions are forced toward the copper plate, which is giving off electrons, and bubbles of hydrogen form on the surface of the copper plate.

The conversion of hydrogen ions to hydrogen bubbles is called *polarization,* and the effect is anything but beneficial. When a large current is drawn from the cell, the bubbles form rapidly and nearly completely cover the copper plate. The tiny bubbles remove much of the plate from contact with the acid solution, greatly reducing the number of electrons given up by the copper. From an electrical standpoint, this has the effect of reducing the maximum current the cell is capable of delivering. If the load conducts for an extended period of time, the polarization will make the cell able to deliver less and less current, until the load can no longer function properly. If, however, the cell is disconnected for 24 hours, the gas bubbles will dissipate, and if there is enough copper left in the cell, it will appear as good as new. The cell has "bounced back," to quote one cell manufacturer.

Certain chemicals, called *depolarizers,* can be added to the electrolyte solution. A depolarizer combines chemically with the hydrogen to form water (H_2O), which does not hinder cell operation. Thus, a depolarizer is added to retard the formation of bubbles, and to speed their dissolution if formed.

A polarized cell has a reduced voltage output as well as a reduced current output. Such a cell is said to have an increased *internal resistance.* That is, a cell with no polarization has a very low resistance from plate to plate. This is, actually, the resistance of the electrolyte itself plus the plates and connections, and can be as small as a fraction of an ohm. But, as polarization develops, the internal resistance rises, and in severe cases *most of the voltage produced by the cell is dropped inside the cell,* leaving little for external loads. This is understandable, since the internal path of current is in series with the load, and any voltage dropped *inside* the cell detracts from that available at the load.

In the foregoing discussion of cell action, the conversion of hydrogen ions into hydrogen atoms at the copper plate is, of course, caused by the gaining of electrons. This process is known as *reduction.* At the zinc plate, the reverse is occurring—the zinc atoms lose electrons; this process is known as *oxidation.* Thus, in such a cell (and nearly all others) reduction and oxidation occur simultaneously, furnishing the energy required for continuous current flow.

One further point regarding electrical cells: During periods of nonuse, chemical action slowly continues, due primarily to impurities in the metals and to temperature factors. Such action, known as *local action,* eventually depletes the materials and the cell becomes useless. Manufacturers of cells and batteries go to great lengths to insure that the materials used are as pure as is practical. A measure of the time a cell will remain usable is the *shelf life,* which allows the user to estimate the life left in a cell or battery manufactured much earlier. Generally, shelf life is given in months or years; it indicates the period of time that elapses until 50% of the cell's original energy is depleted.

10-2 BATTERY AND CELL CLASSIFICATION

Electrochemical cells are classified by the materials used; by whether or not they are rechargeable; and by their physical size and shape and their terminal voltage and rated current. For example, a particular cell might be described as a carbon-zinc cell, size D. From data provided by the manufacturer, such a cell would be described as suitable for moderate current delivery (500 mA or less) for moderate time periods.

Primary Cells

The carbon-zinc cell just described is classified as a *primary* cell; that is, it cannot be successfully recharged. Now, such a statement must be made with certain reservations. Generally, it is more accurate to state that it is not economically feasible to recharge carbon-zinc cells. This is also true of certain other types. The National Bureau of Standards states that under *very controlled conditions* the carbon-zinc cell *may* be recharged to a useful extent. Generally, however, the carbon-zinc cell is classified as nonrechargeable and is therefore a primary cell.

Secondary Cells

A *secondary* cell is one that is specifically designed to be recharged. An example that everyone is familiar with is the lead-acid battery used in automobiles. In this case, the chemical action is completely reversible, and supplying the battery a reverse current from an external source (a battery charger) causes the battery to return completely to its original condition.

Terminal Voltage

The materials used determine the cell voltage. The elementary cell described earlier has a voltage of 1.1 V per cell; a carbon-zinc cell yields 1.5 V; while a lead-acid battery provides 2.0 V per cell. Note that the amount of material used does not determine voltage; only the kind of materials influences this.

Rated Current

The maximum current that can be drawn from a cell or battery as well as the total current versus time are primarily functions of the *amount* of materials used. It is of importance to realize that maximum instantaneous current and total current delivered over a period of time are two separate characteristics.

Maximum instantaneous current is determined primarily by the kind and amount of materials used, and the rate of polarization and depolarization. The lower the internal resistance, the greater the maximum current. *Caution:* Placing a dead short across one or more cells to measure maximum current may damage the cell. If the cell or battery explodes, the observer may be severely injured. Only in a well-equipped laboratory can this measurement be made.

Total current delivered over a period of time is called the *ampere-hour* (AH) rating, and is the product of amperes and time. For example, if a cell is capable of delivering 250 mA for 10 hr, then

$$AH = amperes \times time = 0.25 \times 10 = 2.5 \ AH.$$

This is a very useful measure of the total energy capability of a cell or battery. The larger the number, the greater the total energy delivered to the load before the unit is depleted.

Series and Parallel Connections

If more voltage is needed than one cell can deliver, two or more cells can be connected in series. Figure 10-2a illustrates a 6-V carbon-zinc battery. Four cells are connected in series, so the terminal voltage is $4 \times 1.5 = 6$ V. Any number of cells can be connected in series; the only limitations are those of economics and physical size. When cells are

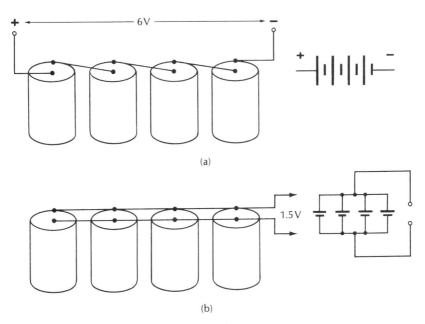

Figure 10-2 (a) Series and (b) parallel connected cells.

connected in series, the voltage is greater, but the total current delivery (AH) is the same as for a single cell.

Connecting cells in parallel (Figure 10-2b) provides more total current delivery to the load (or the same current for a longer time) but terminal voltage is the same as a single cell. Generally, the AH rating is increased by the number of cells. For example, if four 5-AH cells are connected in parallel, the battery rating will be: AH = 4 × 5 = 20 AH.

If a cell goes bad in a series connection, the net result is simply that the terminal voltage drops by the amount that the bad cell decreases. Other cells are not damaged, and the equipment being powered may still function, but at reduced efficiency. However, a bad cell in a parallel connection often continues to draw current from the good ones, rapidly depleting the battery. This occurs even if the load is disconnected. The effect on the overall battery, of course, depends on the kind of trouble in the bad cell. If the cell simply experiences an open circuit internally, it cannot supply energy and it is effectively removed from the circuit.

Dry and Wet Cells

Comparing a flashlight cell with a car battery leads to the conclusion that, physically, they bear no resemblance to each other. The flashlight cell is a *dry* cell, while a cell in the car battery is a *wet* cell. The wet cell contains an electrolyte that is truly in liquid form. This cell must always be operated in the upright position. The dry cell, however, is usually a sealed unit and, while not truly dry, may be operated in any position. The so-called dry cell has an electrolyte that consists of a moist paste, and hence remains in place better than a true liquid. The moist paste, however, functions essentially the same as a liquid as far as the production of energy is concerned.

Cell Sizes

Carbon-zinc cells, as well as many others, are available in a wide variety of sizes. These are given American Standards Association (ASA) designations of N, AAA, AA, C, D, F, G, and 6, in ascending order of size for standard round cells. Some of these are illustrated in Figure 10-3.

10-3 CARBON-ZINC CHARACTERISTICS

The carbon-zinc dry cell, or Leclanche cell, is exemplified by the standard flashlight cell. It is probably the most widely used type today, because of both low cost and reliable performance. Figure 10-4 illustrates a typical method of construction, although others are possible.

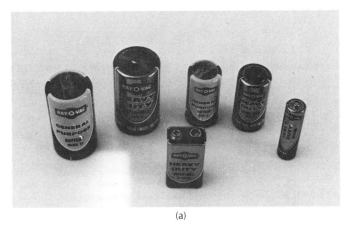

Figure 10-3 (a) An assortment of standard dry cells; (b) a disassembled view of a D-cell. Courtesy of RAY-O-VAC Division, ESB, Inc.

The negative electrode (anode) is the outer can itself, made of zinc. The true positive electrode (cathode) is the moist paste (core mix), consisting of manganese dioxide, acetylene black, and ammonium chloride moistened with a zinc chloride–ammonium chloride electrolyte. The manganese dioxide functions both as the positive plate and as the depolarizer. When the cell is in use, the manganese dioxide loses oxygen, while at the same time the zinc is oxidized. Thus, as the cell is used up, the manganese dioxide becomes less and less active as the cell's positive plate, thus requiring a longer and longer rest period between uses to "bounce back."

The solid carbon rod is connected to the top cap, which is the positive terminal. The rod itself is chemically inert, and its only function is to make a good connection to the paste.

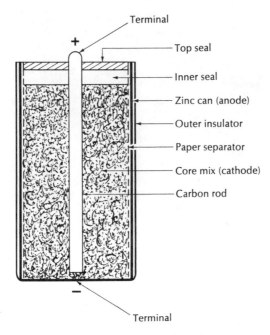

Figure 10-4 Typical construction of a carbon-zinc cell.

Terminal Voltage

As mentioned, carbon-zinc terminal voltage is approximately 1.5 V per cell under open circuit conditions (no load). When the cell or battery is delivering current to a load, the terminal voltage, now called the *working voltage,* drops and continues to drop as long as the load exists. If service is interrupted by removing the load, terminal voltage rises again to 1.5 V per cell. Figure 10-5 illustrates intermittent discharge curves for a typical size D carbon-zinc flashlight cell.

Service Capacity

The total useful energy delivered by a given size of carbon-zinc cell is widely variable, depending as much on manufacturing techniques as on the conditions under which the cell is used. Generally, however, the three rules below are useful in determining proper application.

1. High-current, short-time (hours) discharge results in much less total energy delivery to the load than the cell is normally capable of.

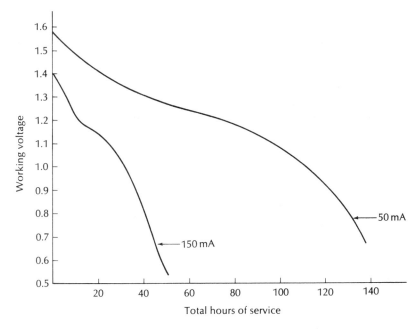

Figure 10-5 Carbon-zinc discharge characteristics; discharge rate is two hours per day at 70°F.

2. Moderate intermittent current delivered over relatively long periods of time (months) results in maximum energy output.
3. Very small current delivered over a very long time (years) results in less-than-normal total energy delivered to the load.

If the load current is too heavy, the electrodes cannot function properly and their reaction rate may be exceeded. On the other hand, if energy is withdrawn too slowly, the time required to deplete the available materials is so long that local action, which reduces the shelf life, expends some percentage of the materials. The carbon-zinc system, therefore, gives maximum service in moderate-current, moderate-time-interval applications. The service capacity is so variable that this system is never rated in ampere hours.

Temperature Effects

Excessive high temperature, either operating or storage, is detrimental to carbon-zinc systems. Temperatures in excess of 125°F may cause catastrophic failure, while use or storage over 70°F reduces the cell life.

At freezing or below, the cells deliver very little energy. Ideal temperatures, both for storage and use, fall between 40°F and 70°F.

Shelf Life

Storage temperature has a large effect on shelf life of the carbon-zinc system, as do the care and expense taken during manufacture. Factors influencing shelf life are storage temperature, vapor-sealing techniques used to prevent the electrolyte from drying out, atmospheric humidity, and local action. To reduce local action, the zinc is often coated with mercury, a process known as amalgamation, to reduce contact with the electrolyte. Shelf life (50% energy left) of the carbon-zinc system varies from several months for cheaply made varieties to two or three years for high-quality units.

10-4 ALKALINE-MANGANESE CHARACTERISTICS

The alkaline-manganese cell system is one of a family of six systems using an alkaline, rather than an acid, electrolyte. These systems are: (1) alkaline-manganese, (2) nickel-cadmium, (3) mercury, (4) silver oxide–zinc, (5) silver oxide–cadmium, and (6) air depolarized. Of these, the first three will be discussed in some detail, and only the basic characteristics of the others will be mentioned, since they are not yet in widespread use.

The alkaline-manganese system is widely available in a variety of cell and battery shapes and sizes. Although originally developed as a primary cell, it is rechargeable to a useful degree *under controlled conditions*. We shall consider it to be a primary cell, however.

The energy-producing reaction in alkaline-manganese cells is basically the same as in carbon-zinc cells. However, the significantly different electrical and physical characteristics are the result of the alkaline electrolyte and of the somewhat different construction. Figure 10-6 illustrates a typical cross-sectional view of the cell. The steel case is in contact with the cathode (which is also the depolarizer). The steel case (often nickel plated) not only provides strength, but because steel does not react with alkalies, there is no adverse reaction that would tend to weaken the case (as happens with some carbon-zinc constructions). The cylindrical cathode (a manganese dioxide mix) is separated from the anode by a porous material that allows full ionic transfer. The anode consists of granulated zinc, usually amalgamated, and the electrolyte, which is composed of potassium hydroxide.

Alkaline cells have reversed polarity; that is, the outer case is positive, as opposed to the carbon-zinc case, which is negative. However,

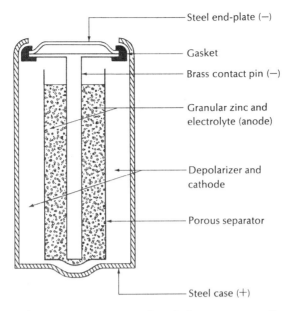

Figure 10-6 Construction of an alkaline-manganese cell.

the external case is arranged to *appear* the same as a carbon-zinc cell, so their application is the same.

Terminal Voltage

A new alkaline-manganese cell has an open-circuit voltage of 1.5 V. However, under nominal load conditions, the terminal voltage is slightly less than that of a carbon-zinc cell. The terminal voltage decreases with continuous discharge, but decreases more slowly than does a carbon-zinc cell. Alkaline-manganese cells are capable of delivering useful energy at, or below, 0.8 V, whereas the carbon-zinc cell is virtually exhausted at 1.0 or 1.1 V.

Service Capacity

Alkaline-manganese cells have 50% to 100% greater capacity than equivalent carbon-zinc cells. In many applications, the greater cost of alkaline-manganese is more than offset by their increased performance. The increase in capacity is due primarily to the more efficient depolarization in the alkaline cell, resulting in higher current delivery for a longer period of time. Also, little if any recovery time is required between uses compared to the carbon-zinc type.

TABLE 10-2 ALKALINE-MANGANESE TEMPERATURE CHARACTERISTICS

Temperature (°F)	Percent of Total Capacity		
	Light drain	Medium drain	Heavy drain
115	100	100	100
70	100	100	100
30	70	40	25
−10	15	10	3

Temperature Effects

The alkaline-manganese cell has good low-temperature characteristics. Table 10-2 shows the percent of total capacity for three different drain rates at four different temperatures.

10-5 MERCURY CELL CHARACTERISTICS

The mercury cell possesses certain electrical characteristics that make it unique. Under light drain conditions, the terminal voltage remains virtually constant until depleted. Mercury cells find very wide application where high-energy content, long life, constant voltage, and dependable service are required. A few examples are given to illustrate the wide range of usage:

Mercury cells, because of their constant terminal voltage, are used as reference cells in electronic laboratory equipment where a known and accurate voltage is required.

They are used in emergency equipment where their very long shelf life still allows ample power after years of standby conditions. Because of their high energy content, they are widely used for electronic wrist watches, where 18 months of service is not unusual for a tiny button cell.

They also find wide use in hearing aids, portable radio communications equipment, and a host of other devices.

Cylindrical mercury cells are constructed with a center post of highly compressed amalgamated zinc particles. This center post is surrounded by an absorbent material that allows access to the electrolyte yet separates anode and cathode structures. Surrounding this separator is the cathode, which consists of a compressed structure of mercuric oxide and conductive graphite. The cathode structure also contains the electrolyte, which is typically a 40% solution of potassium-hydroxide

(KOH) saturated with the zincate ion to prevent dissolving of the zinc anode during periods of nonuse.

As in the case of other alkaline systems, the cell case is made from steel and hence is virtually indestructable. With proper sealing techniques, the caustic electrolyte can be effectively constrained.

Terminal Voltage

The basic EMF produced by a mercury cell is 1.35 V. This is so repeatable and reliable that such cells are frequently used as secondary standards over wide temperature variations. In this application current drain must be kept low.

Service Capacity

The total energy delivery of the mercury system is even greater than the alkaline-manganese. For comparison, alkaline-manganese AA cells are rated at 1.8 AH, while mercury AA cells are rated at 2.3 AH, a nearly 25% increase. Size D cells are rated at 10 AH (alkaline-manganese) and 15 AH (mercury).

Temperature Effects

Low temperature adversely affects the performance of mercury cells. Maximum performance of cylindrical cells occurs at or above 70°F. At 32°F, their efficiency is reduced to about 10% of the 72°F output. Specially constructed mercury cells have efficiencies of 70% at 32°F, and produce 10% of nominal energy at 0°F. New advances are reported that will extend operation to −40°F.

Shelf Life

Mercury cells have a typical shelf life of five to seven years, although some have given reasonable energy output after twelve years under ideal storage conditions.

10-6 LEAD-ACID CHARACTERISTICS

A lead-acid battery, such as is used in automotive applications, is a secondary system, being fully rechargeable many, many times. Such systems are also used in many other applications, providing emergency power for such diverse uses as ship-to-shore communications, diesel-engine starting, emergency lighting, and many others.

The basic characteristics of the lead-acid cell are high output voltage, high current capacity, and reasonably long life. As opposed to the systems described earlier, the lead-acid system does not generate electrical energy—it merely stores it. That is, the cells must first be charged before electrical energy can be delivered. During charge chemical changes occur that, at a later time, will reverse themselves to generate electrical energy.

A typical car battery is constructed as shown in Figure 10-7. Inside each cell, the positive plates are composed of lead dioxide (PbO_2), while the negative plates are sponge lead. Both positive and negative plates are immersed in a dilute sulphuric acid electrolyte. As the cell delivers current, SO_4 combines with the lead in both plates to form

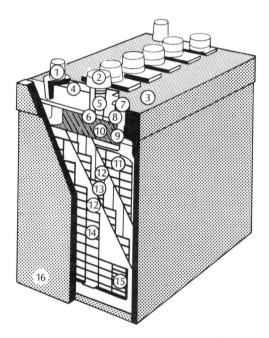

1. Terminal post
2. Vent plug
3. Sealing compound
4. Cell cover
5. Filling tube
6. Electrolyte level mark
7. Inner cell connector
8. Lead insert in cover
9. Plate strap
10. Separator protector
11. Negative plate (Pb)
12. Separator
13. Positive plate (PbO_2)
14. Negative plate with active material removed to show grid
15. Plate grid
16. Container

Figure 10-7 Cutaway drawing of a lead-acid battery.

lead sulphate. At the same time, H_2SO_4 divides into H_2 and SO_4. The H_2 ions combine with oxygen formed at the positive plate to form water. Hence, as the cells are discharged, the acid solution becomes more and more dilute, providing an excellent "state-of-charge" indication. In fact, a device called a *hydrometer* is often used to determine the condition of a storage battery by measuring the proportion of acid to water. This is described below.

To charge a lead-acid battery, an external source is connected so as to force current into the battery in the opposite direction from normal load current. This causes the reverse chemical action, where lead sulphate on both plates is converted to lead dioxide on the negative plate and sponge lead on the positive plate. The electrolyte increases in strength as SO_4 from the plates combines with hydrogen to form H_2SO_4 again.

Terminal Voltage

The nominal terminal voltage for a lead-acid cell is 2 V. However, for 14 to 16 hours after being fully charged, cell voltage may read slightly in excess of 2.1 V. Under load, cell voltage may read from 2.00 to 2.1 V, depending on the state of charge and the degree of load current drawn.

Service Capacity

The ampere-hour capacity of conventional lead-acid batteries is determined primarily by size (which is related to the amount of material available). Most batteries intended for automotive use are rated by the Society of Automotive Engineers (S.A.E.) by the 20-hour rate method. For example, a battery rated at 100 AH will deliver 100 AH/20 = 5 A for 20 hours before being depleted and requiring recharge. Other rating methods exist but are most useful to the manufacturer.

Temperature Effects

Temperatures between 70°F and 100°F provide ideal operating conditions for the lead-acid system. Continuous operation above 110°F shortens overall battery life. Temperatures significantly below 70°F result in lower output. At 32°F, the battery delivers only 65% of nominal energy, while at 0°F it delivers only 40%.

Shelf Life

This rating is not applicable to fully rechargeable systems such as the lead-acid system.

Specific Gravity

The charge condition of a lead-acid cell can be most conveniently checked by measuring the specific gravity of the electrolyte. This can be checked by using a hydrometer, which measures the density, or weight, of the sulphuric acid and the degree of dilution, which is directly proportional to the state of charge. When a battery is discharged, the electrolyte is very dilute; a fully charged battery contains much less water.

The hydrometer consists of a glass syringe that allows some of the electrolyte to be withdrawn from the cell. A calibrated float allows a reading to be made; a dense electrolyte causes the float to ride higher than when it is floating in pure water. A fully charged cell gives a reading (has a specific gravity) of about 1.260; a cell with 75% of capacity left reads 1.225; a cell having 50% of capacity left reads 1.190; a cell having 25% of capacity left reads 1.155; and a cell having 0% of capacity left reads 1.110. By comparison, pure water has a specific gravity of 1.0.

Charging Procedures

The requirements for charging a lead-acid battery are simply met. A voltage source slightly higher in value than the terminal voltage of the battery (≈ 2.5 V per cell), and capable of delivering fairly heavy currents is used. Figure 10-8a illustrates the connection required. Note that current during battery charge flows *into* the negative post and *out* of the positive post. Compare with Figure 10-8b, where the same battery is shown during normal discharge.

Charging current is maintained until a hydrometer reads 1.260 to 1.280. If in doubt as to the full-charge condition, take successive readings at 1-hour intervals. If three readings are made with no change in specific gravity, the battery is fully charged.

10-7 NICKEL-CADMIUM CHARACTERISTICS

Nickel-cadmium, or ni-cad, cells and batteries are probably made in the widest variety of sizes and shapes of any existing system. This is one of the rare systems where *both* dry cells and wet cells are readily available. Tiny *button-type* cells with complete recharging capability are available to power miniature equipment. At the other end of the scale, very large wet-cell types are available that are used, for instance, to start large diesel-locomotive engines, aircraft engines, and other heavy-duty applications.

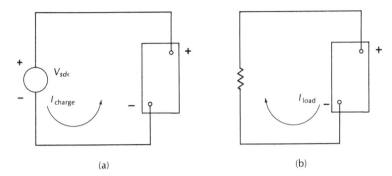

Figure 10-8 Direction of (a) charge and (b) discharge currents in a lead-acid battery.

The primary advantage of the nickel-cadmium system is its complete rechargeability. When constructed as a dry cell, it is capable of at least two hundred recharge cycles and often can be recharged more than a thousand times.

In addition to its rechargeability, the nickel-cadmium system offers certain other benefits. The discharge characteristics (voltage and current) remain nearly constant until the very end of the system's capacity. This can be easily demonstrated with a flashlight equipped with ni-cad cells. As the cells become nearly depleted, the flashlight continues to emit light as though the cells were fully charged. Then, within only a few seconds, the lamp rapidly dims and goes out.

Another advantage of ni-cads is the very large currents that can be delivered in a short period of time, even with small flashlight-size cells. Pulse currents of tens of amperes from flashlight cells are not uncommon. This characteristic is related to the very low internal resistance of the ni-cad.

Additionally, the wet-cell types used, for example, to start helicopter engines can be made much smaller and lighter in weight than lead-acid counterparts, thus saving significant amounts of space and weight. Ni-cads also serve as power supplies in research and communications satellites, where every ounce of weight is critical; the cells are recharged periodically by the energy of the sun acting upon solar cells.

Finally, ni-cads are capable of delivering satisfactory service for very long periods of time. Some sealed cells have been known to function like new, even though they have been in service for twenty years. Ni-cads are used in literally hundreds of rechargeable appliances ranging from shavers to lawn trimmers.

In fabricating ni-cads, sealed cell plates are first constructed of nickel powder sintered upon a fine screen or mesh of nickel. (*Sintering* is the conversion, by means of heat and pressure, of powdered nickel [or other substance] to a solid but highly porous mass.) Two such plates are used; the positive plate is saturated with a nickel salt solution, and the negative plate is saturated with a cadmium salt solution. The two plates are separated by a porous insulator which is saturated with a potassium hydroxide solution. The plates are then assembled by forming them into a tight roll, which is then placed in the case and sealed.

Wet cells using venting techniques to prevent loss of moisture are constructed much like lead-acid batteries and, in fact, appear very similar. The basic characteristics are similar to the sealed dry cells, but differ slightly in charge and discharge characteristics. Whereas sealed cells must be charged slowly, vented wet cells can be fully recharged in ½ hour. Sealed cells will rupture and possibly explode if treated in this fashion. Wet cells also have superior discharge characteristics due to their very low internal resistance.

10-8 OTHER SYSTEMS

Many, many different systems besides those already covered are in use for generating and storing *dc* electrical energy. A few of these systems are the silver oxide–zinc system, the silver oxide–cadmium system, the air-depolarized cell, the Edison cell, and the fuel cell. Additionally, it is possible to generate electrical energy by expending nuclear energy. Lack of space precludes an in-depth discussion of these, and others. However, Table 10-3 lists and compares the major characteristics of each, along with the systems already covered. Other experimental systems are not yet in widespread use, and so are not dealt with herein.

TABLE 10-3 MAJOR ELECTROCHEMICAL SYSTEMS

System	Acid/ Alkaline	Dry/ Wet	Open-Circuit Voltage	Watt-hours /lb	Primary/ Secondary
Carbon-zinc	Acid	Dry	1.5	15	Primary
Alkaline-manganese	Alkaline	Dry	1.5	35	Primary
Mercury	Alkaline	Dry	1.35	45	Primary
Nickel-cadmium	Alkaline	Dry	1.25	15	Secondary
Silver oxide–zinc	Alkaline	Dry	1.5	60	Primary
Silver oxide–cadmium	Alkaline	Dry	1.4	40	Secondary
Air-depolarized	Alkaline	Dry	1.4	53	Primary
Lead-acid	Acid	Wet	2.0	15	Secondary
Nickel-cadmium	Alkaline	Wet	1.25	15	Secondary
Edison (nickel-iron)	Alkaline	Wet	1.2	15	Secondary
Fuel cell	Fuel cells combine hydrogen and oxygen (the opposite of electrolysis), yielding electrical energy as a by-product. They require fuel, often hydrocarbons, to operate.				

One experimental system, however, is worth mentioning, as it seems likely to appear on the market in the near future. This is the *lithium* cell. Lithium is the lightest of all known metals, and in proper combination with other materials produces *ten times* more electrical energy per pound (150 watt-hours per pound) than a carbon-zinc cell. A typical lithium cell designed to power an electronic calculator might measure 23 mm in diameter ($\approx$ 0.9 in.) by 2.5 mm ($\approx$ 0.1 in.) high and weigh only 3.1 grams (0.11 oz). One particular device has a terminal voltage of 2.8 V; the electrolyte in this device is lithium borofluoride solute dissolved in gamma buthyrolactone, an organic material that is relatively inert. The cathode is metallic lithium and the anode is polycarbonmonofluoride. Used with a liquid crystal display, which uses power very economically, it is expected to have a capacity of 140 mAh (milliampere-hours), and may last up to *five years*.

10-9 INTERNAL RESISTANCE OF SOURCES

As has been briefly mentioned, all sources of electricity, whether battery systems or rotating generators, possess internal resistance, R_i, to some degree. The ultimate performance of a source depends to a large degree upon its internal resistance. We shall now investigate the internal resistance of battery systems, but much of what is covered is directly applicable to *any* electrical source.

Figure 10-9 illustrates how the internal resistance of a battery is depicted schematically. The pictorial *(a)* does not show R_i, while the schematic *(b)* illustrates that each cell has some internal resistance. Since the individual cell R_i's are in series, they can be totaled and presented as a single resistance, as in *(c)*. In the case of a battery, the value of internal resistance is usually not constant, since it depends upon such variables as polarization and depolarization, how moist the electrolyte is in a dry cell, and how many hours of use it has had. Since these factors are highly variable, R_i can change in either direction as time passes, whether the cell is in use or not.

R_i and Battery Operation

The ratio of internal resistance to the total resistance, $R_i/(R_{\text{load}} + R_i)$, determines how much the internal resistance influences battery operation. With some cell systems, R_i varies so much, depending upon the application, that it must be determined empirically. That is, the value of R_i must be measured for the particular case at hand.

To determine the R_i of a particular battery empirically, the open-circuit voltage is first measured as indicated in Figure 10-10*a*. Then, a

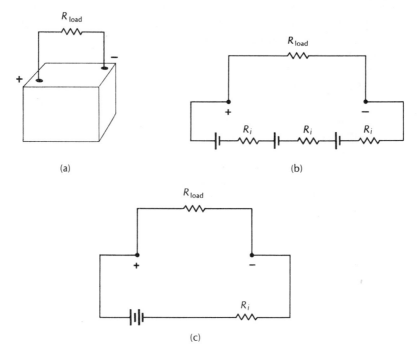

Figure 10-9 Battery internal resistance: (a) pictorial; (b) schematic; (c) total internal resistance schematic.

typical load is connected as shown in Figure 10-10b and the voltage across the load (E_{R_L}) is then measured. Then, the true current through the load is calculated, and the following relationship is solved:

$$R_i = \frac{E_{oc} - E_{R_L}}{I_{R_L}},$$

where E_{oc} is open-circuit voltage. For example, if $E_{oc} = 6.0$ V, $R_L = 1.0$ kΩ, and $E_{R_L} = 5.8$ V, then $I_{R_L} = 5.8/1000 = 5.8$ mA, and

$$R_i = \frac{6.0 - 5.8}{0.0058} = 34.5 \ \Omega.$$

Assume now that this battery is returned to service and checked again later. E_{oc} still is 6.0 V (after a suitable rest period) and the voltage across the load is measured as 4.9 V. Load current is now $4.9/1000 = 4.9$ mA, so

$$R_i = \frac{6.0 - 4.9}{0.0049} = 224.5 \ \Omega.$$

In this instance, as is usually the case with primary systems, the internal resistance increases as the battery is used.

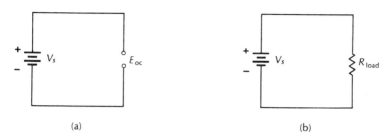

Figure 10-10 Determining internal resistance in a battery.

Note in these two examples that the loaded output voltage is less than the open-circuit voltage. This is true because R_i is in series with the load and total current flows through it, thus dropping a portion of the total available voltage. Any voltage dropped across R_i will, of course, detract from the load voltage, and the greater the value of R_i the lower the voltage across the load. This is the reason that batteries used in heavy-duty service, such as an automobile battery that must deliver perhaps 200 A, are designed to have the lowest possible internal resistance. If such a battery had an internal resistance of even 0.03 Ω, current would be only 133 A instead of 200. This is determined as follows. Assume a 12-V battery. Normal load resistance is

$$R_L = \frac{E}{I} = \frac{12}{200} = 0.06 \ \Omega.$$

With 0.03 Ω of internal resistance, total resistance is $R_L + R_i = 0.06 + 0.03 = 0.09$ Ω. True current, then, is

$$I_L = \frac{E}{R} = \frac{12}{0.09} = 133.3 \ \text{A}.$$

This would no doubt influence the ability of the battery to perform its function properly.

If the internal resistance of a cell or battery is known, the drop across R_i can easily be determined for a given circuit condition. If a 12-V battery is known to have an internal resistance of 200 Ω, and R_L is a 1-kΩ resistor, what is the drop across R_i?

$$E_{R_i} = \frac{R_i}{R_L + R_i} \times (E_{oc}) = \frac{200}{1200} \times 12 = 2 \ \text{V}.$$

The drop across the load is as easily found.

$$E_{R_l} = \frac{R_L}{R_L + R_i} \times (E_{oc}) = \frac{1000}{1200} \times 12 = 10 \ \text{V}.$$

Constant-Voltage Versus Constant-Current Sources

The performance of a source with a variable load determines whether a particular source is classed as a constant-voltage source or a constant-current source.

A constant-voltage source is one where the internal resistance is *very* much (one hundred times or more) smaller than the load resistance. In this instance, the voltage drop across R_i is so small that the voltage across the load remains essentially constant for reasonable changes in load current. Figure 10-11 illustrates a constant-voltage source: $R_i = 0.01 \ \Omega$, while $R_L = 100 \ \Omega$. In this instance, R_i is ten thousand times smaller than R_L. To verify that the load voltage remains nearly constant, the load resistance will be changed from 100 Ω to 50 Ω and the voltage across R_L will be determined for each case. Assume V_s open-circuit voltage is 12 V. For $R_L = 100 \ \Omega$, the voltage across R_L is determined by Ohm's law:

$$I_t = \frac{V_s}{R_i + R_L} = \frac{12}{0.01 + 100} = 0.119988 = 119.998 \text{ mA.}$$
$$E_{R_L} = I_t \times R_L = 0.119988 \times 100 = 11.9988 \text{ V.}$$

If the load resistor is changed to 50 Ω, the voltage will change only very slightly:

$$I_t = \frac{V_s}{R_i + R_L} = \frac{12}{0.01 + 50} = 0.239952 = 239.952 \text{ mA.}$$
$$E_{R_L} = I_t \times R_L = 11.9976 \text{ V.}$$

The voltage across R_L changes by only 0.0012 V, and for all practical purposes the load voltage is 12 V in both cases.

Compare this example of a constant-voltage source with the following example of a constant-current source, where R_i is one hundred or more times R_L. Still referring to Figure 10-11, assume that $V_s = 12$ V, $R_i = 1000 \ \Omega$, and the normal value of $R_L = 5 \ \Omega$. Now the normal voltage across R_L is not the most important factor. In the

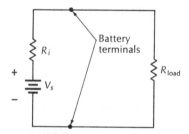

Figure 10-11 A constant current/constant voltage source.

case of a constant-current source, current through the load is of prime interest:

$$I_{R_t} = \frac{V_s}{R_i + R_L} = \frac{12}{1000 + 5} = 0.0119402 = 11.9402 \text{ mA}.$$

Again reducing R_L by 50%:

$$I_{R_t} = \frac{V_s}{R_i + R_L} = \frac{12}{1000 + 2.5} = 0.01197 = 11.97 \text{ mA}.$$

In this example, current changes by only 29.8 μA (0.0000298 A); for all practical purposes current is 11.9 mA in both cases. Note that the voltage or current does not remain absolutely constant, but only relatively so. If the change in voltage or current from one condition to another is 1% or less, the parameter in question is said to be constant.

By far the more useful of the two types of sources is one having constant-voltage output. Nearly all examples used in this book so far have been of this type. However, in later chapters we shall show some examples of how a constant-current source might be useful.

A third major type of source is possible and often desirable. Such a source is neither a constant-voltage nor a constant-current type. It is instead known as a *matched* source, where the load resistance is equal or nearly equal to the internal resistance of the source. This is the required condition for the development of *maximum power dissipation in the load*. When this condition must be met, R_L must equal R_i, and the load resistance is said to be *matched* to the source resistance.

To verify this statement, Figure 10-11 can be used to calculate several values of power dissipation for varying values of R_L. Assume $R_i = 100$ Ω and open-circuit V_s is 12 V; R_L will be assigned values of 40, 60, 80, 100, 120, 140, and 160 Ω. Then current and P_{R_L} will be calculated for each value of R_L.

1. R_L is 40 Ω:

$$I_t = \frac{V_s}{R_i + R_L} = \frac{12}{100 + 40} = 0.0857 = 85.7 \text{ mA};$$
$$P_{R_L} = I^2 R_L = 0.0857^2 \times 40 = 0.2939 \text{ W};$$
$$P_{R_i} = I^2 R_i = 0.0857^2 \times 100 = 0.7347 \text{ W}.$$

2. R_L is 60 Ω:

$$I_t = \frac{V_s}{R_i + R_L} = \frac{12}{100 + 60} = 0.075 = 75 \text{ mA};$$
$$P_{R_L} = I^2 R_L = 0.3375 \text{ W};$$
$$P_{R_i} = 0.5625 \text{ W}.$$

3. R_L is 80 Ω:

$$I_t = \frac{V_s}{R_i + R_L} = \frac{12}{100 + 80} = 0.0667 = 66.7 \text{ mA};$$
$$P_{R_L} = I^2 R_L = 0.3556 \text{ W};$$
$$P_{R_i} = 0.4444 \text{ W}$$

4. R_L is 100 Ω:

$$I_t = \frac{V_s}{R_i + R_i} = \frac{12}{100 + 100} = 0.06 = 60 \text{ mA};$$
$$P_{R_L} = I^2 R_L = 0.36 \text{ W; this is } P_{\max} \text{ for the given circuit.}$$
$$P_{R_i} = 0.36 \text{ W}.$$

5. R_L is 120 Ω:

$$I_t = \frac{V_s}{R_i + R_L} = \frac{12}{100 + 120} = 0.05454 = 54.54 \text{ mA};$$
$$P_{R_L} = I^2 R_L = 0.357 \text{ W};$$
$$P_{R_i} = 0.29752 \text{ W}.$$

6. R_L is 140 Ω:

$$I_t = \frac{V_s}{R_i + R_L} = \frac{12}{100 + 140} = 0.05 = 50 \text{ mA};$$
$$P_{R_L} = I^2 R_L = 0.35 \text{ W};$$
$$P_{R_i} = 0.25 \text{ W}.$$

7. R_L is 160 Ω:

$$I_t = \frac{V_s}{R_i + R_L} = \frac{12}{100 + 160} = 0.04615 = 46.15 \text{ mA};$$
$$P_{R_L} = I^2 R_L = 0.3408 \text{ W};$$
$$P_{R_i} = I^2 R_i = 0.04615^2 (100) = 0.213 \text{ W}.$$

Figure 10-12 illustrates these values graphically. Note that the initial part of the curve rises rather rapidly, while the curve to the right of $P_{\max}$ decreases somewhat more slowly. This is simply due to the fact that power must increase from zero to maximum in the area of the graph representing 0 to 100 Ω. Beyond $P_{\max}$, the load resistance can increase to, perhaps, several thousand Ω, with power dissipation decreasing slowly as R_L is increased in value.

In the area to the left of $P_{\max}$, the internal resistance of the source dissipates more power than the load, but is maximum at $R_L = 0$ Ω. To determine this value, set $R_L = 0.0$ Ω.

$$I = \frac{V_s}{R_i} = \frac{12}{100} = 0.12 = 120 \text{ mA};$$
$$P_{R_i} = I^2 R_i = 1.44 \text{ W}.$$

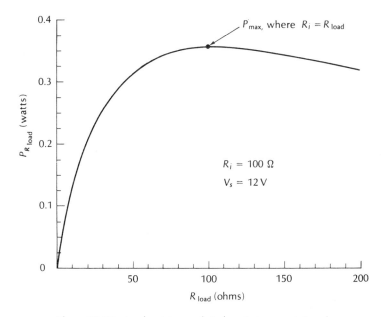

Figure 10-12 Load resistance plotted against source internal resistance.

As R_L increases, the power in R_i decreases continually. Thus, as R_L increases beyond the point where $R_i = R_L$, power dissipation in both R_i and R_L decreases.

SUMMARY

- When certain metals are immersed in a dilute acid solution, they assume a potential with respect to the solution as indicated by the *electromotive series of metals* table.
- When dissimilar metals are immersed in a dilute acid solution, the metal-to-metal voltage becomes the *difference* between their electromotive series numbers.
- Carbon-zinc cells have a terminal voltage of 1.5 V.
- Alkaline-manganese cells have a terminal voltage of 1.5 V.
- Lead-acid cells have a terminal voltage of 2.0 V.
- Terminal voltage of series-connected cells is the *sum* of the cell voltages.
- Terminal voltage of parallel-connected cells is the *same* as that of any one cell.
- AH capacity of series-connected cells is the *same* as one cell.

- AH capacity of parallel-connected cells is the *sum* of the AH capacities.
- Primary cells are those that are essentially not rechargeable.
- Secondary cells are rechargeable.
- Secondary cells are recharged by forcing current through them in the opposite direction to load current.
- Maximum power dissipation in the load occurs when $R_i = R_L$.

QUESTIONS

1. Where certain metals are immersed in dilute acid, the action occurring (is, is not) known as separation of charges.
2. True or false: When zinc is immersed in dilute sulphuric acid, it very quickly dissolves completely.
3. The terminal voltage of a copper-zinc cell (is, is not) 2.0 V.
4. The terminal voltage of a carbon-zinc cell (is, is not) 1.5 V.
5. True or false: Cells are connected in series to increase the current capacity.
6. Cells (are, are not) connected in parallel to increase the current capacity.
7. Cells (are, are not) connected in series to reduce internal resistance.
8. True or false: Cells are connected in parallel to increase the current capacity.
9. True or false: Using carbon-zinc cells, four cells are needed to yield 4.5 V.
10. Using lead-acid cells (five, six, seven) cells are needed to yield 9.0 V.
11. Carbon-zinc cells (are, are not) well suited for high-current, long-time discharges.
12. Alkaline-manganese cells (are, are not) well suited for high-current, long-time discharges.
13. Carbon-zinc and alkaline-manganese cells (are, are not) generally classed as primary cells.
14. Lead-acid and nickel-cadmium cells (are, are not) generally classed as primary cells.
15. A mercury cell (has, has not) about the same energy content as an equivalent size of carbon-zinc cell.
16. An alkaline-manganese cell (has, has not) about the same energy content as an equivalent size carbon-zinc cell.
17. An alkaline-manganese cell (is, is not) useful for its constant-voltage characteristics.
18. A mercury cell (is, is not) useful for its constant-voltage characteristics.

19. A mercury cell (has, does not have) better energy content than a carbon-zinc cell of comparable size.
20. A mercury cell and an alkaline-manganese cell of comparable size (have, do not have) approximately the same energy content.

PROBLEMS

1. A new 12-V battery reads 11.8 volts loaded with a 50-Ω resistor. What is the battery R_i?
2. A new 6-V battery reads 5.0 V when loaded with a 50-Ω resistor. What is the battery R_i?
3. A 12-V battery with an internal resistance of 1-Ω will provide how much voltage across a load of 11 Ω?
4. A 1.5-V carbon-zinc cell will deliver how much voltage to a 500-Ω load resistor if its internal resistance is 100 Ω?
5. A 6-V carbon-zinc battery is known to have an internal resistance of 100 Ω. What will the load voltage be if a 500-Ω resistor is used for R_L?
6. A 4.5-V carbon-zinc battery is known to have an internal resistance of 50 Ω. What will be the voltage drop across R_i if a 200-Ω resistor is used for R_L?
7. A 6-V lead-acid automobile battery has an internal resistance of 0.008 Ω. How much current will flow if the battery is short circuited?
8. A 12-V lead-acid automobile battery has an internal resistance of 0.01 Ω. How much current will flow if the battery is loaded with 0.05 Ω?
9. The terminal voltage of a battery drops from 24 V to 23 V when loaded with a resistor drawing 0.1 A. Determine the value of R_i, R_L, and a matched load.
10. The terminal voltage of a battery drops from 9 V to 7 V when loaded with a resistor drawing 30 mA. Determine the value of R_i, R_L, and a matched load.

CHAPTER **11**

MAGNETISM

Magnets and magnetism have fascinated human beings for centuries. The ability of a magnet to attract and repel other magnets has been investigated and experimented with by some of the world's greatest scientists.

In this chapter we investigate these magnetic forces to the extent that they are important to the work of an electronics technician. The major subject headings to be discussed are listed below.

   11-1   Principles of Magnetism
          Natural magnets
          Magnetic materials
   11-2   Magnetic Fields
   11-3   Magnetic Domains
   11-4   Magnetic Quantities and Units
   11-5   Electromagnetism
   11-6   Magnetic Circuit Examples
   11-7   *B-H* Curves
   11-8   Systems of Measurement

11-1 PRINCIPLES OF MAGNETISM

Although it may not be readily apparent, the effects of electricity and magnetism are closely related. As will be seen, a magnet can exist without concurrent electricity, but electrical current always produces magnetic effects.

Nearly everyone is familiar with the forces of attraction and repulsion exhibited by magnets. Many of the devices in our homes operate due to these forces. Electric motors—in shavers, mixers, fans, and so on—portions of the TV set, and many other devices—rely on these forces for their operation.

Natural Magnets

By definition, a *magnet* is a body of matter that has the property of attracting or repelling magnetic materials and that, if freely suspended, will align itself with the north and south magnetic poles of the earth. Natural magnets, called *lodestones* or *magnetite*, are iron ores that have become magnetized while lying in the earth. The story is told that the early Chinese were familiar with the uses of lodestone as an aid in navigation, although the first recorded use was around 1200 A.D. A modern counterpart, the compass, is still used for this purpose.

A compass needle points to the north and south magnetic poles of the earth because the earth itself is a huge magnet. (Several theories have been advanced to explain why this is true, but none has been proved.) In any case, the compass needle and the earth display a common manifestation of magnetism. They both have what are called poles. A *pole* is defined as one of two related opposites. *Magnetic poles* are the points on the magnetic body where the magnetic force leaves and enters the body and where the force is most intense. Magnetic poles cannot exist singly; every north magnetic pole has a corresponding south magnetic pole and vice versa.

A magnetic pole exists because of magnetic *lines of force,* which are the real entity of magnetism. Magnetic lines of force are illustrated in Figure 11-1. It must be realized that this is only a two-dimensional pictorial representation of the field of force that surrounds the bar magnet. Actually, the field surrounds the magnet in all directions perpendicular to the plane of the magnet. Also, the lines themselves do not actually exist, but the force field is readily illustrated in this manner. The idea of showing the force field as lines of force is derived from the

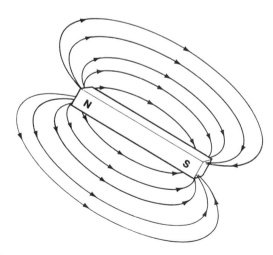

Figure 11-1 Magnetic lines of force surrounding a permanent bar magnet.

pattern generated when iron filings are spread on a thin sheet of cardboard under which is placed a magnet. Such a pattern is shown in Figure 11-2. The iron filings are sprinkled over the card as evenly as possible and the edge of the card is gently tapped to allow the particles to be positioned by the magnetic field. Note the similarity to Figure 11-1. Again, this experiment allows visualization in two dimensions only, but no matter how the magnet is rotated, the same pattern is formed.

The nature of the invisible magnetic field has not yet been defined exactly, but it is known to exhibit certain characteristics of both matter and energy. If, for example, an intense magnetic field is suspended in space and the magnet causing this field is made to disappear *instantly*, the magnetic field converts to particles of matter. Such experiments have been successfully carried out, and they have verified the fact that the magnetic field has mass, even though in minute amount.

Attraction and repulsion between magnets occurs according to specific rules. These rules are:

1. Like poles repel.
2. Unlike poles attract.

For example, a north pole repels another nearby north pole, and a south pole repels another south pole. However, a north pole and a south pole placed close together attract each other.

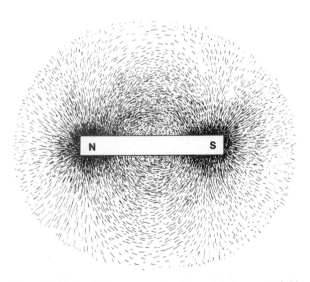

Figure 11-2   Iron fillings assume the shape of the magnetic field.

Similarly, a compass needle points toward the earth's magnetic poles, illustrated in Figure 11-3. Note that the magnetic poles are displaced about 15° from the true geographic poles. By convention, the end of the compass needle that points to the north magnetic pole is called the north pole, so the magnetic pole nearest the true geographical north pole is really a south magnetic pole. In spite of this contradiction, it is nevertheless called the earth's magnetic north pole.

Magnetic Materials

In addition to naturally occurring magnets, certain other materials exhibit magnetic effects. Alloys of iron, steel, cobalt, and nickel possess magnetic characteristics. These materials are known as *ferromagnetic* materials, and they have the ability to easily become strongly magnetized. One widely used alloy, *alnico*, which consists of aluminum, nickel, cobalt, and iron, is capable of very strong magnetic fields.

Nonmagnetic materials are classed as either paramagnetic or diamagnetic. *Paramagnetic* materials (such as aluminum or chromium) can become very slightly magnetized compared to a vacuum. *Diamagnetic*

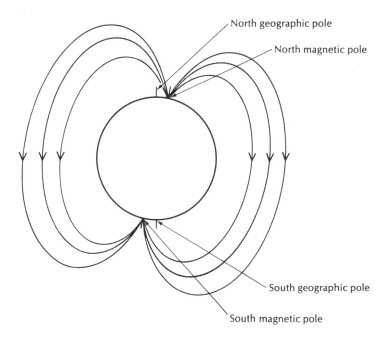

Figure 11-3 Earth's magnetic field.

materials (such as bismuth or copper) are those that exhibit magnetic effects slightly less than that of a vacuum.

11-2 MAGNETIC FIELDS

Magnetic flux is the name given to the total number of lines of force surrounding a magnetized body. The lines are assumed to emerge from the north pole and to reenter the magnet at the south pole. This is indicated in Figure 11-1 by the arrowheads. Placing a small compass alongside a magnet and moving the compass completely around the magnet while observing the compass needle verifies that the magnetic lines of force possess direction. Figure 11-4 illustrates this action. The compass needle aligns itself so that its south pole points toward the magnet's north pole, or the north pole of the compass points toward the magnet's south pole. Comparing Figure 11-4 with Figure 11-1 shows that the compass needle aligns itself *with the lines of force*. It is important to remember that the magnetic lines of force do not exist as lines, nor do they move out of the north pole and into the south pole. The arrowheads in the figure simply imply that magnetic lines of force have *polarity*, or direction; the arrowheads do *not* indicate motion.

The normal magnetic field that surrounds a bar magnet is distorted if magnetic material is brought close. Compare the shape of the field lines in Figure 11-1 with those in Figure 11-5. When two magnets are brought close together, the fields become distorted in a manner that depends on which poles are adjacent. In drawings *a* and *b*, unlike poles are adjacent, and the flux lines emanate from the north pole and flow symmetrically into the south pole.

However, when like poles (N-N; S-S) are placed close together, the flux lines appear as shown in drawings *c* and *d*. The shape of the distorted fields actually suggests a repulsion effect.

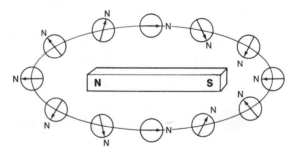

Figure 11-4 A compass is used to show the direction of the lines of force.

Figure 11-6 shows how the field is distorted when a piece of soft iron is brought close to a bar magnet. Because the lines of force can pass through ferromagnetic materials much more easily than through air, the lines concentrate themselves in the soft iron. The iron, in turn, becomes magnetized by *induction*, and as long as it remains in the field it has all the properties of any magnet. When removed from the magnetic field, however, the iron loses its magnetism.

Basically, there are two kinds of magnetic material. These are called *magnetically soft* and *magnetically hard* materials. Pure iron is magnetically soft and thus can easily be magnetized by induction; however, it readily loses its magnetism when removed from the magnetizing force. Most iron alloys are magnetically hard and are somewhat more

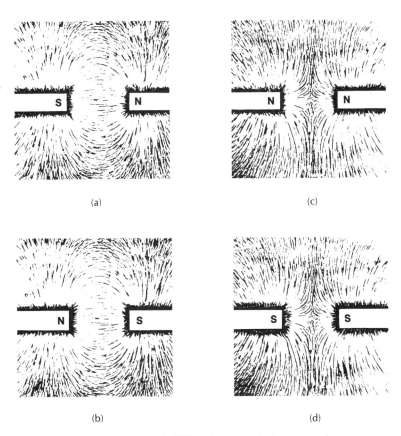

Figure 11-5 Magnetic field distortions caused when a second magnet is placed near the first.

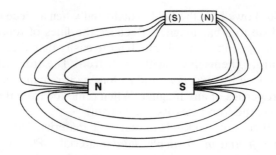

Figure 11-6 Distortion of a magnetic field caused by insertion of a small piece of soft iron into the field.

difficult to magnetize; however, such materials retain magnetism for long periods of time.

The shape of the magnetic field of a straight bar magnet is as shown in Figure 11-1. The strength of the field of a bar magnet is relatively low, due to the long path that the flux must travel through the air. There are several ways to increase the field strength of a magnet, one of which is to form it into a *horseshoe* magnet, illustrated in Figure 11-7a. Bending the magnet so that the north and south poles are closer together gives the flux lines a shorter path to travel, so the field strength is increased. In effect the *air gap* has been reduced, which reduces the length of the path that the flux lines must travel in air. Generally, the shorter the air gap, the more intense the flux.

Figure 11-7b and c illustrate a *pot-core* magnet, which is in effect a circular type of horseshoe magnet. This kind of magnet provides a very dense field and thus a very strong magnetic attraction. The intense field is due in part to the decrease in the air gap, and in part to the magnetic materials used. Figure 11-7d illustrates a design intended to provide a uniform field of force, with all lines of force parallel. This kind of permanent magnet configuration is used primarily in meter movements (and other rotating machinery). A moving coil is mounted in the field, and as the coil is caused to rotate, the flux remains constant wherever the coil positions itself.

11-3 MAGNETIC DOMAINS

Several explanations have been put forward in an attempt to describe the origins of magnetic lines of force. The latest thinking among scientists indicates that the source of magnetic lines of force is the electron. In a normal atom, the electron not only rotates around the nucleus, it

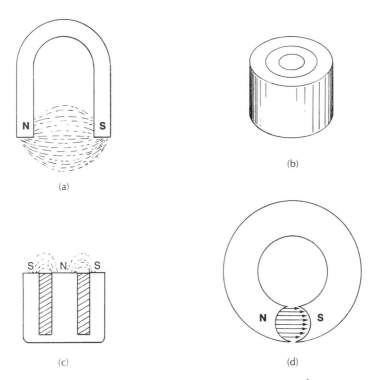

Figure 11-7 (a) A horseshoe magnet; (b) a pot-core magnet; (c) a cross section of the pot-core magnet; (d) a magnet with a uniform field of force, often called a *ring* magnet.

also *spins* in a spiral path along its orbit. It is thought that this spin gives certain materials their magnetic properties.

In nonmagnetic materials, adjacent electrons appear to spin in opposite directions, thus canceling each other's effects. In magnetic materials, however, all electrons tend to spin in the same direction, hence providing magnetic properties. Furthermore, the atoms of magnetic materials tend to form in groups, or *domains*, that have the capability of producing strong magnetic fields.

The domains can be visualized as tiny magnets within the material itself. In soft iron the domains are normally aligned in random fashion, but if the material is placed in a strong magnetic field, the domains line up and the iron becomes magnetized. When the iron is removed from the magnetizing force, however, the domains again become randomly oriented, and the iron is no longer magnetized. Figure 11-8 illustrates this effect.

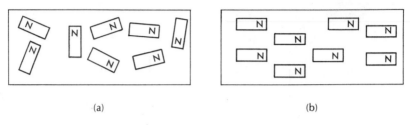

Figure 11-8 (a) Randomly oriented domains; (b) aligned domains in a magnet.

In magnetically hard material, the domains, once aligned by an external force, tend to remain aligned after the force is removed, leaving the material permanently magnetized. The domains, sometimes called *dipoles*, remain permanently aligned unless one of two things happens to demagnetize the material, at least to some degree. (1) If the permanent magnet is repeatedly struck with a hammer, the domains become disoriented and the material becomes demagnetized. (2) If the material is heated to a sufficiently high temperature, called the *Curie point*, the domains become randomly oriented again, and when the material cools, it will have lost its magnetism.

11-4 MAGNETIC QUANTITIES AND UNITS

In order to describe magnetic materials and fields of force in specific detail rather than in general terms, many of the terms used must be defined explicitly or expressed in terms of numerical values. The most important terms and units relating to magnetism are described in the following paragraphs.

Permeability

A material is said to have high *permeability* if it can be easily magnetized or if it is greatly affected by magnetic fields. The permeability of a substance is a measure of the number of lines of force produced by a material with a given magnetizing force, compared to air. Air (or a vacuum) has a permeability of 1. The permeability of diamagnetic materials is slightly less than 1, while that of paramagnetic materials is slightly greater than 1. The symbol for permeability is μ (the Greek letter mu).

Reluctance

Reluctance (ℜ) is the resistance, or opposition, to magnetic lines of force. Magnetic materials have very low reluctance, while nonmagnetic materials have essentially infinitely high reluctance.

Retentivity

Retentivity is a measure of how well a material retains its magnetization after the magnetizing force is removed. Magnetically hard materials such as steel have high retentivity, while magnetically soft iron has low retentivity. How well the domains retain their alignment determines the material's retentivity.

Residual Magnetism

When a ferromagnetic material is subjected to a magnetizing force, a certain field strength is produced. When the magnetizing force is removed, some amount of magnetism remains, and this is known as *residual magnetism*. Magnetically hard materials have a large residual magnetism; magnetically soft materials have little or none.

Magnetic Flux

As mentioned previously, *flux* is the name given to the entire group of lines of force surrounding a magnet. The symbol for flux is ϕ (the Greek letter phi). The stronger the magnetic field, the greater the flux (the more lines of force).

The unit of flux in the cgs (centimeter-gram-second) system* is the maxwell (Mx) named for the Scottish physicist James C. Maxwell (1831–1879). One maxwell is one line of force.

Flux Density

Flux density (B) is a measure of the number of maxwells *per unit area*. In the cgs system the unit in use is one line per square centimeter, called the *gauss* (G). The gauss is named for Karl F. Gauss (1777–1855), a German mathematician. Flux density is a function of the number of lines through a given area, divided by the area:

$$B = \frac{\phi}{A}$$

Figure 11-9 illustrates a magnet having a total flux of 12 lines, or 12 Mx, but the 1-cm² area has a flux density of 3 G, or 3 lines per cm². Note

---

* Systems of measurement are discussed later in this chapter.

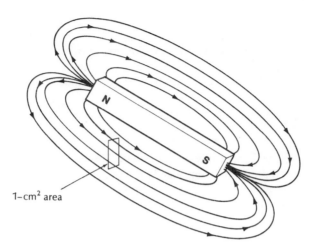

Figure 11-9 Illustrating a flux density of 3 gauss.

that the flux density is much greater at the poles than at the 1-cm² area shown.

11-5 ELECTROMAGNETISM

As briefly mentioned above, electron spin is thought to produce magnetic lines of force. In the case of ferromagnetic materials, these electron spins are such as to reinforce magnetic effects. There is another way to produce magnetic lines of force, without the use of ferromagnetic materials. When an electric current flows in a conductor, *the wire is surrounded by magnetic lines of force*. The strength of the field is directly proportional to the amount of current flowing. Figure 11-10 illustrates this effect. The field is relatively dense close to the wire and decreases in strength away from the wire. At a given distance from the wire the magnetic field is uniform over the length of the wire.

As is true in the case of a permanent magnet, the lines surrounding a current-carrying wire also have direction. Figure 11-11 illustrates the left-hand rule for finding the direction of the flux lines. The thumb of the left hand points in the direction of current flow, and the curled fingers point in the direction of the flux lines. Because the lines of force surround the wire and are continuous, they do not possess polarity as such, only direction. However, if a piece of soft iron is placed close to the wire, it becomes magnetized by induction as shown in Figure 11-12.

Magnetism 247

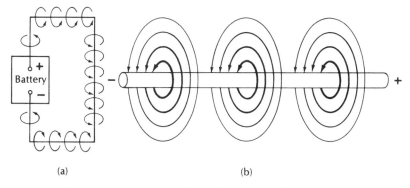

Figure 11-10 (a) Magnetic lines of force surrounding a current-carrying wire; (b) close-up view of small section of wire.

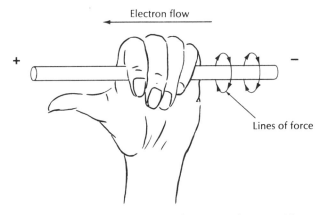

Figure 11-11 The left-hand rule for determining direction of flux lines.

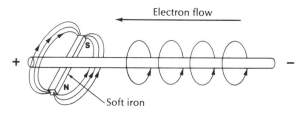

Figure 11-12 Magnetizing soft iron by induction of the field lines surrounding a current-carrying wire.

The iron will exhibit polarity (a north and a south pole), and as long as current flows in the wire it will remain magnetized. When current stops, the lines of force will disappear and the soft iron will become demagnetized. If the applied voltage is reversed, causing the current direction to be reversed, the lines of force will again exist, *but in the opposite direction*. Hence, the magnetic polarity of the soft iron will also reverse.

The magnetic field surrounding a length of straight wire is rather weak for nominal values of current. However, currents on the order of tens or hundreds of thousands of amperes can produce severe stress on the conductors due to the forces of attraction or repulsion. Two conductors placed side by side can easily be physically affected by large currents. This action is exemplified in Figure 11-13, where the left sketch shows adjacent wires having current in the *same* direction (into or out of the page). The wires have a tendency to attract each other with a force dependent upon the value of current. The right drawing illustrates two wires with current flowing in *opposite* directions (the left-hand wire having current flowing into the page and the right-hand wire with current flowing toward the reader). These wires have a tendency to repel each other.

To visualize these forces, it is helpful to consider the magnetic lines of force as if they were similar to tightly stretched rubber. In a few respects the end result is similar. Consider Figure 11-13*a*. If the lines of force are thought of as taut rubber bands, it is easy to see that the two wires will be forced together. In Figure 11-13*b*, if the lines of force are considered to be similar to two inflated balloons that are pressed together, it is easy to understand that the balloons will try to force each other apart.

While this analogy is not completely true, the magnetic lines of force *are* elastic and in theory can be stretched to infinity without breaking.

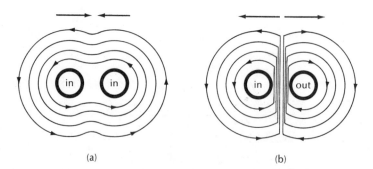

Figure 11-13 End views: (a) Adjacent wires with current flowing in the same direction tend to attract each other; (b) adjacent wires with current flowing in opposite directions tend to repel each other.

The Inductive Coil

As already mentioned, the field intensity surrounding the current-carrying wire is relatively weak. Rather than increase current beyond reasonable levels to increase the magnetic strength, it is more practical to *coil* the wire into a helical shape. Figure 11-14a illustrates a coil of wire wound on a circular insulating form. This has the effect of concentrating the lines of force and forming them into a configuration not unlike that of a bar magnet. A coil with its attendant force field is shown in Figure 11-14b. This field exists as long as current flows. Note that the coil has north and south poles and that the intensity of the field is greater than if the wire were not coiled. The field surrounding each loop

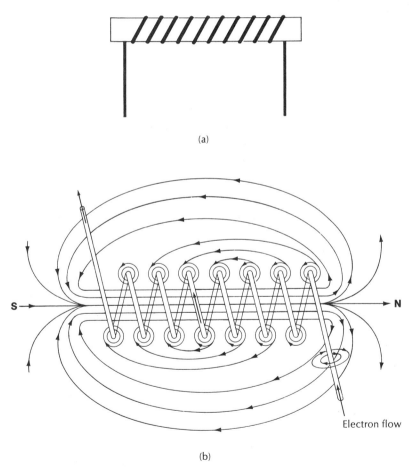

Figure 11-14 (a) A coil of wire wound on a cylindrical insulating form; (b) magnetic lines of force surrounding the coil when current is flowing.

of wire aids the field of the adjacent loop in such a way as to intensify the overall flux.

The left-hand rule can be modified and used to determine the polarity of the energized coil. The fingers of the left hand are placed around the coil so that they point in the direction of the electron flow. The thumb then points in the direction of the north pole, as in Figure 11-15.

Electromagnets

To increase the magnetic strength of a coil of wire, it can be wound on a soft iron *core*. Such a device is an *electromagnet*, which has many uses. An electromagnet is illustrated in Figure 11-16. The usefulness of an electromagnet stems from the fact that the magnetism can be switched on and off with the current flow. When no current is flowing, the core is not strongly magnetized (remember that soft iron has low retentivity).

Magnetizing Force

The field strength of an electromagnet is determined by four major factors. These are (1) the amount of current, (2) the number of turns on the coil, (3) the magnetic qualities of the core, and (4) the length of the coil and core. For a coil with an air core (that is, a coil wound on a plastic or cardboard form), the field strength is a function of the *ampere-turns*, the product of amperes times the number of turns on the coil, which is abbreviated AT.

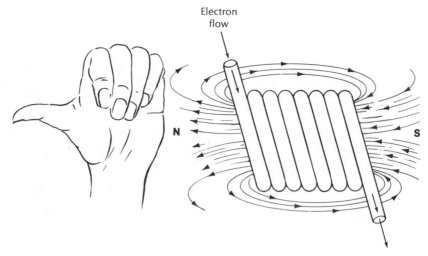

Figure 11-15 The left-hand rule used to determine the north pole of an electromagnet.

To illustrate ampere-turns, Figure 11-17 shows four variations of air-core coils and current. Figure 11-17a represents a 50-turn coil with 1 A flowing through it. In this instance AT = 1 × 50 = 50, and the coil is said to have a magnetizing force, or *magnetomotive force* (mmf), of 50 AT. Note that Figure 11-17b represents a coil having 100 turns with 0.5

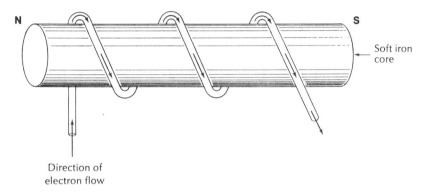

Direction of electron flow

Figure 11-16 An electromagnet made by winding a coil around a soft iron core.

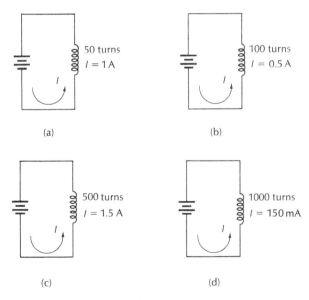

Figure 11-17 Four coils, each with a different number of turns and with different current values, used to illustrate how ampere-turns are calculated.

A flowing through it. Its magnetizing force is 0.5 × 100 = 50 AT, which is the same as the previous example. Drawing *c* has an ampere-turn value of 1.5 × 500 = 750 AT, while drawing *d* has a value of 0.15 × 1000 = 150 AT.

In the cgs system the unit of mmf is the *gilbert*, named for the Englishman William Gilbert (1540–1603), an early investigator of magnetism. One gilbert equals 0.7958 AT and one AT equals 1.257 gilbert. The abbreviation for the gilbert is Gb.

Field Strength

Magnetomotive force, in terms of ampere-turns or gilberts, determines the *available force that can produce flux*. Actually, the amount of flux produced by a given force depends greatly upon the magnetic properties of the material in the space encompassed by the field. A coil with a given ampere-turn force yields far fewer lines of force with an air core than with an iron core.

Thus, field intensity (H) is measured in terms of mmf per unit length. The cgs unit of field intensity is the *oersted*, named for H. C. Oersted (1777–1851), the Danish discoverer of electromagnetism. One oersted equals one gilbert per centimeter. In formula form we have

$$H \text{ (in oersteds)} = \frac{0.4\pi \times \text{ampere-turns}}{l}$$

where l = length in centimeters and where 0.4π is derived from the formula for the surface area of a sphere, $4\pi r^2$.

Note the introduction of length to the relationship. It is, of course, logical to assume that a tightly wound coil will produce more flux density than one in which the turns are widely spread. Hence, the length of the coil is important in determining the total field strength.

Permeability

In the previous section, the permeability of a material was defined as a measure of the degree to which the material is more or less efficient than air in concentrating magnetic lines of force. Air has a μ of 1, while good-quality soft iron has a μ of, perhaps, 5000. This means simply that using such iron as the core of an electromagnet produces 5000 more lines of force than the same coil with an air core.

Field intensity, permeability, and the lines per cm^2 are related as follows:

$$\mu = B/H \quad \text{or} \quad B = \mu H \quad \text{or} \quad H = \frac{B}{\mu}$$

For any coil with an air core, $B = H$, since $\mu = 1$. However, the same coil wound on a soft-iron core with a μ of 650 produces a field density 650 times stronger.

An example of the foregoing relationships is given in Figure 11-18. The flux density in gauss is to be found for the conditions shown. Since $B = \mu H$, and $H = 0.4\pi AT/l$,

$$B = \mu \left(\frac{0.4\pi AT}{l} \right) = 650 \left(\frac{1.257 \times .75 \times 100}{10} \right)$$
$$= 650 \times 9.425 = 6126 \text{ G}.$$

Thus, for the specified conditions, the flux density is 6126 lines per cm^2.

Reluctance and Permeance

As defined earlier, reluctance is the opposition of a material to magnetic lines of force. The symbol for reluctance is $\mathcal{R}$, and it is equal to the magnetizing force divided by flux:

$$\mathcal{R} = \frac{\text{mmf}}{\phi} = \frac{l}{\mu A},$$

where A = area.

Permeance is the reciprocal of reluctance, and is therefore the *ease* with which a material can form a magnetic field with a given mmf. The symbol for permeance is ρ.

Permeability and permeance are related as follows:

$$\mu = \rho \left(\frac{l}{A} \right)$$

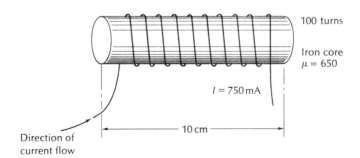

Figure 11-18 An iron-core coil in which the number of gauss is to be determined.

Permeability, then, is directly proportional to permeance and length, but inversely proportional to area. That is, as length increases, permeability increases; but as area increases, permeability decreases. Permeability is therefore a measure of the dimensions of the core material as well as of its magnetic qualities, while permeance is a measure of the magnetic qualities of the material without regard to the amount of core material.

11-6 MAGNETIC CIRCUIT EXAMPLES

To illustrate the application of the foregoing principles, several examples are given.

Example 1. Using as iron-core coil similar to that of Figure 11-18, find the ampere-turns if the coil consists of 1000 turns and the current is 666 mA.

$$AT = \text{amperes} \times \text{turns} = 0.666 \times 1000 = 666.$$

Example 2. A certain iron-core coil is known to have an mmf of 100 AT. What is the mmf in gilberts?

$$\text{mmf} = 1.257 \text{ AT} = 1.257 \times 100 = 125.7 \text{ Gb}.$$

Example 3. A certain iron-core coil is known to have an mmf of 100 Gb. Find the mmf expressed in ampere-turns.

$$\text{mmf} = 0.7958 \text{ Gb} = 0.7958 \times 100 = 79.58 \text{ AT}.$$

Example 4. An air-core coil has a length of 10 cm and consists of 350 turns. It is connected in a circuit of 15-Ω total resistance with an applied voltage of 4.5 V. Find the field strength in oersteds.

Since $H = (0.4\pi AT/l)$, amperes must be calculated first.

$$I = \frac{E}{R} = \frac{4.5}{15} = 0.3 \text{ A}.$$

Field strength can now be found.

$$H = \frac{0.4 \times \pi \times 0.3 \times 350}{10} = 13.2 \text{ Oe}.$$

Example 5. A certain coil has a field density of 250 G and a field strength of 5 H. Find the permeability of the core.

$$\mu = \frac{B}{H} = \frac{250}{5} = 50.$$

Example 6. Refer to Example 4 above. Find the field density in gauss.

1. $B = H$ if $\mu = 1$ (air core) $= 13.2$ G.
2. $B = \mu \left(\dfrac{0.4\pi AT}{1} \right) = 1 \left(\dfrac{0.4\pi \times 0.3 \times 350}{10} \right) = 13.2$ G.

Example 7. Refer to Example 5 above. Find the field density in gauss if the core is changed to one having a permeability of 15.

$$B = \mu H = 15 \times 5 = 75 \text{ G.}$$

11-7 B-H CURVES

One way of describing the magnetic qualities of ferromagnetic materials is to use a *B-H* curve, such as the one shown in Figure 11-19. The curve is a plot of the field density in gauss versus the magnetizing force in oersteds. If the sample of material has never been magnetized, the solid line represents the degree of magnetization for a given magnetizing force. The point of origin *(a)* represents zero magnetizing force and therefore zero flux. As the magnetizing force is increased, the flux increases linearly to a point. However, eventually a point is reached where most of the domains are already aligned, and further increase in

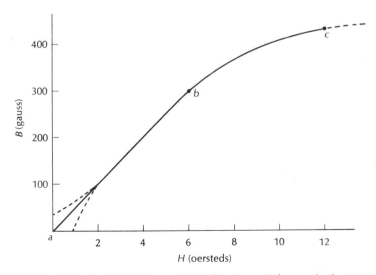

Figure 11-19 A typical *B-H* curve illustrating initial magnetization.

the magnetizing force results in only a small increase in flux. This point, where the increase in flux begins to level off, is called *saturation*. The saturation effect begins to become noticeable at point *b* of the curve, and above this point the increase in magnetization becomes less and less for a given increase in the force.

If an iron-core coil is operated well above the saturation level (in an area where the curve is essentially horizontal, as above point *c*), the iron core contributes little or nothing to a *change* in flux. The coil will perform as an air-core coil for *changes* in magnetizing force. This becomes important when iron-core devices are studied in conjunction with alternating current.

Because the *B-H* curve for a particular sample of iron indicates the degree of flux density for a given mmf, the curve can also be used to find the permeability of the material. For example, in Figure 11-19, if $H = 4$ oersteds, $B = 200$ gauss. Therefore, since $\mu = B/H$, the permeability is $B/H = 200/4 = 50$.

The permeability is normally determined by using values from the linear portion of the curve. The value of μ just determined is a *static* value. However, *dynamic* value is also important. That is, if the mmf is *not* constant but is continually varying, then the result is a dynamic range of values for both B and μ. An example will illustrate this effect.

Example 8. Assume that the mmf is continually varied between 2 and 4 H. *B*, then, varies between 100 and 200 G. The dynamic permeability is then determined as follows:

$$\mu_d = \frac{\Delta B}{\Delta H}$$

where μ_d = dynamic permeability;
ΔB = difference in *B* values, or $B_{max} - B_{min}$;
ΔH = difference in *H* values, or $H_{max} - H_{min}$.
Therefore,

$$\mu_d = \frac{\Delta B}{\Delta H} = \frac{B_{max} - B_{min}}{H_{max} - H_{min}} = \frac{200 - 100}{4 - 2} = \frac{100}{2} = 50.$$

Note that this is exactly the same value as the static permeability. This is due to the fact that the curve is a straight line in this location on the graph.

To illustrate that an iron-core device does not perform well in the saturation region of the curve, let us find μ for the region above the onset of saturation. Values between 8 and 10 oersteds will be used to find the μ_d for this part of the curve.

$$\mu_d = \frac{\Delta B}{\Delta H} = \frac{B_{max} - B_{min}}{H_{max} - H_{min}} = \frac{405 - 370}{10 - 8} = 17.5.$$

Note especially how much lower the permeability is above the saturation point. In the area between 10 and 12 H, μ_d is about 10; if the curve were extended further, the dynamic permeability μ_d would eventually be reduced to *unity,* which is the permeability of air.

Magnetic Hysteresis

The curve of Figure 11-19 is a representation of the magnetic response of the *first* time the material is magnetized only. Because all ferromagnetic materials have some degree of retentivity, there will be some degree of residual magnetism B_R after the first time. This must be accounted for in the *B-H* curve.

Figure 11-20 illustrates the complete curve for all possible combinations of magnetizing force for a typical soft-iron sample. The circuit used to produce this curve is given in Figure 11-21. It allows the current through the iron-core coil to be varied in both amplitude and direction. If the iron has never been magnetized before, the curve between points 1 and 2 describes the magnetic effects. As before, as the magnetizing force is increased, the flux increases linearly to the point where magnetic saturation begins. Now, note that when the magnetizing force is decreased from the maximum value, the curve does not follow its original path, and between points 2 and 3, when $+H$ is decreasing, the total

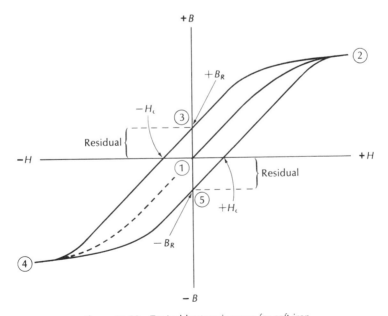

Figure 11-20 Typical hysteresis curve for soft iron.

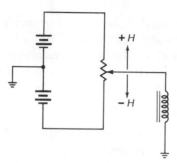

Figure 11-21 The circuit used to yield the hysteresis curve of Figure 11-20. Note the iron-core coil symbol.

flux is greater than when $+H$ was increasing. When the magnetizing force is reduced to zero, point 3, some flux is left. This is the residual magnetism, and the more magnetically hard the core is, the greater the residual magnetism. Conversely, the softer the material, the smaller the residual magnetism.

Now at point 3 the applied force is zero, and there is still a reasonable amount of flux. Once magnetized, a ferromagnetic material will always have some degree of flux present when no magnetizing force exists. Only if a reverse magnetizing force is applied can the flux be reduced to zero (of course, raising the temperature to the Curie point will accomplish this, too). The amount of force required to reduce the flux to zero is called the *coercive force*, symbolized H_c. Coercive force is also measured in oersteds.

Note that $+H$ and $-H$ are simply arbitrary designations denoting opposite directions of current through the coil. If $+H$ represents current in one direction, then $-H$ represents current in the opposite direction.

Continuing the discussion of the hysteresis loop, as current is increased in a direction opposite to that required between points 1, 2, and 3, the magnetic flux increases in the opposite direction along the path between points 3 to 4, until the core begins to saturate, and slightly beyond. As at point 2, point 4 represents a condition of reduced dynamic permeability. As current in the coil is reduced from its maximum value (but still in the same direction) flux again decreases. When current decreases to zero at point 5, residual magnetism is evident, with a value of $-B_R$. As before, the amount of force required to reduce flux to zero is the coercive force H_c. It should be noted that if, at this point, the force is again reduced to zero, the material will return to point 5, indicating that the retentivity is permanent. To maintain flux at zero (just to the right of point 5, or left of point 3) a continuous force must be applied.

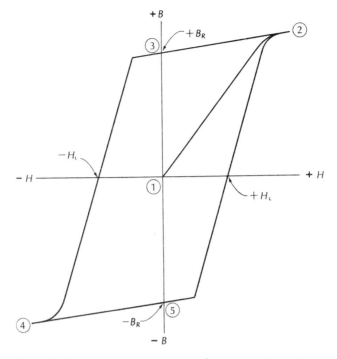

Figure 11-22 Square-loop hysteresis curve for a permanent magnet.

Finally, continuing up the curve from point 5 to point 2, the curve is seen to form a *loop* with a characteristic shape. As magnetization is continued, the core will assume the values described by the loop 2, 3, 4, 5, 2, over and over. Only be demagnetizing (*degaussing*) the core can it be returned to point 1, and then it will again start from the point of origin only once.

Figure 11-20 shows the hysteresis loop for a soft-iron core, where the retentivity is low. A contrasting illustration, Figure 11-22, shows the hysteresis loop for a magnetically hard substance. Note the similar shape but the much greater residual magnetism characteristic of permanent magnets.

Degaussing

As previously mentioned, a magnetically hard substance that has been magnetized is somewhat difficult to demagnetize. One method, previously mentioned, is to raise the temperature of the substance to its Curie point. This is not usually practical, because it often damages the material. The Curie point of iron, for instance, is 770°C; that of cobalt is 1131°C. Obviously, many manufactured articles cannot be subjected to such elevated temperatures. We have also said that severe shock demagnetizes a magnet to some degree. A third method, *degaussing,* is far more practical. In order to degauss a magnet, a gradually diminishing magnetizing force that constantly changes direction is applied to the magnet. Thus, the hysteresis loop is continually formed and reformed in successively smaller loops until it becomes infinitely small. Thus, residual magnetism is minimized.

One method of degaussing involves a coil of wire which consists of many turns and which is energized by alternating current. The energized coil is placed close to the object to be degaussed and *very slowly* moved away. The material becomes magnetized in the two polarities many times per second, and experiences a normal hysteresis loop. As the degaussing coil is slowly moved away, the hysteresis loop of the material becomes smaller and smaller until, when the coil is distant from the object, the hysteresis loop is so small as to be of no consequence. Residual magnetism is therefore minimized.

Several frequently encountered applications of degaussing exist. For example, the recording head of an audio tape recorder must be degaussed regularly to remove static magnetism. This is necessary because audio information to be recorded is alternating current, and the head tends to become permanently magnetized. Using the foregoing technique with a specially designed degaussing tool eliminates this source of audio distortion.

Another example is the degaussing requirement for a color TV picture tube. The metal tube parts can become permanently magnetized, which adversely affects the reproduced color quality. Again, a special degaussing coil is plugged directly into 110-V house current and the tube is degaussed by starting with the coil close to the tube face and gradually moving it farther away while rotating it to cover the entire surface of the tube.

11-8 SYSTEMS OF MEASUREMENT

Up to this point, all the units we have used are part of the system of measurement known as the centimeter-gram-second (cgs) system. There is a second form of the metric system, the meter-kilogram-second (mks) system. The mks system had been developed many years before it was adopted in 1960 through a series of international agreements, and is now known as the *International System of Units*, or *SI*. SI units are being used with more and more frequency; eventually, they will replace all other units. You must therefore become familiar with them. These units are briefly described below. As you study these mks units, note that several of them are closely related to the familiar electrical units, volts and amperes; this close relationship makes mks units very practical.

Flux

The mks unit of flux is the *weber (Wb)*, named for Wilhelm Weber (1804–1890), a German physicist. One weber equals 10^8 lines or maxwells. Since this represents a very large number of magnetic field lines, common values are given in microwebers (μWb). One microweber equals 100 lines or 100 maxwells.

Flux density

The mks unit of flux density is the *tesla (T)*, named in honor of Nikola Tesla (1857–1943), a Yugoslav-born inventor who lived in the United States. One tesla equals one weber per square meter.

MMF

The mks unit of magnetomotive force is *ampere-turns*. One ampere-turn equals 1.257 gilberts.

Field Strength

The mks unit of field strength or intensity is ampere-turns per meter. One ampere-turn per meter equals 0.01257 oersted.

Reluctance

Reluctance expresses the ratio mmf/ϕ. In mks units, this ratio is expressed in ampere-turns per weber.

Permeability

Permeability is a dimensionless (pure) number. When working with mks units, permeability is equal to teslas divided by ampere-turns per meter. Note that with mks units the permeability of air is *not* 1; instead, it is $0.4\pi \times 10^{-6}$.

Permeance

Permeance is the reciprocal of reluctance. In the mks system, permeance equals webers per ampere-turn.

Dynamic Permeability

In the mks system, dynamic permeability (μ_d) is change in flux density (in teslas) divided by the corresponding change in ampere-turns per meter.

In Appendix 6, the International System of Units is discussed in more detail.

SUMMARY

- Natural magnets are called lodestones or magnetite.
- The earth's north and south magnetic poles are displaced about 15° from the true geographic poles.
- A compass needle aligns itself with the earth's magnetic field.
- A magnetic field exhibits properties of both energy and matter.
- Like poles repel, unlike poles attract.
- Magnetic materials include iron, steel, cobalt, and nickel.
- Paramagnetic materials such as aluminum and chromium can be slightly magnetized.
- Diamagnetic materials such as bismuth and copper exhibit magnetic effects slightly less than that of a vacuum.
- A magnetically hard substance retains its magnetism readily; a magnetically soft material loses its magnetism when the magnetizing force is removed.
- Magnetic domains are thought to be caused by electron *spin,* and are similar to tiny magnets throughout the volume of material.
- Permeability (μ) is a measure of the number of lines of force produced by a given magnetizing force, compared to air.
- Reluctance ($\mathcal{R}$) is opposition to magnetic lines.
- Retentivity is a measure of the flux remaining after the magnetizing force is removed.
- One maxwell is one line of force.
- One gauss (G) is one line per cm^2.
- A wire carrying a current has a magnetic field surrounding it.
- If a wire is coiled, its magnetic field is intensified.
- Winding a coil on an iron core greatly increases the field strength of the coil for a given value of current.
- One gilbert (Gb) equals 0.7958 AT.
- Field strength is measured in oersteds (H).

- Permeance is the reciprocal of reluctance.
- A *B-H* curve describes the magnetic qualities of a ferromagnetic material.
- The area encompassed by a *B-H* curve is a measure of the material's hysteresis.

QUESTIONS

1. A natural magnet is called either _____ or _____.
2. The material called magnetite is a natural _____.
3. Briefly describe a ferromagnetic material.
4. Briefly describe a paramagnetic material.
5. Briefly describe a diamagnetic material.
6. The _____ of a magnet is the area of greatest flux concentration.
7. Unlike magnetic poles _____.
8. Like magnetic poles _____.
9. The name given to the total lines of force surrounding a magnet is _____.
10. A(n) _____ is a magnet requiring external magnetizing force.
11. True or false: A permanent magnet requires no external magnetizing force.
12. A magnetically soft material has (high, low) retentivity.
13. True or false: A magnetically hard material has low retentivity.
14. A north pole (attracts, repels) another north pole.
15. True or false: A south pole attracts another south pole.
16. A south pole (attracts, repels) a north pole.
17. The most practical way of demagnetizing a permanent magnet is to _____ it.
18. A magnetic domain is like a tiny _____.
19. To become magnetized, an electromagnet requires _____.
20. The strength of an electromagnet depends upon _____ and _____.

PROBLEMS

1. Refer to Figure 11-18. The coil consists of 1500 turns, and total circuit current is 150 mA. Find the ampere-turns.
2. Refer to Figure 11-18. The coil consists of 200 turns, and total circuit current is 1.125 A. Find the ampere-turns.
3. An iron-core coil has 400 turns and is energized with 250 mA. Find the mmf in terms of gilberts.
4. An iron-core coil has 2500 turns and is energized with 0.375 A. Find the mmf in terms of gilberts.

5. An iron-core coil is energized with an mmf of 250 Gb. Find the ampere-turns.
6. An iron-core coil is energized with an mmf of 1500 Gb. Find the ampere-turns.
7. An air-core coil is 5 cm long, and consists of 135 turns. It is energized by 6.0 V and the total resistance in the circuit is 24 Ω. Find the field strength in oersteds.
8. An air-core coil is 12.5 cm long, and consists of 750 turns. It is energized by 12.0 V and the total resistance in the circuit is 6 Ω. Find the field strength in oersteds.
9. Refer to Problem 7. A 5-cm iron core of the correct diameter is inserted into the coil. The core has a permeability of 167. Determine the new field strength in oersteds.
10. Refer to Problem 8. A 12.5-cm iron core of the correct diameter is inserted into the coil. The core has a permeability of 75. Determine the new field strength in oersteds.
11. An energized coil produces a field density of 250 maxwells/cm^2 with a field strength of 5 H. Find the core's permeability.
12. An energized iron-core coil produces a field density of 1000 G. The μ of the core is 125. Find the field strength in oersteds.

CHAPTER **12**

ELECTROMAGNETIC INDUCTION

Early in the nineteenth century the relationship between electrical current and magnetism was ascertained. First it was discovered that when current flows in a wire a compass needle is influenced in such a way that the needle tends to align itself perpendicular to the plane of the wire, suggesting that magnetic lines of force surround the wire when current flows. When the current is interrupted, the compass needle again aligns itself with the earth's magnetic flux.

Shortly after this discovery, a closely related phenomenon was observed. If a coil of wire is connected to a sensitive voltmeter or ammeter, and if the coil is caused to move in a strong magnetic field, *current is generated and flows in the coil.* It was quickly realized that with the expenditure of kinetic energy (energy of motion), electrical power could be produced. Thus began the utilization of electrical energy for the benefit of mankind.

In this chapter we investigate the motor action that provides motive power to operate electrical equipment. Additionally, we investigate the fundamentals of the generation of electrical power by means of electromagnetic induction. The major subject headings in this chapter follow.

12-1   Motor Action
12-2   Induced Current
12-3   Induced Voltage
12-4   Induced Voltage Across a Coil
12-5   Self-Induction
12-6   Generator Action
12-7   Magnetic Devices

12-1 MOTOR ACTION

The simple attraction and repulsion of permanent magnets is an elementary example of *motor action*. It is difficult, however, to devise a mechanical device that can use this principle alone to perform useful work. Of greater practical interest is the motor action produced by electromagnets, since the current that energizes the device can easily

266 Chapter 12

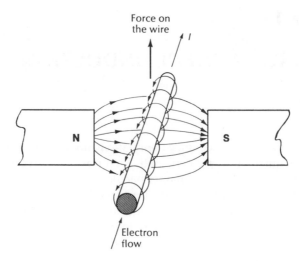

Figure 12-1 Motor action resulting from the interreaction between two magnetic fields.

be controlled to start and stop the device at specific times. In order to grasp the fundamentals of motor action, the interaction between two magnetic fields will be briefly reviewed and then amplified.

Recall from the previous chapter, and especially from Figure 11-13, that two separate magnetic fields tend to attract if the directions of their lines of force are opposite, and to repel if their lines of force have the same direction. Therefore, in Figure 12-1, the wire has a tendency to be forced in an *upward* direction, assuming the mass of the magnet is much greater than that of the wire. This is true since on the lower side of the wire, the flux surrounding the wire and the flux of the magnet are in the same direction, producing a repulsion effect. However, on the upper side of the wire, the two fields are in opposite directions, so that one attracts the other. The force of repulsion is stronger than the force of attraction, and because the wire moves more readily than the magnet, the wire tends to be pushed upward from the strong to the weak field. The actual force applied to the wire is a function of the two field strengths. Hence, for a given magnet, the force is directly proportional to the amount of current in the wire—the more current, the greater the force. Figure 12-2 illustrates the field directions more clearly, with an end view of the wire. Note the concentration of flux lines below the wire.

If the current is made to change directions, then the wire will be forced in the downward direction, as shown in Figure 12-3. In this instance, the two fields above the wire are in the same direction, while

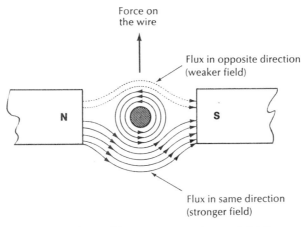

Figure 12-2 End view of the arrangement and magnetic field interreaction shown in Figure 12-1.

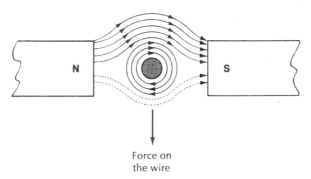

Figure 12-3 End view of the arrangement shown in Figure 12-1; the force on the wire is in the opposite direction, hence the current is reversed.

those below the wire are opposing. The wire therefore is impelled downward.

Using a simple wire results in a very small force, unless the current is made very large. To increase the force, a many-turn coil is used, as illustrated in Figure 12-4. The field produced by the coil is much more intense, and when current is caused to flow the coil tends to rotate on its axis if current direction produces polarities opposite to that of the magnet. As shown in Figure 12-4, if electron current enters the wire at A and leaves at B, the left side of the coil becomes a north pole while the right side becomes a south pole. Because the coil is pivoted at the center, it tends to rotate. The direction of rotation depends on the exact starting alignment of the coil, which will rotate one-half turn, at which point the poles will then be unlike and further

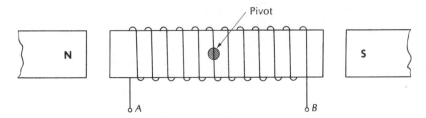

Figure 12-4 Components to illustrate motor action

motion will be impossible. Each magnet pole will now attract the nearest end of the coil.

If, at this time, the current were suddenly reversed, the coil would make another half rotation. By using a device called a *commutator*, our simple coil and magnet become an elementary dc motor, as shown in Figure 12-5. The commutator is, in effect, an automatic switching device that transfers current direction at the instant the attracting poles are adjacent. Inertia rotates the coil some number of degrees beyond this point, but by now the commutator has switched current direction,

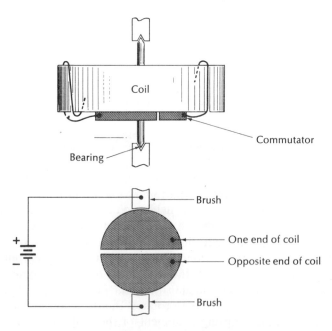

Figure 12-5 The basic dc motor.

and the adjacent poles again repel. The commutator rotates with the coil and makes contact with the source voltage through the fixed *brushes,* which slide easily on the commutator surface.

The simple dc motor just described does function, but it develops little useful power. Part of the time the coil is simply coasting, while it travels parallel to the magnet's field. The only time a large torque is produced is just after the current switches direction, when the coil ends are closest to the magnet and are being either repelled or attracted. To overcome this deficiency, a practical dc motor has several coils wound on an armature, with each coil connected to a multisection commutator. This arrangement ensures that one of the coils is always in the strongest field, and as its rotation takes it out of the strongest field, the commutator switches in the next coil, which is just approaching the strongest field area. Thus, power is developed continuously.

Besides dc motors, other devices use the motor principle. For example, the meter movement described earlier uses this principle to produce the deflection necessary to indicate voltage or current values on the meter face. Other devices that make use of the motor principle are described at the end of this chapter.

12-2 INDUCED CURRENT

Electrical current is generated in commercial quantities by a *generator.* A generator converts mechanical energy (provided by steam or water turbines) to electrical energy by the basic process of *induction.*

As mentioned above, a current is induced in a wire if the wire is caused to move in a magnetic field in such a way that the motion of the wire cuts across the flux lines. (The wire, of course, must be part of a complete circuit in order for current to flow.) Figure 12-6 illustrates this basic action, along with the left-hand rule applied to determine current direction. The wire, which is part of a complete circuit, is caused to move upward, cutting across the flux lines at a right angle to the lines. The generated electron flow is toward the reader, as suggested by the left hand shown in the figure. The amount of electrical energy produced is a function of the number of lines cut per second, in addition to other factors.

It must be emphasized that if neither the magnet nor the wire is in motion, no current is induced. Furthermore, if the wire is moved parallel to the lines of force, no current is generated, since no lines of force are cut. Finally, the induced current flows toward the reader as shown whether the wire is moved in the upward direction, *or* the magnet is moved in a downward direction. These facts, of course, lead one to ask if the magnitude of electrical energy that is induced in the wire is

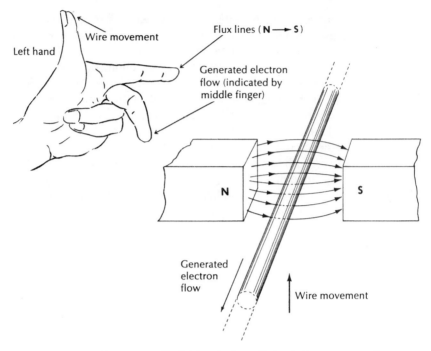

Figure 12-6 The left-hand rule applied to basic generator action.

partially determined by the angle at which the lines are cut. This is certainly true, and as will be discussed in detail at a later point, this has a definite bearing on the formation of a commercially produced ac voltage.

A very important relationship regarding the generation of a voltage and current by induction is called *Lenz's law*. This law states that, in general, when a wire is forced through a magnetic field and an induced current is generated, *a counterforce* is produced that opposes the original force. That is, it takes much more energy to move the coil through the flux if a current flows, due to induction, than if a straight piece of wire (open-circuit) is used. The greater the current flow generated, the greater the opposition. The opposition is due to the field produced by the current, which opposes the original motion. Lenz's law must hold true, or it would be possible to generate large amounts of electrical energy without expending equivalent kinetic energy to do so. This would be "getting something for nothing," which is impossible in the physical world.

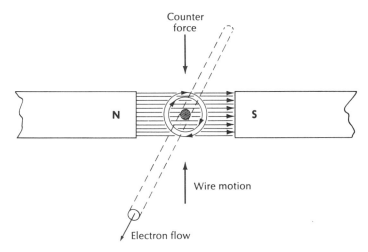

Figure 12-7 The generation of counterforce.

To visualize how this counterforce is produced, refer to Figure 12-7. Assume that the wire is being moved upward as shown. Applying the three-finger left-hand rule, induced current is flowing toward the reader. However, this induced current itself produces a field of force surrounding the wire, with the direction as shown. The magnet's lines of force and the lines created by the induced current are in the same direction *above* the wire, and in the opposite direction *below* the wire. Thus, these fields react to try to force the moving wire downward. Hence, as the wire is moved upward and current is induced in it, the current creates a magnetic field that opposes the original motion. The faster the wire is moved upward, the greater the induced current, and the greater the counterforce. The work expended in overcoming the counterforce is the work necessary to generate the resulting current flow. If the wire is moved in the downward direction, Figure 12-8 illustrates the occurrence.

12-3 INDUCED VOLTAGE

Because there must be voltage before current can flow, it is possible to think of the electrical energy induced in a wire as a voltage, rather than a current. By substituting a voltmeter for an ammeter in the circuit, the induced voltage can be read and interpreted. The amplitude of the induced voltage is determined by the same factors that determine the

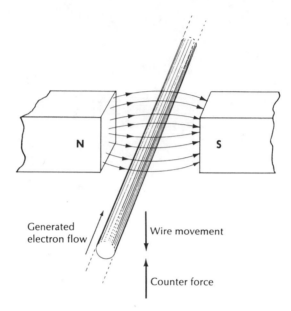

Figure 12-8 The direction of motion is opposite to that shown in Figure 12-6.

current—the rate of cutting (lines per second) and the angle at which the lines are cut.

Note that the lines-per-second rate of cutting can be increased by increasing the magnet strength, by increasing the number of conductors, or by increasing the speed of cutting. Any one or all of these results in a greater induced voltage or current. This relationship is called *Faraday's law,* named after Michael Faraday (1791–1867), an English physicist and chemist. Faraday's law states that the magnitude of a generated voltage is directly proportional to magnet strength, the number of turns on the coil, and the speed (or rate) of cutting.

12-4 INDUCED VOLTAGE ACROSS A COIL

To generate a reasonably large voltage by induction with a reasonable degree of efficiency, a coil with many turns can be used. Such an arrangement is shown in Figure 12-9. If the coil has several hundred turns and if the magnet has reasonable field strength, the induced voltage can easily be read on a standard voltmeter. As the magnet is moved toward the coil, the voltmeter will indicate in one direction, and when the direction of motion is changed, the meter will indicate in the opposite direction. If the magnet is continuously moved back and

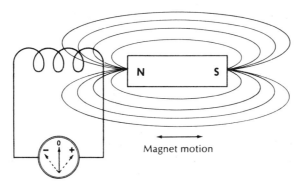

Figure 12-9 Basic generator action.

forth, a continuously alternating voltage will be generated, and the needle will swing back and forth in step with the motion of the magnet.

If the field strength of the magnet is known and the rate of cutting can be estimated, the induced voltage can be calculated by the formula

$$e = N \frac{d\phi}{dt}$$

where e = instantaneous voltage;
     N = number of turns on a coil;
     $d\phi$ = number of lines being cut in webers (10^8 maxwells), during the interval dt; and
     dt = interval of time during which the lines are being cut.

For example, given a coil having 200 turns, with the number of lines of flux being cut equal to 0.65 webers per second and the interval being 1 second, what is the voltage generated?

$$e = N \frac{d\phi}{dt} = 200 \left(\frac{0.65}{1} \right) = 130 \text{ V}.$$

Note that the rate of cutting of 0.65 Wb/s can be generated in several ways: 0.65 Wb in 1 second (0.65/1 = 0.65) or 1.3 Wb in 2 seconds (1.3/2 = 0.65) or 3.25 Wb in 5 seconds (3.25/5 = 0.65). A greater number of flux lines cutting the coil in a correspondingly longer period of time results in the same rate of cutting. The rate of cutting depends on the number of flux lines cutting the coil per unit time; this rate can be increased either by increasing the field strength or by speeding up the motion. For a given coil, the induced voltage is also dependent upon these two factors.

12-5 SELF-INDUCTION

The preceding discussion concerns the generation of voltage and current by the motion between a conductor and a magnetic flux. Another very important way in which EMF is generated has to do with the *self-induction* of a conductor or a coil when energized with a current. Consider Figure 12-10a, which shows a source of current, a wire, and a series switch. On the assumption that such an arrangement will not damage the battery, let us see what the action is when the open switch is closed, then opened again.

With the switch open, as shown, no current flows, of course. When the switch is closed, current begins to flow but cannot be at its maximum value (E/R) instantly. Some amount of time *must* elapse before current attains maximum value. The reason for this is related to the production of the magnetic field that surrounds the wire when current flows. Because the magnetic field must go from zero to a value determined by maximum current, the lines of force must "grow" outward from the center of the wire, as in Figure 12-10b. They extend farther and farther out into space as current increases. When current stops increasing and remains at a static level, the lines of force stop expanding and also remain stationary. When the switch is closed, then, the current increases from zero to maximum, and the magnetic lines of force grow outward away from the wire, as suggested in Figure 12-11. During this time, the lines of force are in motion, and are cutting across the wire. While the lines of force are in motion and cutting across the wire, a voltage *that opposes the applied voltage* is induced in the wire. The effect of this is to retard the increase of current and to prevent the instantaneous increase of current.

In a circuit such as the one shown in Figure 12-10a, the generation of the induced voltage is minimal. To increase this effect, a coil can be

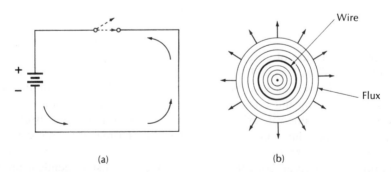

Figure 12-10 A circuit to illustrate the growth of lines of force.

formed; the greater the number of turns, the greater the self-induction. An iron core increases the effect to an even greater extent. Figure 12-12 shows a circuit to be used to further explain self-induction. The foregoing description is valid for this illustration, but the coil concentrates the flux and thus makes it easier to visualize circuit action. As before, when the switch is open, no electrical or magnetic effects are apparent in the circuit. Note that the resistor shown is actually the resistance of the wire in the coil, which is the only limiting factor for direct current. When the switch is closed, current flows and begins to increase, causing the magnetic field to increase and to cut across the turns of the coil. At this first instant, a voltage is induced. This voltage is called the *counterelectromotive force* (CEMF), which is *equal and opposite in polarity to the applied voltage*. Hence, current can only increase slowly. A graph of the circuit response clearly illustrates this circuit action, as

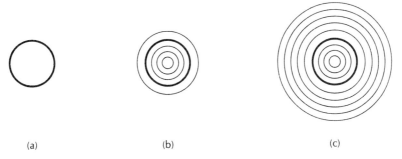

Figure 12-11 Cross-sectional view of a wire with (a) no current flowing, (b) small current flowing, and (c) large current flowing. As the current increases, magnetic lines "grow" outward from the center of the wire.

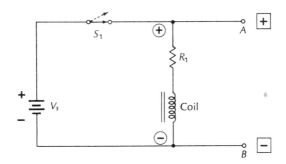

Figure 12-12 A circuit used to investigate self-induction as the switch closes.

shown in Figure 12-13. On this graph we have plotted both current and voltage, to show what happens at given instants of time relative to the induced voltage and the resulting circuit current. The ordinate of the graph is scaled in terms of percent of maximum, to avoid confusing volts and amperes. The abscissa is scaled in units of time, usually seconds.

The graph illustrates that, at the instant of switch closing, the induced voltage is equal and opposite to the applied voltage and the net voltage is zero. For example, if the applied voltage is 10 V, the induced voltage is also 10 V, but of opposite polarity. Hence, $10 - 10 = 0.0$ V, and at this first instant, current is zero.

Now, as some small part of a second goes by, the induced voltage begins to decrease slightly due to the fact that the *rate* of flux lines cutting the coil begins to decrease. That is, the rate of change is maximum at the first instant, because even though the initial current is small, it represents a *very large change* from zero. A smaller rate of change means a smaller induced voltage, and the net circuit voltage increases. If the induced voltage has dropped from 100% to, perhaps, 80% the net circuit voltage is now $100 - 80 = 20\%$ of maximum. Using the preceding example, the net voltage is therefore 2 V ($10 - 8 = 2$ V), and a current is now flowing of a value to satisfy Ohm's law, with $E = 2$ V.

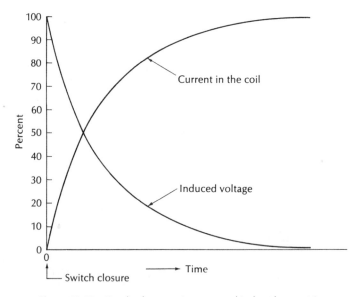

Figure 12-13 Graph of current increase and induced current in a coil energized by dc.

As time progresses, the rate of change of the flux becomes smaller and smaller, and so does the induced voltage. The net circuit voltage grows, until finally it reaches the full applied-voltage value. From this point onward, current remains at its 100% value, determined only by Ohm's law relationships, as long as the switch remains closed. The magnetic field has expanded from the coil and now remains stationary; the rate of change is zero, as is the *induced* voltage.

Referring again to Figure 12-12, note the polarity signs shown. At the instant of switch closing, the full applied voltage appears across the coil as a normal voltage drop across the dc resistance of the coil. This voltage polarity is shown enclosed in circles. Also shown is the induced voltage polarity, enclosed in squares. It must be emphasized that the encircled voltage is simply a *drop* across the dc resistance of the coil. However, the induced voltage must be considered to be a *source,* and as such it influences current flow, as does any source. The induced voltage literally opposes current that would flow due to the applied voltage. At the instant of switch closing, the induced voltage has an amplitude equal to the source voltage, but opposite in polarity. As drawn in Fig. 12-12, the source V_s tries to force current in a counterclockwise direction, while the induced voltage tries to force current in a clockwise direction. At the first instant then, both sources are equal and no current flows. After a period of time, perhaps a tiny fraction of a second to several seconds, the induced voltage becomes zero and full current flows in the coil.

After current has reached maximum, the circuit acts simply as a conventional dc circuit, with current limited only by the dc resistance of the coil, for as long as the switch remains closed. However, when the switch is opened, the magnetic field must *collapse back into the coil.* As it does, another pulse of induced voltage exists, but now the *direction of flux motion is opposite,* and hence the polarity of induced voltage is opposite to the first case. This is illustrated in Figure 12-14, where the polarity of induced voltage at points *A* and *B* is clearly shown.

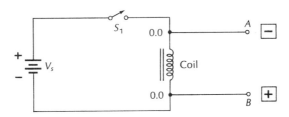

Figure 12-14 A circuit used to investigate self-induction as the switch opens.

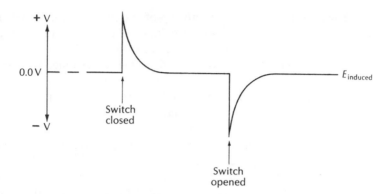

Figure 12-15 Graph of induced voltage as the switch is first closed, then opened.

Figure 12-15 shows the same thing by plotting the amplitude of induced voltage (or CEMF) against time. Such a graph, showing how voltage or current varies as a function of time, is often called a *waveform,* a subject we shall discuss in detail later.

One further point regarding self-induction must be mentioned. The voltage generated across the coil as the switch is opened is shown in Figure 12-15 as having the same amplitude as when the switch is closed. This may or may not be true. In practice, this *spike* of induced voltage produced as the circuit is *opened* is frequently much larger in amplitude than the applied voltage. Many devices make use of this basic principle, a high-voltage automobile ignition system being just one example. The breaker point is simply an elaborate switch that alternately closes and opens as the engine rotates. At exactly the correct time the point opens and the magnetic field in the ignition coil collapses, converting the 12 V from the battery (or alternator) to perhaps 40,000 V. This large voltage is used at the spark plugs to ignite the fuel.

Another device that makes use of the properties of induced voltage is the high-voltage supply in a television set. An arrangement similar to an ignition system converts the low-voltage supply (approximately 12 V for a solid-state receiver, approximately 400 V for a vacuum-tube receiver) to approximately 25,000 V. In a TV receiver, the mechanical switch of the automobile ignition is replaced by a transistor or vacuum tube, but the basic principle remains the same.

12-6 GENERATOR ACTION

A practical generator is shown in simplified form in Figure 12-16. A generator such as this produces *alternating current,* which, as the name

Electromagnetic Induction 279

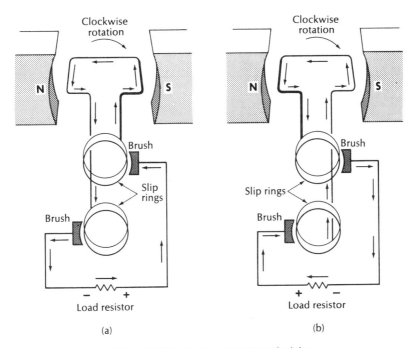

Figure 12-16 Basic ac generator principles.

implies, has values of voltage and current that are constantly changing, or alternating, between positive and negative. The figure illustrates a single turn of wire arranged to be rotated in the magnetic field of a permanent magnet. In actual practice, of course, many turns are used, but the single turn arrangement allows us to visualize the action more easily. The coil of wire is drawn so that one side is represented by a heavy line, while the other is represented by a lighter line. This allows a particular part of the coil to be rotated and to be easily identified. The two ends of the coil are connected to *slip rings,* which also rotate as the coil rotates. The slip rings are connected to the resistive load by means of brushes, which are held against the rings by spring pressure.

Now, to visualize the generation of alternating current, assume that the coil shown in Figure 12-16 is rotating clockwise, as indicated by the arrow. The heavy wire is rotating downward, while the other side of the coil is rotating upward. By applying the left-hand rule to both sides, current must be flowing as indicated, causing the load resistor to exhibit the polarity shown in the figure. As the coil rotates further, the heavy wire is now on the left side (Figure 12-16*b*), so *current through the load has reversed,* and the polarity across the load resistor has also reversed. Obviously, current is now flowing in the opposite direction. As the coil continues to rotate, driven by some external force, current

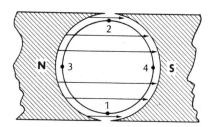

Figure 12-17 Output voltage and current depend on the angle at which the lines of force are cut.

continues to alternate in direction, and voltage polarity changes twice with each revolution of the coil.

With a drawing such as Figure 12-16, only two positions can be shown, and it is necessary to consider the generation of voltage during one complete revolution, or *cycle*. Figure 12-17 illustrates one method of analyzing the occurrences during one cycle of operation. It is apparent that no voltage is generated when a wire is traversing either points 1 or 2, since the wire's motion is in the same direction as the flux, and hence no flux lines can be cut.

However, at points 3 and 4, the wires are traversing the areas where maximum cutting of lines occurs, and therefore maximum voltage is generated. In between the points of minimum and maximum voltage generation, the number of lines of force being cut per unit time varies in a smooth and continuous manner. This is best exemplified by a waveform, or graph, of the voltage or current generated by one complete revolution. Such a waveform is illustrated in Figure 12-18, where one cycle of voltage is shown for one complete revolution.

The drawing assumes that the rotating coil starts in the 1–2 plane of Figure 12-17. Therefore, as rotation starts, zero voltage is generated, because the coil is traveling parallel with the flux. As the coil rotates past the 1–2 plane, an angle is formed between the coil motion and the flux lines. The angle becomes greater and the generated voltage rises from zero, and as the coil passes the 3–4 plane, maximum voltage is generated. As the coil rotates further toward the 2–1 plane, generated voltage decreases, returning to zero as the coil again reaches the vertical plane. As the coil passes the 2–1 plane, the voltage changes polarity, since the lines are now being cut in the opposite direction. Again, the voltage builds up to a maximum in this reverse direction, and again falls to zero as the coil comes back to the starting point. In order for the waveform generated to be as perfect as that in the illustration, the coil *must* rotate at a constant speed.

The waveform in Figure 12-18 is called a *sine wave,* and is typical of nearly all ac voltages. The curve is identical to a plot of all values of the

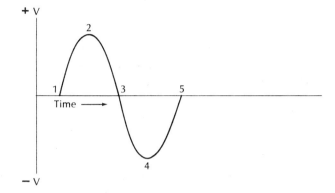

Figure 12-18 One cycle of a sine wave.

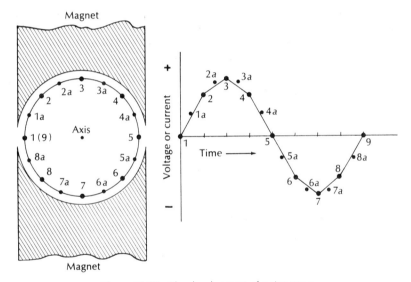

Figure 12-19 The development of a sine wave.

sine of the angles between 0° and 360°. As shown, point 1 represents 0°, point 2 represents 90°, point 3 represents 180°, point 4 represents 270°, and point 5 represents 0° or 360°. Much of the remainder of this book is concerned with alternating voltages and currents having the shape of a sine wave.

To aid in visualizing how a sine wave is developed, refer to Figure 12-19. Shown in the illustration is an ac generator similar to those discussed previously. The rotating coil, or *armature,* is shown, and several points are identified around the coil circumference. Each point represents one wire during one complete revolution. At each point, the voltage is directly proportional to the *angle of cutting* of the lines of force, which determines the number of lines cut per unit time. The points are projected to the right to form a graph. Since the points are evenly spaced in terms of rotational speed and are plotted at equal increments of time on the graph, the result is a very close representation of a sine wave. If, instead of only eight increments, many more were used, the shape would resemble even more closely a true sine wave. This is suggested by the intermediate dots (1*a*, 2*a*, and so on), which are omitted from the graphed line to avoid confusion. When an infinite number of points is used, the shape becomes a true sine wave.

A dc generator is produced when a commutator is used instead of slip rings to connect to the rotating armature. Such a device is shown in Figure 12-20. The basic function is very similar to the ac generator previously explained, except that the current through the load never changes direction. The electrical energy thus produced, however, is not a smooth, unvarying direct current. The waveform produced is shown in Figure 12-21 and is called *pulsating* dc. By proper positioning of the commutator, the switch in polarity can be made to occur just as the generated voltage falls to zero, minimizing both waveform distortion and sparking (arcing) at the brushes. In Chapter 19 we shall discover how to smooth out pulsating dc into a true dc.

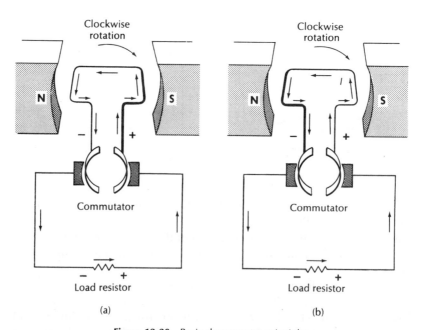

Figure 12-20 Basic dc generator principles.

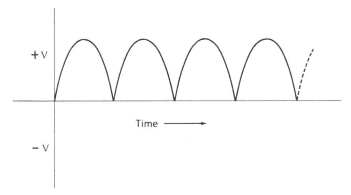

Figure 12-21 A pulsating dc waveform.

12-7 MAGNETIC DEVICES

To acquaint you with a few practical aspects of magnetic devices, we present several examples. In each case, we explain how the device functions and what its major applications are.

Loudspeakers

A loudspeaker accepts electrical energy that represents *audio signals* (low-frequency ac signals that alternate in step with audible sound) and converts the electrical energy to sound waves. Basically, the loudspeaker is a very simple device consisting of a powerful permanent magnet and a lightweight wire coil, to which is attached a large paper cone. When audio signals are sent through the coil (called the *voice coil*), the current produces a varying flux that interacts with the field of the permanent magnet. The net result is that the paper cone is forced to move in step with the audio signals, forcing the air in contact with it to produce sound waves.

Figure 12-22 illustrates a typical loudspeaker. The permanent magnet has a recess into which the voice coil is inserted with barely enough clearance to avoid rubbing the walls of the recess. This keeps the field of the permanent magnet as strong as possible, to obtain maximum cone movement for minimum voice-coil current.

Relays

A relay is simply a switch that can be operated at a remote location by a switch closure at the control point. A relay consists of an electromag-

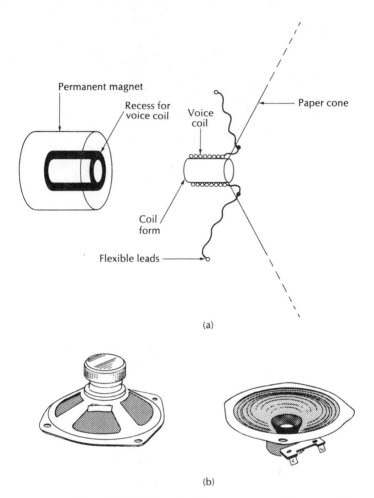

Figure 12-22 (a) Loudspeaker construction and (b) a typical loudspeaker.

net arranged so that the switching contacts are actuated by the electromagnet when it is energized. Figure 12-23 illustrates a typical relay (part *a*), along with several contact configurations (parts *b* and *c*). Form A contacts are single pole and are normally open; when the relay coil is energized, they close. Form B contacts are also single pole and are normally closed; they open when the relay coil is energized. Form C contacts have three segments (double pole) and operate as make-open-break contacts. That is, the normally closed contacts break when the coil is energized, and for a split second while the contacts are

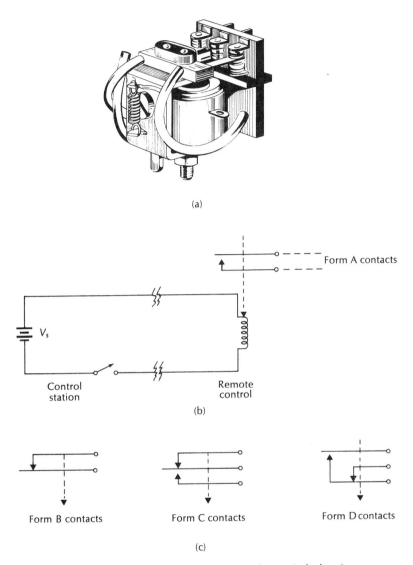

Figure 12-23 (a) Typical relay construction; (b) a typical relay circuit; and (c) form B, C and D contacts.

moving, the circuit is open. Then, the lower contacts close and make the second connection. Finally, form D is a double-pole make-before-break configuration. The upper and lower contacts make when the coil is energized, and the pressure upon the lower contact forces a break with the center one. Hence, at no time does the circuit connection open.

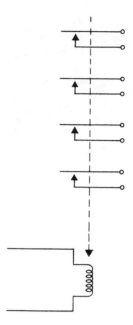

Figure 12-24 A relay with stacked contacts.

In practical relays, there may be many sets of contacts "stacked" as illustrated in Figure 12-24. Shown is a relay with a stack of four form A contacts. With an arrangement such as this, four separate circuits can be controlled from a single switch.

Ignition Coil

A typical automobile ignition coil is shown schematically in Figure 12-25. When the breaker points close, current flows in the lower part of the coil, and a strong magnetic field is built up around the coil. When the breaker points open, the magnetic field collapses rapidly and a large induced voltage is generated. This voltage is made even larger due to the additional turns on the coil that, by *transformer* action, increase the voltage over the value given by a simple coil. (See Chapter 14 for a thorough investigation of transformers.)

Other Magnetic Devices

It is not possible to cover all the different types of magnetic devices. Many will be dealt with in subsequent studies. However, it is interesting to enumerate some of these devices, even if only in passing. Audio

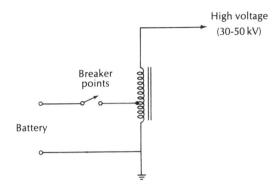

Figure 12-25 The schematic diagram of an automobile ignition coil.

and video tape recorders utilize the magnetic properties of a recording tape, a thin plastic tape coated with a magnetic film that becomes permanently magnetized by the information to be recorded. Nearly all TV programs are first recorded on tape and then rebroadcast. Tape recorders can be quite complex, especially those used to record full-color TV programs.

Many latching devices are operated magnetically by a *solenoid*, which is simply a coil with a *movable* iron core. The iron core is often spring loaded to rest slightly outside the coil. When the coil is energized, the iron core is drawn inside the coil, and the thrust thus developed is used to activate some other device.

A simple doorbell is another device that operates on magnetic principles. An electromagnet attracts a strip of spring steel and as the strip is moved it breaks the connection to the coil. The strip then returns to the rest position, which again makes a connection for the coil, which attracts the strip, and so on, over and over again, as long as current flows. A ball attached to the end of the strip strikes a bell every time the strip is attracted, so the doorbell rings as long as the coil is energized.

SUMMARY

- The repulsion or attraction of magnets is a simple exhibition of motor action.
- A current-carrying wire tends to move if placed in a strong magnetic field. Motion is away from lines in the same direction and toward lines in opposite directions.
- If a wire that is part of a closed circuit is moved through a magnetic field perpendicular to the lines of force, a current is induced in the wire.

- A commutator automatically reverses current direction in a coil so that continuous rotation of the coil is possible when the coil is placed in a magnetic field, thus producing motor action.
- An open coil of wire moved in such a way that it cuts across a magnetic field exhibits an induced voltage across the open as long as magnetic lines are being cut. If the coil's direction of motion is reversed, the polarity of induced voltage is also reversed.
- Self-induced voltage in a coil that is moving in a magnetic field is determined by $e = N(d\phi/dt)$.
- Counterelectromotive force (CEMF) opposes the applied voltage.
- CEMF prevents current from increasing from zero to its maximum value instantly.
- An ac generator delivers its energy by means of slip rings and brushes.
- An ac generator produces a waveform called a sine wave.
- By fitting a commutator to a generator, instead of slip rings, a dc generator is produced.

QUESTIONS

1. Refer to Figure 12-1. What will happen if only the poles of the magnet are reversed?
2. Refer to Figure 12-1. What will happen if only the current direction is changed?
3. Refer to Figure 12-1. What will happen if the polarity of the magnet *and* the current direction are changed?
4. Refer to Figure 12-6. In what direction will current flow (toward the reader; away from the reader) if the wire is moved as shown, but the polarity of the magnet is changed?
5. Refer to Figure 12-6. In what direction will current flow (toward the reader; away from the reader) if the wire is moved from left to right as drawn?
6. Refer to Figure 12-11. Why does current increase more or less slowly toward maximum instead of instantly?
7. Refer to Figure 12-11. What is the difference between the polarity signs shown enclosed in circles and those in squares?
8. Refer to Figure 12-13. How are the polarity signs in squares opposite to those in Figure 12-11?
9. In Figure 12-15, what purpose do the brushes serve?
10. Briefly explain the meaning of *sine wave*.

PROBLEMS

1. A coil having 500 turns is used in conjunction with a flux change of 1 Wb/s. What is the induced (or generated) voltage?

2. A coil having 350 turns is used in conjunction with a flux change of 500 μWb/s. What is the induced voltage?
3. A coil having 1500 turns is used in conjunction with a flux change of 1,500,000 Mx per second. What is the induced voltage?
4. A coil having 350 turns is used in conjunction with a flux change of 1,500,000 Mx per second. What is the induced voltage?
5. A coil is used in a circuit with a changing flux of 2,000,000 Mx per second. The induced voltage is 35 V. How many turns does the coil have?
6. A coil is used in a circuit with a changing flux of 750,000 Mx per second. The induced voltage is 1.0 V. How many turns does the coil have?
7. A 500-turn coil is used in a circuit with an induced voltage of 20 V. What is the rate of change of flux in webers? in maxwells?
8. A 1000-turn coil is used in a circuit with an induced voltage of 95 V. What is the rate of change of flux in webers? in maxwells?

CHAPTER 13

ALTERNATING CURRENT AND VOLTAGE

Nearly all heavy-duty industrial and household equipment operates on alternating current. Generally, alternating current is more economical to transport over long distances than direct current, so ac is used throughout the world as the standard for electrical power. Dc is used for this purpose only rarely, although with technological advances, dc power transmission is becoming more common.

In addition to ac power and the equipment driven by it, an altogether separate classification of ac exists. Alternating voltage and currents that are *not* used to power equipment are known as *signal* voltages or currents. Among these are the audio signals found in a phonograph or stereo tape player, for example. Also, radio and TV signals, for the most part, consist of ac voltages and currents. Two characteristics help to differentiate between power ac and signal ac.

- First is application: If the end use of a voltage or current is simply to provide some kind of motive power, such as for a lamp or motor, the energy is definitely classed as power. If the application calls for application to a *transducer* of some sort, perhaps a TV picture tube or loudspeaker, then it is classed as a signal.
- Second is frequency: If only one frequency is concerned, the ac is, in all probability, power. If the ac voltage or current contains more than one frequency, or consists of a varying frequency, then it is classed as a signal.

While there are a few exceptions to these rules, they hold true in nearly all cases.

In this chapter, we investigate the fundamentals of alternating current, both power and signal, leading ultimately to a number of practical applications. The major subject headings in this chapter are listed below.

13-1 Transporting Ac
13-2 The Sine Wave
13-3 Conversion Factors
13-4 Simple Ac Circuit

13-5 Instantaneous Values
13-6 Signal Ac
13-7 Phase Angles
13-8 Nonsinusoidal Waveforms

13-1 TRANSPORTING AC

Electricity is usually generated some distance from the place where it is to be used. It must, therefore, be transported over wires that are often hundreds of miles long. To avoid severe losses due to the voltage drop in the wire, the energy is transported at very high voltage and relatively low current values, typically 33 to 500 kV at 400 to 3000 A. To see why this is so, consider the following example.

If the energy being transported is rated at 220 kV at 4.545 A, the total power being transported is 4.545 A × 220 kV = 1 MW. Assume that the total resistance of the wire being used is 500 Ω. The voltage drop is $I \times R$ = 4.545 A × 500 Ω = 2273 V. While this seems like a large voltage drop, it actually represents a very small percentage of 220 kV. The voltage delivered to the load is 217,727 V, a loss of only about 1%.

To compare this with the loss that occurs at a lower voltage, assume that 1 MW is to be transported using 50 kV at 20 A (20 A × 50 kV = 1 MW). However, in this case, the drop in the wire is 20 A × 500 Ω = 10,000 V. This represents a 20% loss, a totally unacceptable value, since only 40,000 V is left to supply the load. Obviously, the higher the voltage, and therefore the smaller the current for a given power value, the more efficient the transmission.

In order to attain voltages on the order of 220 kV or higher, a device called a *transformer* is used. A transformer, which we discuss in detail in the following chapter, is a device used to step up (increase) or step down (reduce) voltages and currents, which it does very efficiently. Thus, ac power is generated at convenient voltage levels, on the order of 5 kV, then is immediately stepped up, through a series of transformers, to the much higher transmission level. At the load end, another series of transformers is used to step down the voltage for use by industry and in the home. A typical transmission system is shown in simplified format in Figure 13-1.

13-2 THE SINE WAVE

Electrical power as generated usually has the shape of a sine wave. Because a sine wave is constantly changing, a new vocabulary is required to describe the voltage and current values in typical circuitry.

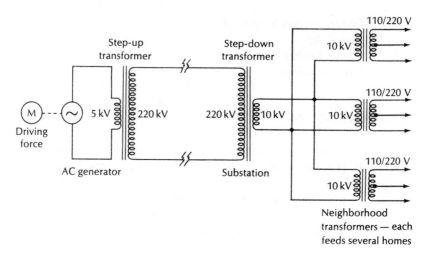

Figure 13-1 A typical ac power distribution system (simplified).

A sine wave is shown in Figure 13-2, along with the names of the various values used to describe numerical quantities. Each of these is identified and described in detail in the following paragraphs. The waveform shown is considered to be a voltage waveform, but everything said about it also holds true for current.

Because the amplitude is constantly changing, there is no single value for voltage or current. Four different values are used to describe amplitude, depending upon how the information is to be used. It is important to label voltage values carefully to avoid mistakes. The *peak* value is the voltage (or current) attained at the maximum excursion away from the base line (0.0 volts). Such a voltage is always labeled peak volts (E_{pk}). The peak value can represent either the positive peak or the negative peak.

When the total excursion must be specified, the *peak-to-peak* value (E_{pk-pk}) is used. This is the difference between the maximum positive peak and the maximum negative peak; for a *symmetrical* waveform, it is twice the peak value. (A symmetrical waveform is one in which the positive and negative peaks have the same shape and amplitude, although opposite signs.)

The *rms*, or *effective*, value (E_{rms}) is the value that is indicated on an ac voltmeter. Effective voltage is always 0.707 of peak voltage. For example, assume the peak value is 170 V. The rms value is

$$E_{rms} = 0.707 \times E_{pk} = 0.707 \times 170 = 120 \text{ V.}$$

The factor 0.707 is derived from $1/\sqrt{2} = 1/1.414 = 0.707$. The term *rms* means *root-mean-square*, which is a method used to derive the numeri-

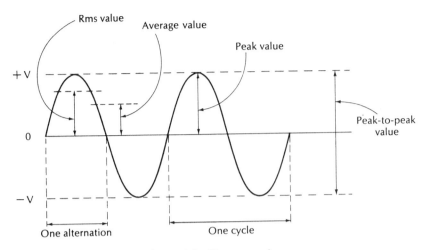

Figure 13-2 Sine wave values.

cal value for a given instance. To derive this, *all* instantaneous values for either peak are squared, then added together and divided by the number of instantaneous values (which gives an average, or mean) and the square root is extracted. This process is illustrated in Table 13-1. The computation in the table is not completely accurate, since only ten values are used. Accuracy would be slightly increased if, for example, a value for *every degree* were to be used. This, however, would require 90 calculations, and is too ponderous to illustrate herein for the very small error involved.

TABLE 13-1 DERIVATION OF RMS VALUE OF A SINE WAVE

Angle	sine θ	sine θ^2
1. 0°	0	0
2. 10	0.1736481	0.030153689
3. 20	0.3420201	0.1169777
4. 30	0.5	0.25
5. 40	0.6427876	0.4131759
6. 50	0.7660444	0.5868241
7. 60	0.8660254	0.75
8. 70	0.9396926	0.8830222
9. 80	0.9848077	0.9698463
10. 90	1.0000	1.000

Sum: 5
Average (mean): 5/10 = 0.5
$\sqrt{\text{Mean}}$: $\sqrt{.5}$ = 0.7071067

The process begins with finding the sine of every 10°-increment between 0° and 90°. These are given in the second column of the table (θ is the Greek letter theta). Each of these is squared, and these values appear in the third column. The values in the third column are then added, giving a sum of 5. Because there are ten increments, the sum is divided by 10 to give the average (mean) of the squared values, 0.5. Finally, the square root of the mean is taken to yield 0.7071067, which is in agreement with $1/\sqrt{2} = 0.7071067$.

Returning to Figure 13-2, the fourth way in which voltage is specified is the average value (E_{av}). This is the simple arithmetic average of all of the instantaneous values for 90° or 180°. If the ten values of sine θ in Table 13-1 are added and divided by 10 to obtain an average, the number arrived at is 0.6215, which is close to the true average value. If *all* instantaneous values are used (each degree for a full 180°), rather than just the ten in Table 13-1, the true average value is 0.636 times the peak value. Note that if the average for a *complete* cycle is determined, the computed average is *zero*, since the positive and negative values cancel. Hence, the average value is determined for a half-cycle period. As long as the waveform is symmetrical, only the first 90°, or quarter cycle, is required, since between 90° and 180° the curve is a mirror image of the 0° to 90° portion. Thus, doubling the increments and dividing by twice the number yields the same result.

Also shown in Figure 13-2 is the interval of time labeled *1 cycle*. This is a very important parameter, since it determines the frequency of the generated waveform. *Frequency* is defined as the number of events per unit time. Since a cycle represents one complete sequence of events, electrical frequency is measured in *cycles per second*. The unit for cycles per second is the hertz (Hz). The interval of time encompassed by one cycle is called the *period,* which is measured in seconds (s), milliseconds (ms), microseconds (μs), or whatever unit is appropriate.

Frequency and period are related to each other reciprocally. For example, in the United States the standard frequency for power is 60 Hz. That is, there are sixty complete cycles generated in a time period of one second. The period *(P)* for a 60-Hz waveform is determined as follows:

$$P = \frac{1}{F} = \frac{1}{60} = 0.1667 \text{ s} = 16.67 \text{ ms},$$

where P = period in seconds, and
F = frequency in Hertz.

Conversely, if the period of a waveform is known, its frequency can easily be found.

$$F = \frac{1}{P} = \frac{1}{0.1667} = 60 \text{ Hz}.$$

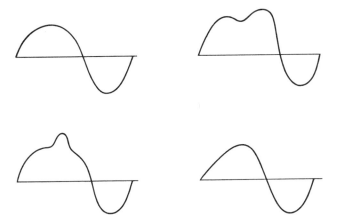

Figure 13-3 Sine wave distortions.

Finally, one *alternation*, commonly called a half-cycle, is shown in Figure 13-2. Note that the positive half-cycle is, except for polarity, exactly the same as the negative half-cycle. That is, the *area* under the curve for the positive half-cycle is the same as the *area* under the curve of the negative half-cycle. If this is not true, the waveform is said to be distorted. Several distorted waveforms are shown in Figure 13-3.

13-3 CONVERSION FACTORS

The peak, rms, and average values of a sine wave (or sinusoidal wave) are all used when analyzing ac circuits. Several examples follow that illustrate the application of these values in typical problems.

Example 1. Given a sinusoidal waveform having a positive peak of 100 V, find the rms value and the average value.

$$E_{rms} = 0.707 \times E_{pk} = 0.707 \times 100 = 70.7 \text{ V}.$$
$$E_{av} = 0.636 \times E_{pk} = 0.636 \times 100 = 63.6 \text{ V}.$$

Example 2. Given a sinusoidal waveform having an rms value of 115 V, find the peak value and the average value.

$$E_{pk} = 1.414 \times E_{rms} = \frac{E_{rms}}{0.707} = 162.7 \text{ V}.$$
$$E_{av} = \frac{0.636}{0.707} \times E_{rms} = 0.636 \times E_{pk} = 103.5 \text{ V}.$$

Note: $1/0.707 = 1.414$; see Table 13-2 for all sinusoid conversion factors.

TABLE 13-2 CONVERSION FACTORS FOR AC SINUSOIDS

Given (x)	To Find	Multiply (x) by
E_{pk}	E_{rms}	0.707
E_{pk}	E_{ave}	0.636
E_{rms}	E_{pk}	$1/0.707 = 1.414$
E_{rms}	E_{ave}	$\frac{0.636}{0.707} = 0.8995756 \cong 0.8996$
E_{ave}	E_{pk}	$1/0.636 = 1.572327 \cong 1.572$
E_{ave}	E_{rms}	$\frac{0.707}{0.636} = 1.1116352 \cong 1.112$

Example 3. Given a sinusoidal waveform having an average value of 76.32 V, find the peak value and the rms value.

$$E_{pk} = \frac{1}{0.636} \times 76.32 = 120 \text{ V}.$$

$$E_{rms} = \frac{0.707}{0.636} \times 76.32 = 84.84 \text{ V}.$$

Example 4. Given a sinusoidal waveform having a frequency of 1000 Hz (cycles per second), find the period.

$$P = \frac{1}{F} = \frac{1}{1000} = 0.001 \text{ s} = 1 \text{ ms}.$$

Example 5. Given a sinusoidal waveform having a period of 1 μs, find the frequency.

$$F = \frac{1}{P} = \frac{1}{0.000001} = 1{,}000{,}000 \text{ Hz} = 1 \text{ MHz}.$$

13-4 SIMPLE AC CIRCUIT

To illustrate a similarity between ac and dc circuits, consider Figure 13-4. A simple incandescent lamp is connected to a source of 120-V alternating current. The lamp is rated at 100 W, and this determines the amount of current that will be drawn, as well as the amount of light. An ac voltmeter is shown measuring the voltage across the lamp, and ignoring the fact that the ac voltmeter works in a somewhat different manner than a dc voltmeter, the measurement is similar to the measurement of a similar dc circuit. If a VOM is used, the selector switch is

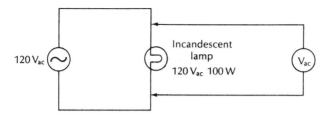

Figure 13-4 A simple ac circuit.

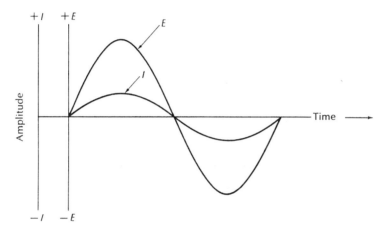

Figure 13-5 Voltage and current relationships in a resistive ac circuit.

placed on ac volts with a range that will accommodate 120 V, and the reading is taken using the ac scales. If the source is producing 120 V, the meter reads 120 V_{rms}.

Nearly all ac voltmeters read effective voltage (E_{rms}) because the effective value of ac corresponds directly to dc. That is, a 120-V_{dc} battery connected to the same lamp will dissipate *exactly* the same power (and produce the same amount of light) as a 120-V_{rms} ac voltage.

Now, the voltage drop across the lamp is 120 V_{ac}, and the current through the lamp is easily found. Since $P = I \times E$, $I = P/E = 100/120 = 0.833$ A. This current value is also in terms of rms, because the voltage is rms. The voltage and current relationship is illustrated in Figure 13-5. The voltage waveform is shown larger simply because its numerical value is usually larger. As in this illustration, one scale must be used for current values, and another scale for voltage, whenever the

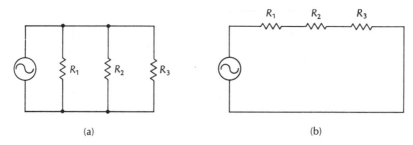

Figure 13-6 Ac circuits with (a) loads in parallel and (b) loads in series.

two values are plotted together. Such an illustration, however, helps visualize the *phase* relationship between voltage and current. Because the lamp is purely resistive, voltage and current rise and fall in step with one another, and so are said to be *in phase*. We shall later discover that, very often, voltage and current are *not* in phase, or step, and are therefore *out of phase*. This occurs when the load in a circuit is not purely resistive. Such load devices are capacitors and inductors, which will be dealt with in detail later.

The measurement of ac current poses a few problems. Most VOMs have no provision for such measurements. However, many digital multimeters *do* have this capability. Also used to provide this function are such devices as clamp-on ammeters and the oscilloscope, both of which will be covered subsequently.

The lamp in the foregoing example could be replaced with a simple resistor capable of dissipating 100 W. Current and voltage values would remain the same, the only difference being that the resistor would not produce light.

More often than not, power-line loads are connected in parallel. For example, all receptacles in your home are wired in parallel. In certain applications, for example, Christmas tree lights, it is possible to find loads in series. In both parallel and series instances, the circuits are treated exactly as though they were dc circuits. In Figure 13-6, parallel and series ac circuits are shown. The loads in both circuits are shown as resistors, but they could be any purely resistive load, such as an incandescent lamp. Again, the circuits are treated exactly as though they were dc circuits. Voltages are read in the same way, but with the realization that the value is the rms value with most meters.

13-5 INSTANTANEOUS VALUES

Recall from an earlier discussion that an ac generator produces a sine wave because of the *angle* at which the coil of wire cuts the lines of

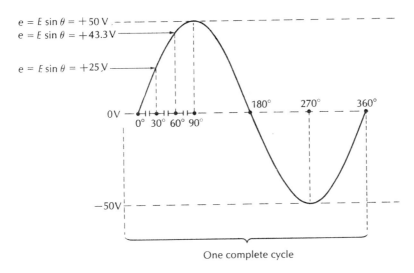

Figure 13-7 Instantaneous values of a sine wave.

force. Since one complete revolution generates one complete cycle, each must encompass 360°. The voltage at any part of the sine wave is a function of the peak voltage attained by the waveform and of the sine of the angle of rotation, which is generally measured from a zero-volt point. Figure 13-7 illustrates the instantaneous value of a sine wave as a function of the angle of rotation attained to the point in question. The point of origin of a graph such as this is usually the point on the waveform where the voltage is zero and is positive-going. Because all cycles are similar, it makes no difference which cycle is graphed, except that the generator should be at full operating speed, so that the period of each cycle is the same.

The time base (zero-volt line) of Figure 13-7 is divided according to degrees of rotation. At this point we are only concerned with the first 90°. The voltage attained by the sine-wave curve at any point in time is determined by the following relationship.

$$e = E \sin \theta,$$

where e = instantaneous voltage at a specified time,
E = peak voltage, and
$\sin \theta$ = the sine of the angle θ. Here, the angle θ is simply the linear distance along the time base starting from 0°.

Example 6. Find the voltage at 30° if peak voltage is 50 V.

$$e = E \sin \theta = 50\text{V} \times \sin 30° = 50\text{V} \times 0.5 = 25 \text{ V}.$$

(The sine of 30° can be obtained from the trigonometric function tables in Appendix C, or from a slide rule, or from an electronic calculator with trig functions.)

As shown in Figure 13-7, the voltage at 30° can be read from the curve directly above the 30° mark on the base line. At 60° the value is much greater.

Example 7. Find the voltage at 60° if peak voltage is 50 V.

$$e = E \sin \theta = 50\text{V} \times \sin 60° = 50\text{V} \times 0.86603 = 43.3 \text{ V.}$$

This value is also indicated on the drawing. At 90°, the curve has attained peak value, or 50 V. The voltage at 90°, therefore, is already known. It could, however, be calculated.

Example 8. Calculate the voltage at 90° if peak voltage is 50 V.

$$e = E \sin \theta = 50 \text{ V} \times \sin 90° = 50 \text{ V} \times 1 = 50 \text{ V.}$$

At first, many people find it confusing to measure time in degrees instead of seconds. It is a simple matter to convert degrees to seconds if the frequency or period of the wave is known.

Example 9. Assume that the frequency of the waveform in Figure 13-7 is 25 kHz, or 25,000 cycles per second. The period of each cycle is therefore

$$P = \frac{1}{F} = \frac{1}{25 \times 10^3} = 4 \times 10^{-5} = 40 \text{ }\mu\text{s.}$$

Hence, 40 μs is the period of time of one cycle. The time interval in seconds from 0° to 30°, then, is simply

$$\frac{30}{360} \times 40 \text{ }\mu\text{s} = 3.333 \text{ }\mu\text{s.}$$

Conversely, the fraction of a period and the full period can be used to determine the angle.

$$\theta = \frac{\text{partial period}}{\text{period}} \times 360° = \frac{3.333 \text{ }\mu\text{s}}{40 \text{ }\mu\text{s}} \times 360° = 30°.$$

13-6 SIGNAL AC

More frequently than not, the electronics technician has occasion to work with alternating voltages and currents that are not intended to supply electrical power. As mentioned in the introduction to this chapter, these are frequently called *signal* voltages and currents. While power-line voltages have fixed frequencies, such as 50, 60, or 400 Hz, signal frequencies range from perhaps one to millions of cycles per second. Signal frequencies are broken down into categories for easier classification.

Audio Frequencies

Frequencies from approximately 20 Hz to 20 kHz are called audio frequencies (af), since they can be heard by the human ear as sound when applied to a *transducer* (such as a loudspeaker or headphones). The lower the frequency, the lower the pitch; the higher the frequency, the higher the pitch. In other words, low frequencies provide bass tones and high frequencies provide treble tones. Most people have difficulty hearing tones lower than about 20 Hz or higher than 15 to 20 kHz. On the average, women can hear higher audio frequencies than men. By the same token, dogs can hear much higher frequencies than human beings. Witness the "silent" dog whistle, which is silent only for most people. Loudness is a function of the amplitude of the audio frequency signal—the greater the amplitude, the louder the sound produced by a transducer.

Audio frequencies are subdivided into subsonic, sonic, and supersonic frequencies, according to the response of the average human ear. *Subsonic* sounds are those below the range of human hearing, those below approximately 20 Hz. *Sonic* frequencies are those to which the ear responds readily. *Supersonic* frequencies are those which are too high to hear, but still in or near the audio range, or about 20 kHz and higher. Regardless of frequency, once launched into the atmosphere, the compressions and rarefactions of sound waves travel at a speed of 1130 feet per second (in dry air at 21°C). You already know that this rate of speed produces a very noticeable delay if the source of sound is some distance away from the listener.

Radio Frequencies

Above 30 kHz, it is possible to "launch" radio waves from a suitable radio antenna, so these frequencies are known as radio frequencies (rf). Values from 30 kHz up to 300,000 MHz are used for radio transmissions of all kinds, from AM, FM, and TV broadcasting to radar signals, to extra-high frequency telecommunications to and from outer space. Radio waves travel through the atmosphere (or a vacuum) at a speed of approximately 186,000 miles per second. Thus, for point-to-point communications on Earth, transmission and reception appear to happen instantaneously. Communications between Earth and moon produces a noticeable delay of a few seconds. Since the average distance between Earth and the moon is 239,000 miles, the delay for a round-trip signal is approximately 2.6 seconds.

As we extend our explorations further and further into space, the delay becomes more and more pronounced. For instance, at its closest, Mars is approximately 35 million miles from Earth. The round-trip delay is, therefore,

$$\frac{70,000,000}{11,160,000} = 6\frac{1}{4} \text{ minutes.}$$

Note that we convert the speed per second of radio waves to speed per minute (186,000 × 60 = 11,160,000 mi/min). In August 1977 the United States launched a probe which is scheduled to investigate Jupiter and Saturn. At its closest, Jupiter is 368 million miles from Earth, and Saturn is 745 million miles. Can you see that in your lifetime, this time delay will become a serious problem?

Appendix E shows in more detail the classification of radio frequencies and some typical applications of each signal type.

One important concept relating to radio signals is *wavelength*. When a radio signal is transmitted, the linear distance covered by a single cycle is called the wavelength. Wavelength is determined by the speed of propagation and the frequency of the signal, and is calculated as follows:

$$\lambda = \frac{V}{F},$$

where λ = wavelength (this symbol is the Greek letter lambda),
V = velocity of the wave, and
F = frequency of the wave.

Example 10. What is the wavelength of a radio wave with a frequency of 1 MHz?

$$\lambda = \frac{V}{F} = \frac{1.86 \times 10^5}{1 \times 10^6} = \frac{186,000}{1,000,000} = 0.186 \text{ mi.}$$

Since there are 5280 feet per mile, 0.186 × 5280 = 982 ft.

Example 11. It is often necessary to determine wavelength in meters or in multiples or submultiples of that unit. One inch equals 0.0254 meter or 2.54 centimeters. Therefore, the wavelength in centimeters of the radio wave in Example 10 is calculated as follows:

$$\lambda = \frac{V}{F} = \frac{2.99 \times 10^{10}}{1 \times 10^6} = 29,934 \text{ cm} \cong 30,000 \text{ cm,}$$

where V = velocity of the wave in centimeters, which is determined by 2.54 cm/in. × 12 in./ft × 5280 ft/mi × 186,000 (the velocity of the wave in miles per second.).

In meters, this wavelength is 30,000 cm × 0.01 = 300 m.

Example 12. Determine the length of an antenna that must be ½ wavelength at 27 MHz. Because we want the answer in feet, we must convert 186,000 miles per second to feet per second.

$$V = 186{,}000 \times 5280 = 9.8208 \times 10^8 \text{ feet per second.}$$

Therefore,

$$\lambda = \frac{9.8208 \times 10^8}{27 \times 10^6} \times 36.37 \text{ ft,}$$

and

$$\frac{\lambda}{2} = 18.19 \text{ ft.}$$

13-7 PHASE ANGLES

As briefly mentioned before, it is often the case that voltage and current do not increase and decrease exactly in step with each other. Figure 13-5 illustrates a voltage and current exactly *in phase* with each other. Figure 13-8, on the other hand, illustrates a voltage and current that are *not* in phase. Here the waveforms are displaced by 60°, and the voltage

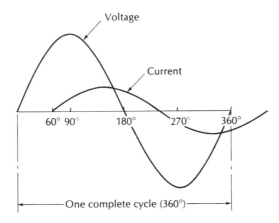

Figure 13-8 Voltage leading the current by 60°.

is said to *lead* (occur first in time) the current. Alternatively, the current can be said to *lag* (occur later in time) the voltage. In either case, the *phase angle,* or *phase difference,* or *phase shift,* between the two quantities is 60°. A voltage or current is caused to lead or lag the other quantity if the current contains either inductance or capacitance or both. These qualities will subsequently be discussed in detail.

The amount of lead or lag may be virtually anything from 0° (no phase shift) through 180° (completely out of phase) to 360° (no phase shift). Also, a comparison may be made between two voltages, or two currents. One very popular electronic circuit provides two voltages that are 180° out of phase with each other. Figure 13-9 illustrates this in simplified form. The circuit, represented by the square box labeled *amplifier*, has a single sine-wave input. By virtue of its internal configuration, it produces as output two 180°-out-of-phase sine waves.

Radian Measurement

When performing angular measurement it is often desirable to use the *radian* unit of measurement instead of degrees. The radian (rad) is

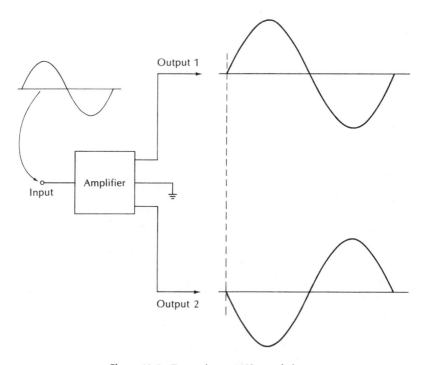

Figure 13-9 Two voltages 180° out of phase.

equal to 57.296° ≅ 57.3°. The radian is derived from the arc on the circumference of a circle having a length equal to the length of the *radius* of the circle. This is illustrated in Figure 13-10. The circumference of a circle encompasses 360°, or 2π rad. Hence, one cycle is the equivalent of 2π rad, 180° is π rad, 90° is $\pi/2$ rad, and so on, as shown in Figure 13-11.

Vectors and Phasors

The two waveforms shown in Figure 13-8 have a phase difference of 60°. The drawing shows one way of specifying voltages or currents having a phase difference. There is, however, an alternate way, the *phasor* diagram. As the first step toward understanding phasors, we

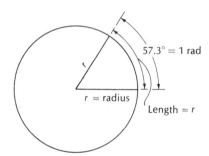

Figure 13-10 Deriving the radian.

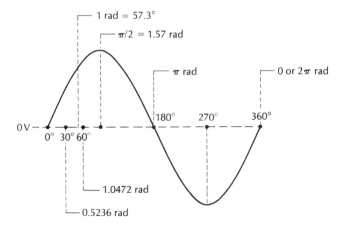

Figure 13-11 Relationship of degrees and radians.

illustrate a *vector* drawing in Figure 13-12. At the bottom of the figure, you see a boat propelled at 3 mph directly across a river. The current, however, is traveling at a rate of 4 mph. Therefore, the boat has *two* forces acting on it—its own propulsive force of 3 mph and the force of the current at 4 mph. To determine the true speed of the boat and its true direction, a vector diagram can be drawn, as indicated at the top of Figure 13-12. Each line is a *vector line,* which contains information concerning *both* amplitude and direction. The three-unit line represents the direction of the boat and its speed (three units = 3 mph). The vertical line is the vector of the river current, which is traveling at a right angle to the boat at a speed of 4 mph (four units = 4 mph). To find

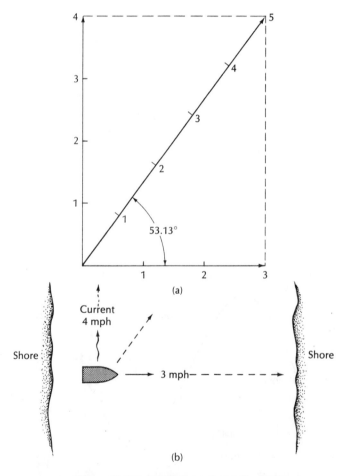

Figure 13-12 Developing a vector diagram.

the true speed *and* direction of the boat, a *parallelogram* is constructed as shown by the dashed lines. A line drawn from the point of origin to the intersection of the dashed lines specifies the true direction and speed of the boat. Note that the boat's speed is 5 mph, indicated by the five-unit line length. The boat's true direction is at an angle of 53.13° away from its intended direction, so that it will strike the far shore some distance downstream from the spot it set out for.

A very similar graphical construction can be arranged for the waveforms in Figure 13-8. A phasor is a graphical line, or construction, having the dimensions of both amplitude *and* time (or phase) difference. Comparing a vector and a phasor, both convey information as to amplitude, but the second dimension of a vector is *direction,* while the second dimension of a phasor is *time,* or *phase,* difference. Figure 13-13 shows the phasor diagram for the waveforms in Figure 13-8. The horizontal phasor represents the voltage waveform and is the *reference* phasor. Because phasors are rotated in a counterclockwise (CCW) direction, the current phasor is shown rotated 300° in that direction (or 60° CW), indicating that it *lags* the voltage.

13-8 NONSINUSOIDAL WAVEFORMS

Certain applications of alternating current require a sinusoidal waveform. Ac power distribution systems and audio-frequency and radio-frequency amplifiers are among these. Many other ac applications, however, utilize voltage and current waveforms that are *not* sinusoidal. These waveforms are called *pulse* waveforms, and a pulse is often defined as a waveform having any shape *except* that of a sinusoid. Several pulse waveforms are illustrated in Figure 13-14.

Note that these are ac waveforms in the strictest sense, since each is centered around a zero-volt base line. Hence, both positive and negative voltage polarities exist, and so the current reverses direction on each alternation (half cycle).

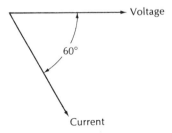

Figure 13-13 This vector diagram represents the voltage and current graphed in Figure 13-8.

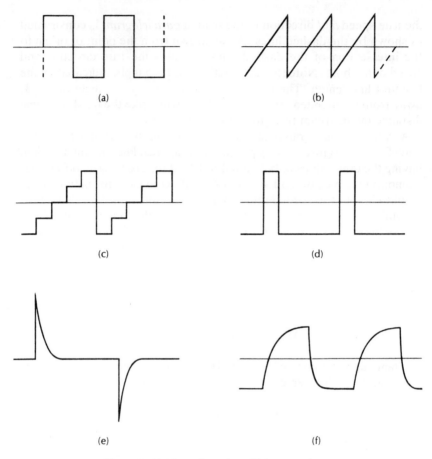

Figure 13-14 Several nonsinusoidal ac waveforms.

Figure 13-14*a* is a symmetrical *square wave* pulse train. Note the equal excursions above and below the zero-volt base line, as well as the equal time interval for each half cycle. For this waveform, as well as the others illustrated, the peak and peak-to-peak values are the only ones of interest. That is, the rms and average values have little, if any, application with nonsinusoidal waveforms. If these values are required, they are different for each waveform, and so must be determined separately for each instance. For example, the square wave shown in Figure 13-14*a* has rms and average values that are exactly the same, and both are equal to the peak value for a half-cycle period. The factors for converting one value of voltage to another, which are discussed in Section 13-3, apply *only* to sine waves.

Figure 13-14*b* is a *sawtooth* waveform, while *c* shows a *stairstep,* or *staircase,* waveform. Figure 13-14*d* shows a nonsymmetrical square wave, *e* is a *differentiated* waveform, and *f* is an *integrated* waveform.

Such waveforms are encountered primarily in *digital* equipment and, more often than not, appear as *signal* voltages rather than power waveforms. We shall later discuss some of these waveforms in detail, for two of them are necessary to understanding the function of inductors and capacitors. In fact, Figure 13-14*e* has already been covered briefly, in the discussion regarding the operation of an iron-core coil when a series switch is closed, then opened.

Many of the terms used to describe sine waves are applicable to nonsinusoidal waveforms. Peak and peak-to-peak values describe the same quantities as they do in reference to sine waves. One cycle of a repetitive waveform still refers to one complete sequence of events, such as base line to base line in a positive-going direction. Frequency (often called pulse repetition rate) and cycles per second (Hz) also have the same meaning as with sine waves, as do period and half period.

SUMMARY

- Alternating current is widely used for power distribution systems.
- Ac power is transported over long distances at very high voltages and at relatively small currents to reduce losses.
- A transformer is used to step voltages and currents up or down.
- Voltage and current waveforms of ac power are sinusoidal in shape.
- The instantaneous value of a sine wave, or sinusoid, can be found from $e = E \sin \theta$.
- Peak value is the maximum value attained.
- Rms value of a sine wave is 0.707 times peak value.
- Average value of a sine wave is 0.636 times peak value.
- Frequency *(F)* is the number of cycles per second.
- The time required for one complete cycle is the period *(P)*.
- Frequency and period are related by $F = 1/P$ and $P = 1/F$.
- A 100-W lamp will function identically if operated on either 120 V_{dc} or 120 V_{rms}.
- Audio frequencies extend from about 20 Hz to 20 kHz.
- Radio frequencies extend from 30 kHz to above 3000 kMHz.
- Radio waves travel at a speed of approximately 186,000 miles per second.
- Wavelength is the linear distance encompassed by one cycle of a radio wave.
- Wavelength (λ) = velocity (V)/frequency (F).
- Ac voltages and currents are frequently out of phase with each other.

- Current can either lead or lag the voltage.
- One radian equals 57.3°.
- A vector is a line (or quantity) having both amplitude and directional information.
- A phasor is a graphical construction (or quantity) having both amplitude and phase information.
- Alternating waveforms having shapes other than that of a sinusoid are called pulses.

QUESTIONS

1. A lamp rated at 120 V_{ac} at 40 W is energized. What value of current is drawn?
2. A lamp rated at 120 V_{ac} at 60 W is energized. What value of current is drawn?
3. An incandescent lamp rated at 120 V_{ac} is known to draw 0.1667 A. What is its wattage?
4. An incandescent lamp rated at 120 V_{ac} is known to draw 4.167 A. What is its wattage?
5. Briefly define a *vector*.
6. Briefly define a *phasor*.
7. Given the peak value of a sinusoid, what is the instantaneous value?
8. What is the range of audio frequencies?
9. Given a frequency of 1 MHz, what classification would this fall into?
10. A radian consists of how many degrees?
11. 180° equals how many rads?
12. A sine wave has a period of 60 μs. What is the angle if 25 μs has elapsed?
13. A sine wave has a partial period of 1 ms; at this point in time the angle is 90°. What is the full period? What is the frequency?

PROBLEMS

1. An ac generator provides energy for a load that draws 10 A at 120 V_{ac}. The distance between generator and load is 3 miles. The total cable resistance is 10 Ω. Find the required generator voltage and current in rms values.
2. Refer to problem 1. Find the peak voltage and current for the generator.
3. Refer to problem 1. Find the generator voltage and current in terms of average values.

4. An ac generator provides energy for a load that draws 2.3 A at 120 V_{ac}. The distance between generator and load is 10 miles. The total cable resistance is 13 Ω. Find the required generator voltage and current in rms values.
5. Refer to problem 4. Find the generator voltage and current in terms of peak values.
6. Refer to problem 4. Find the average voltage and current required for the generator.
7. Find the period of a waveform that has a frequency of 500 Hz.
8. Find the period of a waveform that has a frequency of 1.2 kHz.
9. Find the period of a waveform that has a frequency of 15 kHz.
10. Find the period of a waveform that has a frequency of 1.5 MHz.
11. Refer to Figure 13-4. A similar circuit is to be constructed, but using two 120-V, 100-W lamps connected in series. Find the required voltage.
12. Refer to Figure 13-4. A similar circuit is to be constructed, but using three 120-V, 100-W lamps in parallel. Find the required voltage and current.
13. A 60-Hz, 120-V_{ac} waveform is to be analyzed. Find the instantaneous value of voltage after 3 ms has elapsed on the positive excursion after crossing the zero-volt line.
14. A 400-Hz, 50-V_{ac} waveform is to be analyzed. Find the instantaneous value of voltage at 45°.
15. A certain broadcast antenna is to be ½ wavelength. The transmitter frequency is 560 kHz. Find the length of the antenna in feet.
16. A radio transmitter broadcasts on a frequency of 108 MHz. Find the wavelength in feet.
17. Two waveforms differ in phase by 114.59°. How many radians does this represent?
18. Two waveforms differ in phase by 160°. How many radians does this represent?
19. A portion of a sine wave is known to be 2 rads. How many degrees is this?
20. Two sine waves differ in phase by π rad. How many degrees is this?

CHAPTER 14

INDUCTANCE AND INDUCTIVE DEVICES

Inductance is the ability of a conductor to produce induced voltage when the current changes value. The unit of measure of inductance is the *henry (H)*, named for Joseph Henry (1797–1878). The amount of inductance depends upon many factors, and varies from millionths of a henry (μH) to hundreds of henrys. Devices exhibiting inductance include smoothing chokes; power, rf, and af transformers; and motors and generators. In this chapter, we study the physical characteristics of some of these devices, as well as their electrical characteristics. Fundamentally, an inductor is simply a coil of wire with either an air core or an iron core. However, the applications of inductors are many and varied.

In Chapter 12 you studied some of the effects of self-inductance. In this chapter, we apply these principles and add to them the effects of inductance in both dc and ac circuits. One of these effects is the phase difference between voltage and current that occurs when an inductor is placed in an ac circuit. One result of this phase difference is the *opposition* to alternating current produced by the counterelectromotive force. This opposition is called *inductive reactance*, which is similar in some ways to resistance, but which, as we shall see, differs from resistance in important respects.

The subjects to be discussed are listed below.

14-1 Inductance in Ac Circuits
14-2 Circuit Examples
14-3 Inductors in Series and Parallel
14-4 Transformers
14-5 Inductance in Dc Circuits

14-1 INDUCTANCE IN AC CIRCUITS

When a coil is placed in an ac circuit, it might be expected to react somewhat differently from an ordinary resistance, owing to its active opposition to a changing current. This is exactly true, and the circuit action is quite different from purely resistive circuit action.

One difference is the counterelectromotive force (CEMF) that is generated when current changes in a coil of wire, or inductor. The ability to produce CEMF when current changes is indicative of the electrical property called *inductance*. Inductance is symbolized by the letter L, and is measured in henrys. One henry (H) is that amount of inductance that causes a CEMF of 1.0 V to be induced if the rate-of-change of current is one ampere per second. For example, consider a coil or wire that has a linearly changing current of 0.25 A per second. That is, the current starts from zero and at the end of one second the current is 0.25 A, at the end of two seconds the current is 0.50 A, and so on. The CEMF is measured and found to be 1.0 V. What is the value of inductance in henrys?

$$L = \frac{E_L}{di/dt} = \frac{1}{0.25/1} = 4 \text{ H},$$

where L = inductance in henrys,
E_L = induced voltage (CEMF),
di = amount of change of current, and
dt = time during which the current is changing.

The foregoing formula states that the inductance in henrys is determined by the induced voltage (E_L), or CEMF, divided by the *rate of change of current, di/dt*. As before, the d represents delta (sometimes Δ is used) and means "a small change in. . . ." Hence, in this particular coil, the rate of change of current is 250 mA per second, which yields a CEMF of 1.0 V. The coil therefore has a value of inductance of 4 henrys.

By rearranging the formula for inductance, a useful variation is formed:

$$E_L = L \frac{di}{dt}$$

This expression allows us to determine the CEMF if the inductance and the rate of change of current are known. For example, a 14-mH coil is energized with a current that changes linearly (in a straight line) from 2 mA to 7 mA in 30 μs. Determine the CEMF induced across the coil.

$$E_L = L \frac{di}{dt} = 0.014 \times \frac{7 \text{ mA} - 2 \text{ mA}}{30 \times 10^{-6}} = 2.33 \text{ V}.$$

In order for the induced voltage to have a constant value, the current must increase or decrease linearly. A graph of such circuit action produces the waveforms shown in Figure 14-1. If the current increases and decreases linearly the induced CEMF will have the shape of a square wave. Such waveforms, however, are seldom encountered. Usually,

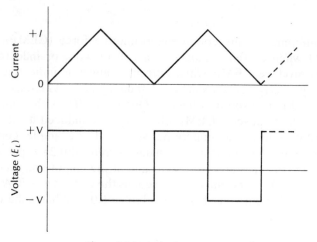

Figure 14-1 Inductive current waveforms.

the property of inductance is used in alternating current circuits, and the circuit effects are distinctive.

Characteristics of Inductive Ac Circuits

One characteristic of inductance in ac circuits is that the induced voltage is also sinusoidal. This is true only with a sinusoidal current. That is, a sinusoidal current is the only waveshape that produces the same shape of induced voltage. Any other current waveshape produces a CEMF that does not resemble the waveshape of the current. Another important characteristic is phase difference—voltage and current are *not* in phase with each other, as illustrated in Figure 14-2a. Here, the ac generator voltage and the current through the inductor are shown to be 90° apart in time. The current is said to *lag* the voltage by 90°. (We could say that the voltage leads the current by 90°.) The circuit that generates these waveforms is shown in part *b* of the figure.

The voltage leads the current because it depends on the *rate of change of current* through the coil. Figure 14-2a illustrates this fact. Note the current (*i*) waveform: its rate of change is maximum *as it crosses the zero base line*. Hence, induced voltage is maximum as current approaches and crosses zero. On the other hand, as the current peaks, the rate of change is minimum and so induced voltage is zero.

To illustrate the concept of rate of change, study Figure 14-2c. The waveform of current shown represents the current of Figure 14-2a. As the current waveform crosses the base line, the rate of change *(di/dt)* is

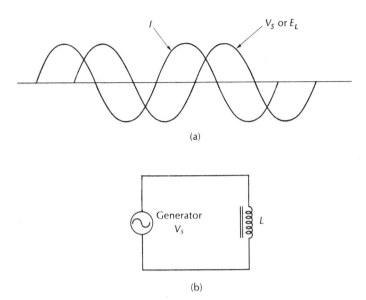

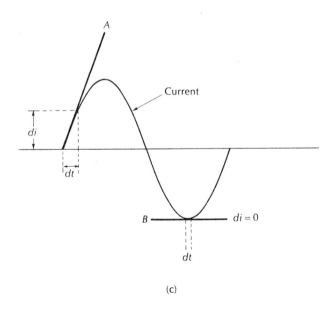

Figure 14-2 (a) The 90° phase difference between voltage and current in a purely inductive circuit; (b) the circuit that produces these waveforms; (c) illustrating the rate of change.

greatest. Note line A, which is drawn tangent to the waveform in this area. Line A is nearly parallel to the waveform, which shows a relatively large current change in a small amount of time. During this time, then, induced voltage is large.

However, line B is drawn tangent to the negative peak, and for a short time, the change in current is essentially *zero*. Hence, induced voltage is zero. Thus, near the peak current values the induced voltage is low (or zero at the very peak); and at the zero-crossing points, the current *change* is maximum. So CEMF is maximum, causing the 90° phase difference shown in Figure 14-2a.

The foregoing discussion is true only if the inductor has pure inductance with *no* resistance. Obviously, it is not possible to wind a copper-wire coil with zero resistance. All practical inductors have some degree of resistance, and the effect of this resistance is to *reduce the angle of lag*. That is, the inductance of the coil tends to make the current lag the voltage by 90°; while the dc resistance of the coil tends to reduce the angle of lag to less than 90°. The relative proportion of inductance to resistance determines the actual angle of lag. The more the resistance, the smaller the angle—if there is *no* inductance but only pure resistance, the angle is 0°.

In order to determine the angle of lag, we apply trigonometric relationships, which will also allow us to determine the total opposition to current offered by the inductance and resistance, as well as the voltage across each element. The following discussion of trigonometry is of sufficient depth to allow you to work all problems in this book. A summary of trigonometric functions and relationships as they concern your studies appears in Appendix 3, in case you feel that you need a brief review.

Because an inductor tends to cause a current lag of 90°, while circuit resistance tends to cancel this effect, we are very concerned with the study of angles, specifically the *right triangle* (a triangle containing a right angle), since this accurately represents the electrical quantities found in a simple ac circuit having the property of inductance.

As briefly mentioned in the previous chapter, the right triangle can be used to solve vector or phasor problems where the quantities have a directional or angular difference of 90°. Recall the earlier discussion regarding vectors and phasors, which are representations of quantities having both magnitude and either direction (vector) or phase difference (phasor). The phasor is very useful in helping to determine electrical quantities that occur in circuits such as those you are currently studying.

Consider Figure 14-3a, which shows a simple series circuit having an inductor and a resistance connected to a source of 60-Hz power. In order to determine the electrical quantities in a circuit such as this, it must be understood that the voltage drops across the inductor and the

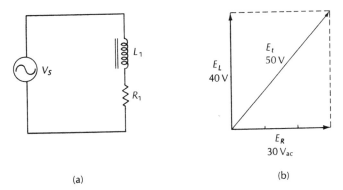

Figure 14-3 (a) An ac circuit and (b) the phase relationship for this circuit.

resistor are *not* in phase. This can be illustrated by the use of a phasor diagram, Figure 14-3b. The voltage across the inductor (assuming negligible resistance) is 40 V, while the voltage across the resistance is 30 V. If this were a dc circuit, you would determine the total applied voltage to be 40 + 30 = 70 V. *This is not true in the ac circuit shown.* The actual applied voltage is 50 V. To see why this is so and to determine the phase difference between the two voltage drops, we must apply the concepts of trigonometry.

Trigonometric Concepts

A right triangle is shown in Figure 14-4. The angle in which we are interested is the one labeled θ (theta). In most circuit analysis, θ represents an unknown quantity. Two sides of the triangle are labeled the *side adjacent* (to the angle θ) and the *side opposite*. The line connecting the ends of the two sides is the *hypotenuse*. The Greek mathematician Pythagoras, who lived in the sixth century B.C., discovered the relationship between the length of the sides and the length of the hypotenuse. This relationship has come to be known as the *Pythagorean theorem*, which states that the square of the hypotenuse is equal to the sum of the squares of the other two sides. In formula form we have

$$\text{hypotenuse}^2 = \text{side adjacent}^2 + \text{side opposite}^2,$$

or

$$c^2 = a^2 + b^2.$$

As an example, in Figure 14-4 the adjacent side *(a)* is 3 units long, the opposite side *(b)* is 4 units long, and the hypotenuse *(c)* is 5 units long. Substituting these values in the equation gives

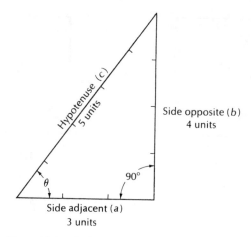

Figure 14-4 The basis of trigonometric relationships.

$$c^2 = a^2 + b^2$$
$$5^2 = 3^2 + 4^2$$
$$25 = 9 + 16$$
$$25 = 25$$

Now, seldom are all three quantities known. Often, the hypotenuse is the unknown, and the equation can be rearranged so that we can solve for c:

$$c = \sqrt{a^2 + b^2}.$$

Again using the triangle in Figure 14-4, assume that the hypotenuse is unknown. Then

$$c = \sqrt{a^2 + b^2} = \sqrt{3^2 + 4^2} = \sqrt{9 + 16} = \sqrt{25} = 5.$$

If the hypotenuse is known and one of the other sides is not, the basic equation may again be adjusted as follows:

$$a = \sqrt{c^2 - b^2},$$
$$b = \sqrt{c^2 - a^2}.$$

In electrical circuits, a, b, and c represent the various values of the circuit itself. The a and b sides can represent any two related values having phasor quantities that operate at a 90° phase difference. When this is done, c represents the result of combining the a and b values. The angle θ is the actual phase difference, often the difference in phase between two voltages, or between a voltage and a current.

Given any two sides, several additional relationships are used to find the phase angle θ. The *ratio* of any two sides gives information relating

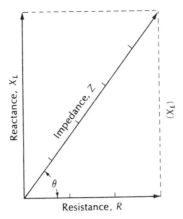

Figure 14-5 Phasor relationships in an inductive circuit.

to the angle θ. These ratios are given the names *sine* (sin), *cosine* (cos), and *tangent* (tan). The ratios are

$$\sin \theta = \frac{b}{c},$$

$$\cos \theta = \frac{a}{c},$$

$$\tan \theta = \frac{b}{a}.$$

For the triangle in Figure 14-4, these ratios are

$$\sin \theta = \frac{b}{c} = \frac{4}{5} = 0.8,$$

$$\cos \theta = \frac{a}{c} = \frac{3}{5} = 0.6,$$

$$\tan \theta = \frac{b}{a} = \frac{4}{3} = 1.33.$$

These ratios allow us to find the angle θ. Refer to the trigonometric function tables in the Appendix. Under the column headed sin θ, the nearest value to 0.8 is .799, opposite which is given the angle of 53°. Hence, the unknown angle is approximately 53° (actually 53.130102°). Note that the same angle can also be read from either the cosine or tangent column, opposite the appropriate ratio.

Now, to relate these relationships to an electrical circuit, the sides of the triangle are renamed to agree with the quantities found in the circuit. Figure 14-5 shows the same right triangle given in Figure 14-4, but relabeled to agree with the electrical circuit given in Figure 14-3a. The only difference between this drawing and that of Figure 14-4 is that the line representing X_L is moved from the right-hand side to the left, so that all quantities start from the same point. The side adjacent to the angle is labeled resistance, and its length is proportional to the magnitude of resistance. The side opposite is labeled *reactance,* specifically *inductive reactance,* X_L. Inductive reactance is the opposition to alternating current offered by the inductor, and is caused by the CEMF of the coil. Like resistance, X_L is measured in ohms. Inductive reactance is drawn at a 90° angle to the resistance simply because of the 90° phase shift caused by the inductor.

The hypotenuse is shown as *impedance, Z,* which is the *total combined opposition* to alternating current offered by the total circuit. As before, the angle θ defines the phase difference between applied voltage and circuit current. The magnitude of inductive reactance is determined by the frequency of the applied ac voltage and by the amount of inductance (in henrys) of the coil. This relationship is

$$X_L = 2\pi FL,$$

where X_L = inductive reactance (in ohms),
2π = 6.283,
F = frequency in hertz, and
L = inductance in henrys.

Refer to Figure 14-6. Using this circuit as an example, we shall assign values to the components to illustrate how these ac relationships are determined. The source is 120 V_{ac}, 60 Hz. The resistor has a value of 30 Ω, and the inductor has a value of 0.1061 H. First, the inductive reactance will be determined.

$$X_L = 2\pi FL = 6.283 \times 60 \times 0.1061 = 39.99 \cong 40 \ \Omega.$$

The resistor has a resistance of 30 Ω, and the inductor has a reactance of 40 Ω; these two quantities *cannot* be simply added together, because their effects in the circuit operate at a 90° angle to each other. By using the phasor diagram shown in Figure 14-6b, their combined effects are seen to be 50 Ω. In place of a phasor diagram, we could apply the Pythagorean theorem to determine the impedance.

$$Z = \sqrt{R^2 + X_L^2} = \sqrt{30^2 + 40^2} = \sqrt{900 + 1600} = \sqrt{2500} = 50 \ \Omega.$$

To determine the phase angle between the applied voltage and current, any of the following relationships can be used, depending on which two factors are known.

$$\sin \theta = \frac{X_L}{Z} = \frac{40}{50} = 0.8,$$

$$\cos \theta = \frac{R}{Z} = \frac{30}{50} = 0.6,$$

$$\tan \theta = \frac{X_L}{R} = \frac{40}{30} = 1.333.$$

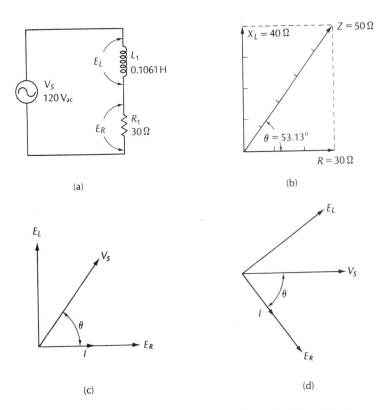

Figure 14-6 (a) The ac relationships in this ac circuit are given by (b) the phasor diagram. (c) Composite phasor for the five quantities shown, with total current as the reference, (d) Composite phasor with V_s as reference.

Referring to the trig tables reveals that the angle for this circuit, using any of the three ratios, is approximately 53°.

In this instance, 53° is the angular difference between the voltage across the inductor and the current through it. Because this is a series circuit, the current at any instant has the same value throughout, and so can be thought of as a reference for all angular measurements. Thus, the voltage across the inductor *leads* the current by 90°, since, in an inductive circuit, current lags. Also, the voltage across the inductor leads the voltage across the resistor, also by 90° because the voltage across the resistor is proportional to, and in phase with, the current through it. Figure 14-6c is a composite phasor diagram illustrating the phase relationships between the five quantities shown. The angle between I_t and E_R is zero, while the angle between I_t (or E_R) and E_L is 90°. The angle between V_s and I_t (or E_r) is θ (53° in the numerical example). Figure 14-6d illustrates the same phase relationships if V_s is considered to be reference.

The amount of current drawn by this circuit is determined by the voltage and the impedance. Note that the following relationship is equivalent to the Ohm's law relationship for dc circuits. Assume $V_s = 120 \text{ V}_{ac}$. Then

$$I = \frac{V_s}{Z} = \frac{120}{50} = 2.4 \text{ A}.$$

Because this is a simple series circuit, the current is the same, at any instant, measured anywhere in the circuit. However, the magnitude of the voltage drops across the resistor and inductor depends on the opposition of each device. Furthermore, these voltage drops are *not* in phase with each other.

To find the voltage drop across each component, total circuit current and R or X_L values are used:

$$E_R = I \times R = 2.4 \times 30 = 72 \text{ V},$$
$$E_L = I \times X_L = 2.4 \times 40 = 96 \text{ V}.$$

Note that you cannot add these values to determine total voltage: 72 + 96 does not equal 120 V_{ac}. Because there is a *phase difference* between the two voltage drops, they cannot be simply added. While it is true that a voltmeter would indeed measure 96 V across the coil and 72 V across the resistor, it is equally true that the same meter will read 120 V if placed across *both* components. Figure 14-7 illustrates this phase difference. Analyzing these waveforms will allow us to make a general statement regarding the reason why the voltage across each component appears to be greater than one might think. Note the portion of the curves just to the right of 180°. While E_{L_1} is still in the positive region, E_{R_1} is already into the negative region. Hence their sum at this instantaneous value is much less than when both are positive or negative together. This occurs every half-cycle, and thus the *total* voltage is much less than the simple sum of the voltage drops. Note that the greater the angular difference between voltage and current, the greater the difference between the arithmetic sum and the true total voltage.

The applied voltage in such a circuit is related to E_R and E_L as follows:

$$V_s = \sqrt{E_R^2 + E_L^2}.$$

By rearranging this equation, either of the other two values may be solved for:

$$E_R = \sqrt{V_s^2 - E_L^2}$$

and

$$E_L = \sqrt{V_s^2 - E_R^2}.$$

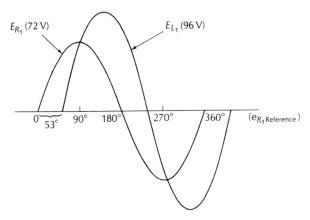

Figure 14-7 The graph of phase difference between E_{R_1} and E_{L_1} for the circuit in Figure 14-6.

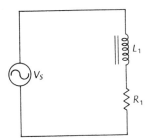

Figure 14-8 A circuit for the examples in Section 14-2.

14-2 CIRCUIT EXAMPLES

Several circuit examples follow to illustrate the preceding relationships and to show their application. In each case, refer to the circuit in Figure 14-8.

Example 1. $L_1 = 0.1591$ H, $R = 30\ \Omega$, and $V_s = 120\ V_{ac}$, 60 Hz. Find the values of X_L, Z, I_t, and θ.

$$X_L = 2\pi FL = 6.283 \times 60 \times 0.1591 = 59.98 \cong 60\ \Omega.$$
$$Z = \sqrt{X_L^2 + R^2} = \sqrt{60^2 + 30^2} = 67\ \Omega.$$
$$I_t = \frac{V_s}{Z} = \frac{120}{67} = 1.79\ A.$$
$$\sin\theta = \frac{X_L}{Z} = \frac{60}{67} = 0.896,$$
$$\text{arc}\sin\theta = \sin^{-1}\theta = 63.576°;$$

or

$$\cos\theta = \frac{R}{Z} = \frac{30}{67} = 0.448,$$
$$\text{arc } \cos\theta = \cos^{-1}\theta = 63.4°;$$

or

$$\tan\theta = \frac{X_L}{R} = \frac{60}{30} = 2.0,$$
$$\text{arc } \tan\theta = \tan^{-1}\theta = 63.434°.$$

Note that *arc sin* means "the angle whose sine is. . . ." Also, $\sin^{-1}$ is simply an abbreviation for arc sin, and similarly with the other functions. (See Appendix C for further information.) Also note that the slight difference in answers above is due to round-off errors.

Example 2. $L_1 = 0.1591$ H, $R = 30\ \Omega$, and $V_s = 120\ V_{ac}$, 400 Hz. Find the values of X_L, Z, I_t, θ.

$$X_L = 2\pi FL = 6.283 \times 400 \times 0.1591 = 399.9 \cong 400\ \Omega.$$
$$Z = \sqrt{X_L^2 + R^2} = \sqrt{400^2 + 30^2} = 401.1\ \Omega.$$
$$I_t = \frac{V_s}{Z} = \frac{120}{401.1} = 0.299 \cong 0.3\ A.$$
$$\sin\theta = \frac{X_L}{Z} = \frac{400}{401} = 0.9975$$
$$\text{arc } \sin\theta = 85.95 \cong 86°.$$

Note that inductive reactance increases as a function of frequency; and that as inductive reactance increases, total opposition Z is less influenced by R and more influenced by X_L. Also note how close the angle in the preceding example is to 90°. This is, of course, due to the relatively small resistance (30 Ω) compared to reactance (400 Ω). In the next example, observe what occurs when resistance is large and reactance is small.

Example 3. $R_1 = 1000\ \Omega$, $L = 0.00159$ H, and $V_s = 25\ V_{rms}$, 1000 Hz. Find the value of X_L, Z, I_t, θ.

$$X_L = 2\pi FL = 6.283 \times 1000 \times .00159 = 9.99 \cong 10\ \Omega.$$
$$Z = \sqrt{X_L^2 + R^2} = \sqrt{10^2 + 1000^2} = \sqrt{1000100} = 1000.05\ \Omega.$$
$$I_t = \frac{V_s}{Z} = \frac{25}{1000.05} = 0.025\ A.$$
$$\sin\theta = \frac{X_L}{Z} = \frac{10}{1000.05} = 0.009999.$$
$$\text{arc } \sin\theta = 0.573°.$$

In this example, the amount of inductive reactance is very small compared to the resistance, and it therefore does not influence the circuit to any significant extent. While the amount of inductance is quite low, it may become significant *if* the frequency is increased. The next example illustrates this point.

Example 4. $R_1 = 1000 \ \Omega$, $L = 0.001$ H, and $V_s = 25$ V, 159 kHz. Find the value of X_L, Z, I_t, θ.

$$X_L = 2\pi FL = 6.28 \times (1.59 \times 10^5) \times (1 \times 10^{-3})$$
$$= 999 \cong 1000 \ \Omega.$$
$$Z = \sqrt{X_L^2 + R^2} = \sqrt{1000^2 + 1000^2} = 1414 \ \Omega.$$
$$I_t = \frac{V_s}{Z} = \frac{25}{1414} = 0.0177 \text{ A}.$$
$$\sin \theta = \frac{X_L}{Z} = \frac{1000}{1414} = .7072136$$
$$\text{arc} \sin \theta = 45°$$

In this example, the resistance and inductance values are nearly the same as in the preceding one, yet the inductive effects are *much* larger. This, of course, is due to the much higher frequency.

Example 5. $R_1 = 1000 \ \Omega$, $Z = 2000 \ \Omega$, and $I_t = 0.011$ A. Find the value of X_L and V_s.

Since $Z^2 = X_L^2 + R^2$, then $X_L^2 = Z^2 - R^2$ and

$$X_L = \sqrt{Z^2 - R^2} = \sqrt{2000^2 - 1000^2} = 1732 \ \Omega.$$
$$V_s = I_t \times Z \left(\text{since } E = I \times R, \text{ and } I_t = \frac{V_s}{Z} \right) = 0.011 \times 1732 = 19.05 \text{ V}.$$

14-3 INDUCTORS IN SERIES AND PARALLEL

When inductors are connected in series or parallel, whether or not they are treated in much the same way as resistors depends on whether any flux linkage exists between inductors. If any flux from one coil cuts across a second coil, the amount of inductance can be changed. Figure 14-9 illustrates several such arrangements that assume *no* flux linkage.

The first two parts of the figure illustrate two series arrangements. If no flux linkage exists, the inductances are summed as are resistances.

$$L_t = L_1 + L_2 + L_3 \ldots$$

This is true because the series ac current is the same in each coil, and therefore the induced voltage is dependent upon the total number of turns and the combined core characteristics.

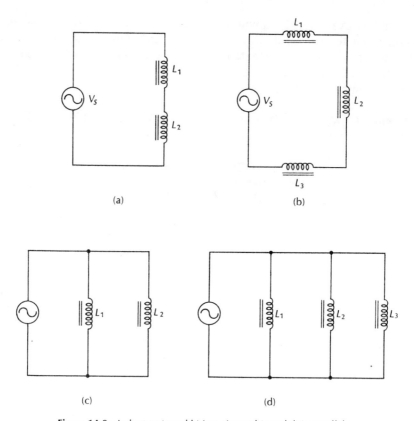

Figure 14-9 Inductors (a and b) in series and (c and d) in parallel.

When inductors are placed in parallel, the total inductance is computed by the familiar reciprocal formula. In Figure 14-9c and d,

$$L_t = \frac{1}{1/L_1 + 1/L_2 + 1/L_3 \ldots},$$

or

$$L_t = \frac{L_1 \times L_2}{L_1 + L_2}$$

for a circuit containing two inductors in parallel.

In addition to the value of inductance, reactance must often be considered. When inductors are connected in series, their inductive reactance is simply the sum of the individual reactances.

$$X_{L_t} = X_{L_1} + X_{L_2} \ldots$$

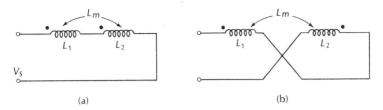

Figure 14-10 Determining mutual inductance.

When inductors are placed in parallel, the total reactance is computed by the same relationship that is used for resistors in parallel.

$$X_{L_t} = \frac{1}{1/X_{L_1} + 1/X_{L_2} \ldots},$$

or

$$X_{L_t} = \frac{X_{L_1} \times X_{L_2}}{X_{L_1} + X_{L_2}}$$

for a circuit containing two inductors. As is the case with resistance, the total reactance is less than the smallest individual value.

The situation is slightly more complicated when flux from one coil *links* with flux from another. The actual linkage is termed *mutual inductance*, L_m. To determine the mutual inductance in two coils such as shown in Figure 14-10, use the following procedure. In part *a*, the coils shown are in series-aiding, which means that the flux of each aids the other (or is in the same direction). That is, at any instant of time, both left ends are north or south simultaneously, as indicated by the dots.

While connected in series-aiding, the total inductance is measured and the total recorded; this value is symbolized $L_{t(\text{aid})}$. Next the connections to each coil are reversed, as shown in Figure 14-10*b*, and the total inductance is measured again. This value is indicated as $L_{t(\text{opp})}$. The mutual inductance can then be determined by the following relationship:

$$L_m = \frac{L_{t(\text{aid})} - L_{t(\text{opp})}}{4}.$$

Once the mutual inductance is found, the total *effective* inductance can be calculated.

$$L_t = L_1 + L_2 \pm 2L_m.$$

Note that a positive L_m indicates a value that *increases* total inductance, while a negative L_m *decreases* total inductance.

Now, once L_m is given a value, the *coefficient of coupling (k)* can be determined. This factor is a measure of the number of lines of force linking the two coils compared to the total flux developed by one coil. If *all* the flux lines from one coil link the other, $k = 1.0$. If none link, $k = 0.0$. To determine k, use the following relationship:

$$k = \frac{L_m}{L_1 L_2}.$$

A few examples are given below to illustrate the application of these relationships.

Example 6. Two coils are to be placed in a series circuit similar to Figure 14-10. Each is a 750-μH coil. When connected in series-aiding, the total inductance is 2000 μH; and in series-opposing, the total inductance measures 1000 μH. Find the value of L_m and k.

$$L_m = \frac{L_{t(\text{aid})} - L_{t(\text{opp})}}{4} = \frac{2000\ \mu\text{H} - 1000\ \mu\text{H}}{4} = 250\ \mu\text{H}.$$

$$k = \frac{L_m}{\sqrt{L_1 L_2}} = \frac{250\ \mu\text{H}}{\sqrt{(750\ \mu\text{H})(750\ \mu\text{H})}} = 0.333.$$

Example 7. Two coils are to be placed in a series circuit similar to Figure 14-10. L_1 is a 750-μH coil and L_2 is a 1000-μH coil. In series-aiding, L_t is measured as 2000 μH; in series-opposing L_t is 1500 μH. Find L_m and k.

$$L_m = \frac{L_{t(\text{aid})} - L_{t(\text{opp})}}{4} = \frac{2000 - 1500}{4} = 125\ \mu\text{H}.$$

$$k = \frac{L_m}{\sqrt{L_1 L_2}} = \frac{125\ \mu\text{H}}{\sqrt{(750\ \mu\text{H})(1000\ \mu\text{H})}} = 0.144.$$

14-4 TRANSFORMERS

A device designed to have the greatest possible mutual inductance is the *transformer*. Briefly mentioned earlier, a transformer is used mainly to step up (or down) *alternating voltage* or *current*. A transformer consists of two or more windings (coils) generally wound upon an iron core, although not necessarily so. Such iron-core devices are shown in Figure 14-11 for two configurations, along with the schematic diagrams. Figure 14-11a, b, and c illustrate a simple two-winding transformer. If this transformer is a step-up transformer, drawing b is applicable, while c shows a step-down configuration. Unless otherwise specified, a transformer is a step-up or step-down transformer according to its action on the *voltage*. As you will find, a transformer that steps up voltage will step down current, and vice versa. (The number of turns shown on the coils in the schematic is only relative and does not repre-

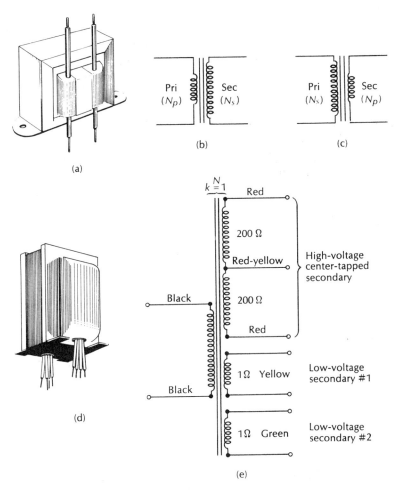

Figure 14-11 (a) A typical iron-core power transformer; (b) the schematic for a step-up transformer; (c) the schematic for a step-down transformer; (d) a typical transformer with multiple secondary windings; and (e) the schematic for the transformer pictured in (d).

sent the true number of turns on any coil.) Figure 14-11d and e show a typical power transformer with multiple secondary windings, each of which delivers power to a separate circuit.

The coil that is *driven* by a source is called the *primary* winding, while the coil that drives the load is known as the *secondary* winding. Whether the transformer is step up or step down depends on the number of turns on each winding. Because of the iron core, the coefficient of coupling k is unity for many iron-core devices, and only the number of turns influences the output voltage (for a fixed input voltage).

If both primary and secondary windings have the same number of turns, the output voltage is the same as the input voltage. That is, if 120 V_{ac}, 60 Hz is impressed on the primary, the secondary voltage will also be 120 V_{ac}, 60 Hz. However, if the *turns ratio* (the number of turns on the secondary winding divided by the number of turns on the primary winding [N_s/N_p]) is not unity, output voltage is different from input.

Example 8. A transformer has 1000 turns on the primary and 400 turns on the secondary. What is the turns ratio?

$$\text{turns ratio} = \frac{N_s}{N_p} = \frac{400}{1000} = 0.4.$$

This means simply that the transformer secondary voltage is 4/10 as large as the primary voltage, or that the primary voltage is 2.5 times larger than the secondary. If 120 V_{ac} is applied to the primary, the secondary voltage will be $120 \times 0.4 = 48$ V. Thus, this device is a step-down transformer. An example of a step-up transformer follows.

Example 9. The number of turns on the primary of a transformer is 400, and the number of turns on the secondary is 1000. Therefore, the turns ratio is $N_s/N_p = 1000/400 = 2.5$. In this case, the secondary voltage is 2.5 times larger than input voltage; if 120 V_{ac} is supplied to the primary, secondary voltage is $120 \times 2.5 = 300$ V. The above relationship can be expressed by the following proportions:

$$E_s = \frac{E_p N_s}{N_p}$$
$$N_p = \frac{E_p N_s}{E_s}$$
$$E_p = \frac{E_s N_p}{N_s}$$
$$N_s = \frac{E_s N_p}{E_p}$$

Example 10. An iron-core transformer has 500 turns on the primary and 1200 turns on the secondary. The primary voltage is 120 V_{ac} at 0.5 A. What is the secondary voltage?

$$E_s = \frac{E_p N_s}{N_p} = \frac{120 \times 1200}{500} = 288 \text{ V}.$$

Current values are equally important, and they can also be derived from this proportion:

$$\frac{E_s}{E_p} = \frac{N_s}{N_p} = \frac{I_p}{I_s}$$

With this formula, knowing either the voltage ratio or the turns ratio allows secondary current to be found if primary current is known, or vice versa. For example, what is the secondary current in the preceding example?

$$\frac{E_s}{E_p} = \frac{I_p}{I_s} \quad \text{or} \quad \frac{N_s}{N_p} = \frac{I_p}{I_s}$$

and cross-multiplying gives

$$E_s I_s = E_p I_p.$$

Then dividing both terms by secondary voltage yields

$$I_s = \frac{E_p I_p}{E_s} = \frac{120 \times 0.5}{288} = 0.208 \text{ A}.$$

Since the voltage is stepped up, the current must be stepped down. This rule is of vital importance. To verify it, remember that the amount of electrical power dissipated by the secondary load must be the same as the power delivered by the primary (excluding any efficiency losses). Figure 14-12 illustrates this point, using the same transformer as before, but with a 576-Ω resistive load connected to the secondary. It is this load that determines the current drawn by the transformer from the source. Secondary voltage has already been determined to be 288 V. The load resistor will therefore draw some value of current.

$$I_{R_L} = \frac{E_{R_L}}{R_L} = \frac{288}{576} = 0.5 \text{ A}.$$

This secondary current (0.5 A) must draw some value of primary current consistent with the turns ratio:

$$I_p = \frac{N_s I_s}{N_p} = \frac{1200 \times 0.5}{500} = 1.2 \text{ A}.$$

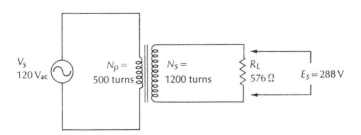

Figure 14-12 A transformer with a resistive load.

Now, secondary power and primary power must be equal. That is, $E_s \times I_s = E_p \times I_p$. To prove this, use the preceding example and simply substitute values:

$$E_s \times I_s = E_p \times I_p$$
$$288 \times 0.5 = 120 \times 1.2 \text{ and } 144 = 144.$$

If secondary power were greater than primary power, the transformer would be producing something for nothing, which, of course, is not possible.

Transformer Efficiency

The preceding example, based on Figure 14-12, assumes a transformer that is 100% efficient. That is, power out *exactly* matches power in. While transformer efficiencies are quite good, perhaps from 80% to 98%, certain losses do occur that detract from perfect operation. These losses are of four types: (1) I^2R losses, (2) eddy-current losses, (3) hysteresis losses, and (4) flux losses.

I^2R losses are those caused by the copper wire in each winding. Each coil usually has a very appreciable resistance due to the long length of wire used in the winding. Hence, to minimize such losses, the largest size of wire is used that is consistent with size and weight requirements, and maximum rated current in each winding. I^2R losses are evidenced as heat, and are one reason why transformers become warm during operation. In fact, temperature rise is a measure of whether a transformer has excessive losses or not. When a transformer becomes too warm, the insulating paper or plastic used between windings begins to deteriorate, as does the enamel coating on the wire itself. This can lead to shorts within the transformer.

Eddy currents are electrical currents induced in the iron core itself. Since the varying magnetic field cuts across the iron core, the field induces currents in the core. If the core were made of solid iron, these currents would be quite large, which would result in excessive losses. To reduce eddy currents, the core is *laminated,* made of thin slices or sheets of core material. This has the effect of causing the resistance of the core material to increase by reducing the cross-sectional area, thus reducing the eddy currents and the consequent losses. Also, certain alloys possess inherently higher resistance, and so reduce eddy currents still further. Figure 14-13 illustrates a laminated iron core. The thin sheet-like shape is evident, but not apparent in the drawing is the fact that each lamination is coated with an oxide film, so as to reduce contact between laminations. This constrains current flow to each lamination, since each is effectively insulated from adjacent laminations.

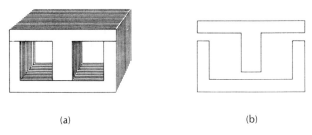

Figure 14-13 Iron-core lamination construction.

Hysteresis losses are those caused by the hysteresis of the core material itself. Recall from an earlier discussion that energy must be expended to remagnetize a material in successively opposite directions. The area enclosed by the hysteresis loop is a direct measure of the hysteresis loss. Hysteresis loss is usually described as a loss due to friction of the magnetic dipoles as they are continually switched end for end. Good core material, then, has as small a coercive force (residual magnetism) as possible.

The fourth loss listed above is that of loss of flux. Any coil or transformer that is operated in the vicinity of any metallic or conducting surface will suffer some flux linkage with the surface. The energy required to magnetize the material, or to induce a current therein, is a loss. All four types of losses occur in iron-core devices, but only I^2R and flux losses occur in devices having an air core.

Eddy currents and hysteresis losses are both apparent as heat, although it is the core itself, rather than the copper wire winding, that becomes heated. When I^2R losses, eddy-current losses and hysteresis losses are considered, the net effect is to raise the temperature of the device, *and* to reduce both output voltage and current. In fact, total loss can be determined by measuring both currents and voltages (primary and secondary) and applying the following relationship:

$$\% \text{ eff} = \frac{P_{\text{out}}}{P_{\text{in}}} \times 100,$$

where P_{out} = power output and
P_{in} = power input.

Example 11. A transformer energized with 120 V_{ac} at 0.5 A delivers to the secondary load a voltage of 238 V at 0.24 A. What is the efficiency? What is the power loss?

$$\% \text{ eff} = \frac{P_{\text{out}}}{P_{\text{in}}} \times 100 = \frac{57.12}{60} \times 100 = 95.2\%$$

power loss = power in − power out = 60 − 57.12 = 2.88 W.

Thus, 2.88 W represents the *total* power loss and is the sum of the I^2R, eddy-current, and hysteresis losses. The transformer will then produce heat according to the 2.88 watts being expended.

Transformer Types

Transformers are constructed in a wide variety of sizes and shapes, as well as core material. Those illustrated thus far have been of the iron-core type, but you will encounter other types as well. The end use determines the physical characteristics of a given transformer. To be able to handle large amounts of power, both the core and the windings must be correspondingly large; hence physical size is a good measure of the designed power level. Two factors determine the size of a particular unit: the allowed temperature rise, and the possibility of *core saturation*. If the current level rises excessively, the core may saturate. A saturated core will fail to produce changes in flux of a magnitude sufficient to provide good transformer action. When this happens, the output power levels off *and* the sinusoidal waveshape becomes distorted. For a given amount of power delivered to the load, the more massive the core and winding structure, the more efficient the operation is, and the less heat lost.

Transformers without iron cores are widely used in signal-type circuits, not so much for transfer of power, but to transfer signals efficiently and to step signal voltage or current up or down. A typical air-core transformer is shown in Figure 14-14. The primary and secondary are wound separately on a cardboard or plastic form, with the coil ends soldered to terminal lugs. Such a device might have a coefficient of coupling of 0.1 to 0.2, so the transfer of energy is not 100% efficient. However, at the radio frequencies normally used, perhaps in the megahertz range, transfer of energy is adequate.

To increase the efficiency of RF signal transformers, a core made of powdered iron or ferrite is often used as illustrated in Fig. 14-15. A core suitable for 60-Hz power-line use cannot be used at significantly higher frequencies, since core losses would be much too high. A *powdered-iron* core consists of particles of granulated iron compressed into a solid mass and held together with a binding agent. Because the individual particles are so small, eddy currents are reduced to very low values; hence the losses are correspondingly small. If the individual particles are coated with an oxide, losses are smaller still. A powdered-iron core can be effectively applied in applications having frequencies in the megahertz region.

Ferrites are ceramic materials having excellent magnetic qualities. Since these materials are also *insulators,* eddy-current losses are reduced to essentially zero; hence they are useful with *very* high frequencies.

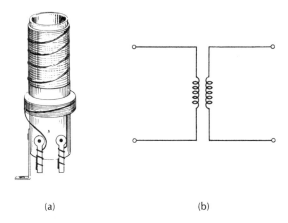

Figure 14-14 (a) A typical air-core transformer and (b) the schematic.

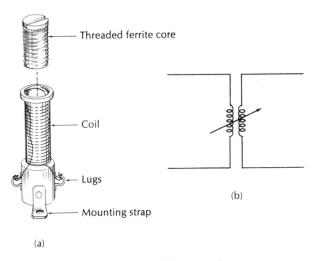

Figure 14-15 A variable RF transformer.

Powdered-iron or ferrite cores are often used with transformers appearing much like the one shown in Figure 14-15. The core is placed inside the hollow coil form; due to the threading moulded as part of the core, the core can be screwed in or out to change the total inductance offered by the coil.

An *autotransformer* is a special kind of transformer having step-up or step-down qualities but *only one coil*. Such a device is shown schematically in Figure 14-16. Drawing *a* is a step-up transformer, since the

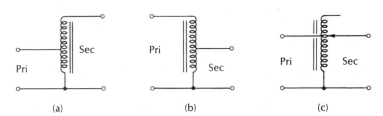

Figure 14-16 The schematics for (a) a step-up transformer, (b) a step-down transformer, and (c) a variable output autotransformer.

secondary has more turns than the primary, while *b* represents a step-down transformer. Drawing *c* is a variable autotransformer. This device is useful whenever the ac line voltage must be set to an exact value.

The autotransformer has one distinct disadvantage over transformers with separate windings. In a conventional transformer, *there is no physical connection from input to output.* This affords a large degree of protection from electrical shock when using 60-Hz power lines. The autotransformer provides *no such protection* and so is more hazardous to use.

Impedance Matching

Another use to which transformers are often put is that of *impedance matching*. Remember that maximum power is delivered to a load when the load resistance equals the source resistance. It is equally true when considering source impedances. For example, if an ac voltage source has an internal resistance of 500 Ω and must drive a load of 25 Ω with maximum power transfer, a transformer is used. Because of the turns ratio, this can be accomplished if the correct turns ratio is used. The following relationship allows the correct turns ratio to be found:

$$\frac{N_p}{N_s} = \sqrt{\frac{R_{\text{source}}}{R_{\text{load}}}} = \sqrt{\frac{500}{25}} \cong 4.5.$$

Thus, if the primary has about 4.5 times as many turns as the secondary, maximum transfer of power will occur with these circuit values.

14-5 INDUCTANCE IN DC CIRCUITS

Recall from an earlier discussion that when an inductor is used in a dc circuit, the *effects* of inductance are evident only when current is changing. This usually occurs only when the power is switched on or off. Then, as current is building up, it increases from zero to maximum rather slowly, and if its increasing values are plotted on a graph, they

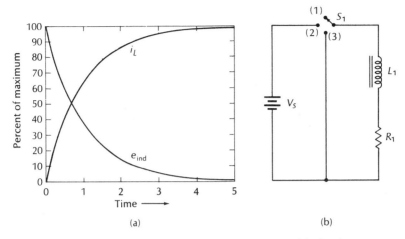

Figure 14-17 (a) The universal time constant curve and (b) the circuit used to generate the curve.

assume a characteristic shape. Such a graph is shown in Figure 14-17, along with the circuit used to generate the curve.

Briefly reviewing the circuit action from which the curve is derived, it is first assumed that the switch is open and hence that no current is flowing. When the switch is closed (position 2), current begins to flow; at this first instant the induced voltage across the coil equals the applied voltage V_s. The current, then, is zero, but it begins immediately to rise as the induced voltage begins to fall. After a period of time during which the current continues to increase more or less gradually, it rises to essentially its maximum value, limited now by V_s/R_1. During this time the magnetic field of force is expanding outward, becoming more dense (stronger) as current increases. In fact, the lines of force cutting across the coil produce the induced EMF.

For as long as the switch remains closed after current attains its maximum value, the circuit acts as a simple dc circuit. There are no inductive effects in the circuit, and the magnetic field is stationary, surrounding the coil in all directions. Because energy was required to produce the field, the field itself is considered to be a *reservoir* of energy, which will be literally dumped or returned back into the circuit when the switch is opened. The amount of energy stored in the field of an inductor is given by the formula

$$J = \tfrac{1}{2} L \times I^2$$

where J = joules or watt-seconds, a measure of the rate of dissipation of power per unit time,
 L = inductance in henrys, and
 I = current in amperes.

Example 12. A 500-mH coil has a current of 250 mA flowing through it. What is the energy stored in the magnetic field?

$$J = \tfrac{1}{2}LI^2 = 0.5 \times (500 \times 10^{-3}) \times (250 \times 10^{-3})^2$$
$$= 0.0156 \text{ joules}.$$

When the switch is opened, the stored energy is returned to the circuit, often causing an arc (spark) across the switch contacts. By using contact 3 on the switch, instead of contact 1, to interrupt current, sparking is often reduced, since the resistor absorbs some of the energy and limits current to some small value determined by simple Ohm's law (assuming that the switch transfers very quickly from 2 to 3).

Time Constant

The term *time constant* can be applied to the curves of Figure 14-17a. This graph represents the growth of current over some period of time. For a given inductor and resistor, the amount of time required for current to go from a value of zero to maximum will *always be the same*. That is, the *total time* required for this circuit action is determined by the values of the inductor and the resistor, and is divided into increments called time constants.

One time constant t_c is defined as the amount of time required for the current to rise from zero to 63.2% of maximum (actually 63.21206%). Note on the graph the numerals 1 through 5 on the abscissa. Each of these represents one time constant. That is, from 0 to 1 represents one t_c; note that at point 1 the current has risen to 63.2% of maximum. Between 1 and 2 is another time constant, and so on.

During the second t_c, the current rises another 63.2% of *what is left to be attained*. This is true for all succeeding time constants. Thus current can *never* rise to the true value specified by Ohm's law, since at the end of each period of time 36.8% of the balance always remains. However, in practical circuitry, the current rises to 99.3% of Ohm's law maximum after 5 t_c, and we consider this to be maximum for all practical purposes.

To determine a value for one time constant in a particular case, the following formula is used:

$$t_c = \frac{L}{R},$$

where t_c = time constant in seconds,
  L = inductance in henrys, and
  R = resistance in ohms.

If values are given to the circuit in Figure 14-17b, the time constant can easily be computed. If $L = 1.0$ H and $R = 1$ kΩ, then

$$t_c = \frac{L}{R} = \frac{1.0}{1000} = 0.001 \text{ s} = 1 \text{ ms}.$$

In this circuit arrangement, current will rise to 63.2% of maximum in 1 ms. Current will attain 99.3%, its final (maximum) value, in 5 ms, at which time it is considered to be at full value. If V_s is given a value, the actual current can be found instead of simply percent of maximum. After five or more time constants have passed, current will be at an approximate value determined by V_s and R. For the same circuit, assume $V_s = 50\ V_{dc}$.

$$I_{max} = \frac{V_s}{R} = \frac{50}{1000} = 0.05\ A = 50\ mA.$$

Therefore, in 1 t_c, current has risen to 63.2% of 50 mA:

$$I_{1\,t_c} = 0.632 \times 50\ mA = 31.6\ mA.$$

The curve for rise in current is called an *exponential* curve, since it is described mathematically by the exponential equation

$$i = I(1 - \epsilon^{-t/L/R}) = i(1 - \epsilon^{-Rt/L}),$$

where i = instantaneous value of current lying on the curve at time t;
 I = maximum current;
 ϵ = epsilon $\cong$ 2.72, the base of natural logarithms;
 t = the time interval during which the curve has grown to the point of interest; and
 L/R = time constant.

A brief example of the use of this equation, which allows the value of any point on the curve to be calculated, will illustrate its application. Referring to Figure 14-17, assume the maximum value (100%) of the curve is 1.0 A. Find the value of current after 1 time constant (equal to 1 ms) has elapsed.

$$\begin{aligned}i &= I(1 - \epsilon^{-Rt/L}) = 1(1 - \epsilon^{-1000 \times 0.001/1}) \\ &= 1(1 - \epsilon^{-1}) = 1(1 - 2.7183^{-1}) \\ &= 1(1 - 1/2.7183) = 1 - 0.36787 = 0.6321\ A\end{aligned}$$

This, of course, simply verifies that the current rises to 63.2% of maximum in one t_c. We shall return to these principles in a later chapter.

SUMMARY

- Inductance is the measure of the ability of a conductor to produce induced voltage when the current changes value.
- The unit of measure of inductance is the henry (H).
- A coil has an inductance of 1 H if the CEMF is 1 V when current changes linearly at a rate of 1 A/s.
- The CEMF is determined by $E_L = L(di/dt)$.
- In ac circuits, voltage and current are not in phase with each other unless the load is completely resistive.

- In ac circuits, the presence of pure inductance causes the current to lag the voltage by 90°.
- In ac circuits, the presence of both resistance and inductance causes the current to lag the voltage by less than 90°, but more than 0°.
- Inductive reactance X_L is the opposition to current generated by the CEMF, and is measured in ohms.
- Impedance is the total opposition to current in an ac circuit.
- Impedance is found by the formula $Z = \sqrt{R^2 + X_L^2}$.
- Total current in a simple inductive and resistive ac circuit is V_s/Z.
- The arithmetic sum of E_R and E_L does *not* equal the applied voltage, since E_R and E_L are not in phase.
- Inductors in series have a total inductance equal to the sum of the values.
- Inductors in parallel have a total inductance as follows: $X_L = 1/(1/X_{L_1} + 1/X_{L_2} \ldots)$, for as many inductors as there are in parallel.
- Mutual inductance is the total effective inductance when two or more coils experience flux linkage.
- Coefficient of coupling $k = L_m/\sqrt{L_1 L_2}$.
- A transformer consists of two or more coils coupled so that ac energy from one is transferred to the other.
- In a transformer, the primary coil is *driven* by V_s, while the secondary *drives* the load.
- If the secondary coil of a transformer has more turns than the primary, the transformer steps up the voltage and steps down the current.
- If the secondary coil of a transformer has fewer turns than the primary, voltage is stepped down, but current is stepped up.
- In any high-efficiency (iron-core) transformer, $E \times I$ is essentially the same for both primary and secondary.
- Iron-core transformers can have efficiencies to 97 or 98%.
- Losses in transformers occur due to I^2R losses, eddy currents, hysteresis, and flux losses. (Eddy-current and hysteresis losses occur only in iron-core transformers.)
- Iron cores are laminated to reduce losses.
- Transformers are also used to match impedances.
- Time constant is the time interval for current to increase to 63.2% of its final value.
- Time constant $t_c = L/R$.
- The curve describing current rise in an inductive circuit is an exponential curve that follows equation $i = I(1 - \epsilon^{-Rt/L})$.

QUESTIONS

1. Briefly explain why current in an inductive ac circuit is not simply limited to E/R.

2. Define CEMF.
3. The current through a perfect inductance (no resistance) lags the voltage across it by 90°. Why?
4. State the Pythagorean theorem.
5. Briefly describe what is meant by core losses.
6. What causes I^2R losses?
7. Briefly describe hysteresis losses.
8. What is an eddy-current loss?
9. Name three types of core material used to reduce core losses.
10. Briefly define mutual inductance.
11. What is the approximate value of the coefficient of coupling for a transformer using an iron core?
12. Energy stored in a magnetic field can be expressed in units known as joules. What is another name for this unit?

PROBLEMS

1. An inductor has a linear rate of change of current applied that is equal to 1 mA per μs. The induced voltage is 1.0 V. What is the inductor's value?
2. In the previous problem, the rate of change of current is changed to 1 A per s, using the same coil. What is the induced voltage?
3. A 100-mH coil is energized with a current that changes linearly from 20 mA to 30 mA in 50 μs. What is the induced voltage?
4. Refer to Problem 3. The same coil is energized with a current that changes from 10 μA to 100 μA in 90 μs. What is the induced voltage?
5. What is the inductive reactance of a 1.5-H coil energized with 120 V, 60 Hz?
6. What is the inductive reactance of a 1.0-mH coil energized with 30 V, 1 kHz?
7. A 150-mH coil is energized with an audio signal of 12 V at 10kHz. What is its X_L?
8. A 750-μH coil is energized with a 100-kHz signal. What is its X_L?
9. A 1.0 mH coil is energized with a 15-kHz signal and is connected in series with a 188-Ω resistor. What is the phase angle between E_L and E_R?
10. A 75-mH coil is energized with a 50-kHz signal and is connected in series with a 45-Ω resistor. What is the total series impedance? What is the phase angle between E_L and E_R?
11. A 100-μH coil is energized with a 15.9-kHz signal. What is the phase angle if a 10-Ω resistor is placed in series?
12. Refer to the previous problem. If the applied voltage is 7.5 V_{rms}, what is the value of current?

13. Refer to Problem 11. If the applied voltage is 35 milliwatts rms, what is the value of current?
14. A 100-μH coil is energized with a signal frequency of 1.59 MHz. What is the phase angle if a 2000-Ω resistor is placed in series? What is the impedance?
15. An inductor with $X_L = 50$ Ω is in series with a 40-Ω resistor. $V_s = 30$ V and $E_R = 18.74$ V. What is E_L?
16. An inductor with $X_L = 200$ Ω is in series with a 100-Ω resistor. $V_s = 10$ V and $E_L = 8.94$ V. What is E_R?
17. In Problem 15, what is θ?
18. In Problem 16, what is θ?
19. A 10-mH coil is placed in series with a 20-mH coil. What is the total inductance if no flux linkage exists?
20. A 500-μH coil is in series with two coils, each rated at 250 μH. What is the total inductance if $L_M = 0$?
21. Two inductors, each having a value of 300 μH, are placed in parallel. If $L_M = 0$, what is the total inductance?
22. Three inductors, each having a value of 2.7 mH, are placed in parallel. If $L_M = 0$, what is the total inductance?
23. Refer to Problem 19. What is the total reactance if a frequency of 1.0 kHz is applied?
24. Refer to Problem 21. What is the total reactance if a frequency of 1.0 MHz is applied?
25. Two coils are to be placed in a series circuit similar to Figure 14-10a and b. Each is a 0.75-mH coil. When connected in series-aiding, the total inductance is 2.0 mH; in series-opposing, total inductance measures 1.0 mH. Find the value of L_m and k.
26. Two coils are to be placed in a series circuit similar to Figure 14-10a and b. Each is a 350-mH coil. When connected in series-aiding, the total inductance is 1000 mH; in series-opposing, total inductance is 400 mH. Find the values of L_m and k.
27. A power transformer has a 1000-turn primary and a 100-turn secondary. A 48-Ω resistor is connected across the secondary. If 120 V_{ac} is applied to the primary, find the secondary voltage and current.
28. A power transformer has a 1000-turn primary and a 200-turn secondary. A 24-Ω load is connected across the secondary. If 120 V_{ac} is applied to the primary, find the secondary voltage and current.
29. A power transformer draws 100 W from the primary source. Secondary voltage is 12 V and the load draws 7.8 A. What is the percent efficiency?
30. A power transformer primary circuit is measured. The input has a voltage of 250 V_{ac} at 0.25 A. Secondary voltage is 57.5 V at 1.0 A. What is the percent efficiency?

CHAPTER 15

INDUCTIVE CIRCUIT ANALYSIS

In this chapter we continue the study of inductive and resistive circuits. More complex circuit arrangements and greater mathematical depth are presented, and the subject of power dissipation in inductive circuits is discussed. The most common troubles in inductive devices are covered. The major subject headings in this chapter are listed below.

15-1 Complex Numbers
15-2 Power in Inductive Circuits
15-3 Series and Parallel Circuit Examples
15-4 Types of Inductors
15-5 Trouble in Inductors

15-1 COMPLEX NUMBERS

A complex number is one that has a real part and an imaginary part. Such numbers are very useful (in fact necessary) in solving certain problems in ac circuits. To begin to understand complex numbers, we must first define real and imaginary numbers. Recall earlier discussions regarding the phasor relationship between resistance and inductance in ac circuits. Resistance and inductive reactance can be plotted graphically, and when they are, they must be drawn *at a 90° angle* to each other. This is true because the effect of inductance in an ac circuit is such as to cause current to lag voltage by 90° (for a pure inductance). Because of the out-of-phase condition in an ac circuit that contains inductance, normal circuit analysis procedures, such as adding parallel currents to obtain the total current or adding voltage drops in a series circuit, cannot be done by simple arithmetic.

Numbers such as those representing resistor values are called *real* numbers, since they represent either positive or negative values in the usual sense. Figure 15-1 illustrates an example of a set of real numbers. Such numbers can be either *positive* or *negative,* but can have no other value except that of magnitude. The thermometer illustrated shows the Celsius scale, and those numbers above 0 are positive real numbers, while those below 0 are negative real numbers. Real numbers, then, are the numbers that you are accustomed to using. Real numbers have both magnitude and either positive or negative direction.

344 Chapter 15

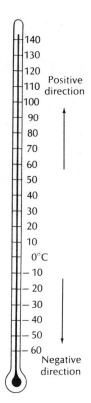

Figure 15-1 The Celsius temperature scale illustrates positive and negative numbers.

Imaginary numbers are not truly imaginary in the sense that they exist only in our minds. However, a name had to be chosen to differentiate these numbers from all real numbers, and *imaginary* is the name originally chosen many years ago. An imaginary number is one that operates at a 90° angle to real numbers. In mathematics an imaginary number is symbolized by the letter *i*, but since this represents current in electronics, we use the letter *j* instead. To further symbolize the imaginary number, the square root of −1 is used ($\sqrt{-1}$); since there is no square root of −1 this makes an ideal symbol for the so-called imaginary numbers. The symbol *j* therefore is the same as $\sqrt{-1}$.

Figure 15-2 illustrates both real and imaginary numbers. The *abscissa* (horizontal coordinate) represents both positive and negative values for real numbers. The *point of origin* is where the two coordinates cross, and values to the right are positive real numbers. Numbers to the left of the point of origin are negative real numbers.

Inductive Circuit Analysis 345

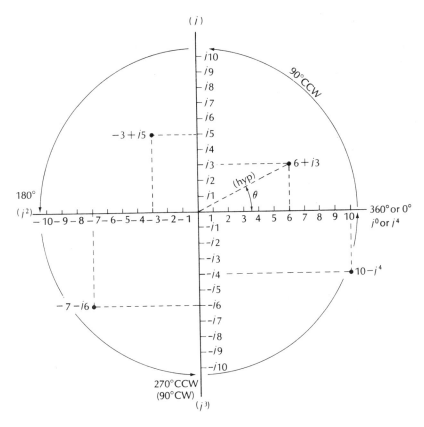

Figure 15-2 The j operator.

The imaginary numbers lie along the *ordinate* (vertical coordinate), and in the upward direction are labeled $j1, j2$, and so on, while in the downward direction are labeled $-j1, -j2$, and so on. Note that positive values of j are rotated 90° counterclockwise in relation to positive real numbers. By the same token, negative values of j are rotated 90° clockwise in relation to positive real numbers.

When a real number and an imaginary number are combined in order to locate any single point on the graph, the result is known as a *complex number*. Note the point in the upper right-hand quadrant labeled $6 + j3$. The location of this point is the combined effect of 6 positive units of real numbers and $+j3$ units of imaginary numbers. By simply writing $6 + j3$, this one point on the graph is specifically identified and at the same time, all other points in all quadrants are excluded.

Note the additional information that is inherent in this one $6 + j3$ statement. Not only are the real and imaginary numbers given mag-

TABLE 15-1

$j = \sqrt{-1} = 90°$ CCW
$j^2 = -1 = (\sqrt{-1})(\sqrt{-1}) = 180°$ CCW (or 180° CW)
$j^3 = (\sqrt{-1})(\sqrt{-1})(\sqrt{-1}) = 270°$ CCW (or 90° CW)
$j^4 = +1 = (\sqrt{-1})(\sqrt{-1})(\sqrt{-1})(\sqrt{-1}) = j° = 0°$ or 360°

nitude, but so is their *combined effect,* as indicated by the diagonal dashed line. In terms of trigonometry, this dashed line is of course the hypotenuse. Furthermore, the angle θ is also implied in the $6 + j3$ statement. Thus, a complex number such as $6 + j3$ gives the lengths of two sides of a right triangle, and implies the values of the hypotenuse and the angle θ.

Substituting the now-familiar electrical quantities for the quantities in the $6 + j3$ expression, the 6 represents resistance, since this is the real part of the complex number. Thus, the resistance is equal to 6 Ω. The $+j3$ must therefore represent inductive reactance having a value of 3 Ω. The hypotenuse represents the impedance offered by a resistance of 6 Ω and an inductive reactance of 3 Ω, and the phase difference between voltage and current is specified by θ. You can now appreciate that a complex number is simply a shorthand method of describing the circuit response in ac circuits, and of condensing a large amount of information into a compact statement.

Several other points are identified in Figure 15-2, one in each quadrant. Each is explained as in the example above, except for the differing angles of rotation from reference. Table 15-1 lists the imaginary values for the entire graph in Figure 15-2. As explained before, j is the equivalent of 90° CCW rotation. Note that j^2 is a *real negative number,* as indicated by the 180° rotation from reference; its value is -1. Also note that j^3 is an imaginary number indicating an angular rotation of 270° CCW from reference, or 90° CW, which is the same thing. Finally j^4 or j^0 is simply $+1$, a real positive number. That is, 360° angular rotation is the same as none. As will be seen in examples to follow, the *j operator* is not difficult to work with once the basic principles are mastered.

Forms of Notation

There are three basic forms used to specify complex notation. These are:

1. Trigonometric notation
2. Rectangular notation
3. Polar notation

Two of these forms have been discussed previously, but the third, *polar* notation, has not. To illustrate each of these, the same information will be written in each of these forms. The now-familiar 3-4-5 right triangle will be used, and the correct statement, or statements, for each will be given. The circuit has 30 Ω of resistance, 40 Ω of inductive reactance, and 50 Ω of impedance at a phase angle of 53.1°.

Trig	Rectangular	Polar
$Z = \sqrt{R^2 + X_L^2} = 50$	$Z = 30 + j40$	$Z = 50 \underline{/53.1°}$
$\theta = \arctan \frac{X}{R} = 53.1°$		where $\angle$ = at an angle of ....

These statements are equivalent, insofar as the information they contain is concerned; only their form is different. The reason for using several forms of notation is simply that some mathematical operations (such as addition and subtraction) are done more simply and accurately with one system than another. You will find your computations are less difficult when you are able to convert one form of notation to another.

The trigonometric form specifies the total circuit impedance and the angle as two separate quantities, Z and θ. Rectangular notation specifies the resistive and reactive quantities that determine the magnitude of impedance, while the R and j quantities allow the angle to be found. Finally, the polar form gives the impedance and the angle as a single entity.

Notation Conversion

It is frequently necessary to change from one form of notation to another. The reason for this is that it is very difficult to multiply or divide using rectangular notation, and it is not possible to add or subtract using polar notation. Hence, to simplify matters, the following rules are necessary:

1. To add or subtract complex notation, use the rectangular form.
2. To multiply or divide complex notation, use the polar form.

Rectangular to polar conversion is simply a matter of evaluating the expression for Z and θ. Again using the familiar 3-4-5 right triangle, we shall convert the rectangular expression $Z = 30 + j40$ to polar form.

$$Z = \sqrt{30^2 + 40^2} = 50 \text{ Ω}; \theta = \arctan \frac{40}{30} = 53.1°,$$

therefore, $Z = 50 \ \Omega/53.1°$. If trig tables are handy, the following equations may be substituted. $Z = R/\cos \theta$, or $Z = X_L/\sin \theta$. (Of course, to use either equation, the angle must first be known.)

Polar to rectangular conversion is accomplished just as easily. We can use the foregoing example to illustrate this process.

Example 1. Convert $50 \ \Omega/53.1°$ to rectangular form.

1. Multiply the impedance by the *cosine* of the angle to obtain the *real* term: $R = 50 \times \cos \theta = 50 \times 0.6 = 30 \ \Omega$.
2. Multiply the impedance by the *sine* of the angle to obtain the *j* term: $X_L = 50 \times \sin \theta = 50 \times 0.8 = 40 \ \Omega$.

Therefore, $Z = 30 + j40$.

Adding and Subtracting Complex Numbers

As previously noted, when adding and subtracting complex numbers, the rectangular form is used. The expressions are simply written as shown below and added or subtracted algebraically.

Example 2. Add the following:

$$\begin{array}{r} 5 + j6 \\ 2 + j5 \\ \hline 7 + j11 \end{array} \qquad \begin{array}{r} 5 + j6 \\ 2 - j5 \\ \hline 7 + j1 \end{array}$$

Example 3. Subtract the following:

$$\begin{array}{r} 5 + j6 \\ 2 + j5 \\ \hline 3 + j1 \end{array} \qquad \begin{array}{r} 5 + j6 \\ 2 - j5 \\ \hline 3 + j11 \end{array}$$

Note that the conventional rules for adding and subtracting positive and negative numbers apply.

Multiplying and Dividing Complex Numbers

There are two simple rules to remember when multiplying or dividing complex numbers. First, however, recall that these expressions must be in polar form.

1. To multiply polar expressions, multiply the real parts and *add the angles*.
2. To divide polar expressions, divide the real parts and *subtract the angles*.

Example 4. Multiply $15\underline{/25°}$ by $5\underline{/10°}$.

$$15\underline{/25°} \times 5\underline{/10°} = 75\underline{/35°}$$

Multiply $50\underline{/30°} \times 5$.

$$5 \times 50\underline{/30°} = 250\underline{/30°}$$

Note: an angle of $0° + 30° = 30°$.

Example 5. Divide $20\underline{/30°}$ by $5\underline{/15°}$.

$$\frac{20\underline{/30°}}{5\underline{/15°}} = 4\underline{/15°}$$

Divide $20\underline{/30°}$ by 4.

$$\frac{20\underline{/30°}}{4} = 5\underline{/30°}$$

Note: $30° - 0° = 30°$.
Divide 20 by $5\underline{/30°}$.

$$\frac{20}{5\underline{/30°}} = 4\underline{/-30°}$$

Note: when the numerator has no angular part, *change the sign* of the angular denominator and use that in the answer.

15-2 POWER IN INDUCTIVE CIRCUITS

When considering power dissipation in alternating-current circuits, the fact that current and voltage are seldom in phase must be taken into account. In dc circuits, the simple product of voltage and current yields the power dissipation. However, when voltage and current are *not* in phase, they cannot be simply multiplied together to obtain the true amount of power. If total circuit current and voltage are simply multiplied, the result is a value called *apparent power*, P_{app}. Apparent power does not give a true picture of circuit action. True power, P_{true}, however, does specify exactly the total power, in watts, delivered to the load.

Consider the circuit shown in Figure 15-3. An inductor is shown in series with a resistor, which might be a separate resistor, or it might represent the dc resistance of the wire in the coil. Any current flowing in the coil must also flow in the resistor, and as in any circuit, current through a resistance dissipates power. Assuming that the coil symbol represents a pure inductance, and the resistance of the wire is therefore shown separately, *the inductance will be found to dissipate no power.* Any power lost in the coil is represented by the resistor symbol shown separately.

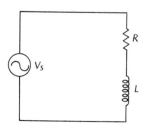

Figure 15-3 A series RL circuit.

This phenomenon can be explained as follows. At some instant in time, circuit current is zero, but will immediately begin to grow in amplitude. During the time current is *increasing*, (90° interval) the CEMF opposes the applied voltage and energy is expended overcoming this CEMF. When current reaches its peak value, it then begins to decrease again. During this 90° interval, a voltage is again induced in the coil, but now *energy is put back into the circuit* in exactly the same quantity that was used to overcome the CEMF. Averaging the power consumed by the inductor gives, therefore, zero watts, as suggested in Figure 15-4a. That is, with equal amounts of "positive" power and "negative" power, the net result is zero. A pure inductance, then, dissipates, or consumes, no power. However, in the practical case, a coil must be wound with wire, and so there is always some resistance. In all practical circuits, then, there is some power consumption.

Figure 15-4b shows the power in a purely resistive circuit (*no inductance*). All power is positive, as evidenced by the curves *above* the graph's abscissa.

The difference between true power and apparent power is best explained by an example.

Example 6. In the circuit shown in Figure 15-3, assume that $V_s = 12$ V, $R = 30\ \Omega$, and $X_L = 40\ \Omega$. To determine circuit current, find the values of Z and θ.

$$Z = \sqrt{R^2 + X_L^2} = \sqrt{30^2 + 40^2} = 50\ \Omega.$$
$$\theta = \arctan \frac{X_L}{R} = \arctan \frac{40}{30} = 53.1°.$$

The value of current, then, is

$$I_t = \frac{V_s}{Z} = \frac{12}{50} = 0.24 = 240\ \text{mA}.$$

Now, apparent power is simply the product of total voltage V_s and total current I_t.

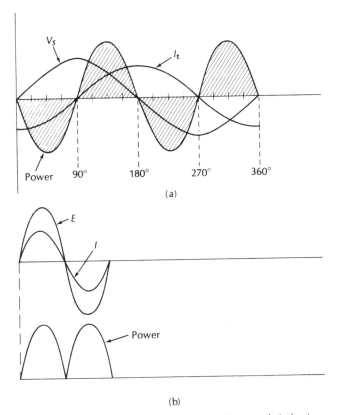

Figure 15-4 Voltage, current, and power (a) in a purely inductive circuit and (b) in a purely resistive circuit.

$$P_{app} = V_s \times I_t = 12 \times 0.24 = 2.88 \text{ W}.$$

Only that part of the apparent power that is dissipated in the resistor is the true value of power dissipation. True power can be determined by one of the following relationships.

$$P = I^2R = 1.728 \text{ W} \cong 1.73 \text{ W}.$$
$$P = \cos \theta(EI) = 1.728 \cong 1.73 \text{ W}.$$
$$P = \frac{R}{Z}(EI) = 1.728 \cong 1.73 \text{ W}.$$

In an inductive circuit, true power is always less than apparent power, because some part of the total is returned to the circuit by the collapsing field of the coil.

Power factor (PF) is a measure of the amount by which true power is less than apparent power and is the ratio of the first to the second.

$$PF = \frac{P_{\text{true}}}{P_{\text{app}}} = \cos \theta.$$

Using the previous example,

$$PF = \frac{P_{\text{true}}}{P_{\text{app}}} = \frac{1.73}{2.88} = 0.6,$$

or,

$$PF = \cos \theta = \cos 53.1 = 0.6.$$

15-3 CIRCUIT EXAMPLES

In order to illustrate the application of the foregoing material, several examples follow that show correlation to actual inductive circuits. You must realize that the Ohm's law relationships between voltage and current apply to ac circuits just as to dc circuits. For example, if in a series dc circuit it is true that $R_t = R_1 + R_2 + R_3$, then in a similar ac circuit $Z_t = Z_1 + Z_2 + Z_3$. By the same token, $Z_t = 1/[(1/Z_1) + (1/Z_2) + (1/Z_3) \ldots]$, which corresponds to $R_t = 1/[(1/R_1) + (1/R_2) + (1/R_3) \ldots]$, for the parallel arrangement. You must remember, how-

TABLE 15-2 COMPARISON OF DC AND AC OHM'S LAW RELATIONSHIPS

DC	AC
BASIC	
$E_T = I_T R_T$	$E_T = I_T Z$
$R_T = \dfrac{E_T}{I_T}$	$Z_T = \dfrac{E_T}{I_T}$
$I_T = \dfrac{E_T}{R_T}$	$I_T = \dfrac{E_T}{Z_T}$
SERIES	
$R_T = R_1 + R_2 + R_3$	$Z_T = Z_1 + Z_2 + Z_3$
$E_{R_1} = I_T R_1$	$E_{Z_1} = I_T Z_1$
$E_{R_2} = I_T R_2$	$E_{Z_2} = I_T Z_2$
$E_{R_3} = I_T R_3$	$E_{Z_3} = I_T Z_3$
$E_T = E_{R_1} + E_{R_2} + E_{R_3} = V_s$	$E_T = E_{Z_1} + E_{Z_2} + E_{Z_3} = V_s$
PARALLEL	
$R_T = \dfrac{1}{1/R_1 + 1/R_2 + 1/R_3}$	$Z_T = \dfrac{1}{1/Z_1 + 1/Z_2 + 1/Z_3}$
$I_{R_1} = V_s/R_1$	$I_{Z_1} = V_s/Z_1$
$I_{R_2} = V_s/R_2$	$I_{Z_2} = V_s/Z_2$
$I_{R_3} = V_s/R_3$	$I_{Z_3} = V_s/Z_3$
$I_T = I_{R_1} + I_{R_2} + I_{R_3}$	$I_T = I_{Z_1} + I_{Z_2} + I_{Z_3}$

Inductive Circuit Analysis 353

ever, that in ac circuits the phasor quantities must be dealt with as explained earlier in this chapter. Nevertheless, the same basic formulas hold true if Z is substituted for R in each. Table 15-2 lists many relationships for both dc and ac circuits, while Figure 15-5 illustrates series and parallel relationships for ac circuits.

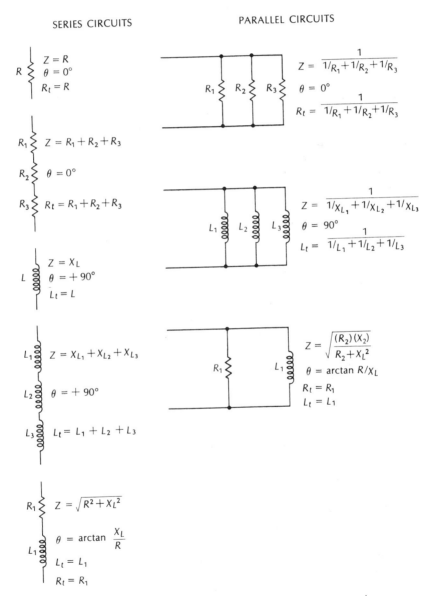

Figure 15-5 Series and parallel configurations of resistive and inductive circuits.

Example 7. Express $250 + j150$ in polar form.

$$Z = \sqrt{250^2 + 150^2} = 291.5 \ \Omega.$$
$$\theta = \arctan \frac{X_L}{R} = \arctan \frac{150}{250} = 30.96° \cong 31°.$$

Therefore, $291.5 / 31°$.

Example 8. Express $300 / 45°$ in rectangular form.

$$\text{Real term} = 300 \times \cos \theta = 300 \times 0.7071 = 212 \ \Omega.$$
$$j \text{ term} = 300 \times \sin \theta = 300 \times 0.7071 = 212 \ \Omega.$$

Therefore, $212 + j212$.

Example 9. Refer to Figure 15-6. $R = 1800 \ \Omega$ and $L = 4.77$ H. Express the impedance in rectangular and polar forms.

$$X_L = 2\pi FL = 6.28 \times 60 \times 4.77 \cong 1800 \ \Omega.$$

Therefore, $Z = 1800 + j1800$ in rectangular form.

$$Z = \sqrt{R^2 + X_L^2} = \sqrt{1800^2 + 1800^2} = 2546 \ \Omega.$$
$$\theta = \arctan \frac{X_L}{R} = \arctan \frac{1800}{1800} = \arctan 1 = 45°.$$

$Z = 2546 / 45°$ in polar form.

Example 10. Refer to Figure 15-6. $R = 100 \ \Omega$ and $L = 0.8$ H. Express the impedance in rectangular and polar form.

$$X_L = 2\pi FL = 6.28 \times 60 \times 0.8 = 302 \ \Omega$$

Therefore, in rectangular form, $Z = 100 + j302$.

$$Z = \sqrt{R^2 + X_L^2} = \sqrt{100^2 + 302^2} = 318 \ \Omega$$
$$\theta = \arctan \frac{X_L}{R} = \arctan \frac{302}{100} = \arctan 3.02$$
$$= 71.7°;$$

in polar form, $Z = 318 / 71.7°$.

Example 11. A certain circuit has a set of values described as $500 / 25°$. Express this statement in rectangular form.

$$R = Z \times \cos \theta = 500 \times \cos 25° = 500 \times 0.906 = 453 \ \Omega.$$
$$X_L = Z \times \sin \theta = 500 \times \sin 25° = 500 \times 0.423 = 211 \ \Omega.$$

Therefore, $Z = 500 + j211$.

Example 12. A certain circuit has a set of values described as $1500 / 65°$. Express this statement in rectangular form.

Inductive Circuit Analysis 355

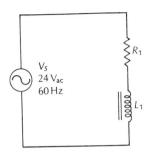

Figure 15-6

$$R = Z \times \cos \theta = 1500 \times \cos 65° = 1500 \times 0.423$$
$$= 634 \ \Omega.$$
$$X_L = Z \times \sin \theta = 1500 \times \sin 65° = 1500 \times 0.906$$
$$= 1360 \ \Omega.$$

Therefore, $Z = 634 + j1360$.

Example 13. Add the following complex numbers.

$$\begin{array}{r} 150 + j150 \\ 200 + j200 \\ \hline 350 + j350 \end{array} \qquad \begin{array}{r} 150 + j150 \\ 200 - j200 \\ \hline 350 - j50 \end{array}$$

Example 14. Subtract the following complex numbers.

$$\begin{array}{r} 150 + j150 \\ 200 + j200 \\ \hline -50 - j50 \end{array} \qquad \begin{array}{r} 150 + j150 \\ 200 - j200 \\ \hline -50 + j350 \end{array}$$

Example 15. Multiply the following complex numbers.

$$(150 \underline{/25°})(50 \underline{/10°})$$

1. $150 \times 50 = 7500$,
2. $(\underline{/25°}) + (\underline{/10°}) = \underline{/35°}$.

Therefore, the product is $7500 \underline{/35°}$.

Example 16. Divide the following complex numbers.

$$\frac{150 \underline{/25°}}{50 \underline{/10°}}$$

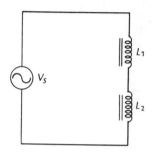

Figure 15-7

1. $\dfrac{150}{50} = 3$,
2. $\underline{/25°} - \underline{/10°} = \underline{/15°}$

Therefore, the quotient is $3\underline{/15°}$.

Example 17. Refer to Figure 15-7. Given that $V_s = 12$ V_{rms} at 1 kHz, find the total reactance if $L_1 = 10$ mH and $L_2 = 200$ mH.

$$X_{L_1} = 2\pi FL = 6.28 \times 1000 \times 0.01 = 62.8 \ \Omega,$$
$$X_{L_2} = 2\pi FL = 6.28 \times 1000 \times 0.02 = 125.6 \ \Omega,$$
$$X_{L_4} = 62.8 + 125.7 = 188.5 \ \Omega.$$

Example 18. Refer to the previous example. Determine the impedance and total circuit phase angle.

As drawn, the circuit assumes zero resistance; hence,

$$Z = X_{L_t} = 188.5 \ \Omega \text{ (note that } Z = \sqrt{0^2 + 188.5^2} = 188.5 \ \Omega).$$
$$\theta = \arctan \dfrac{X_L}{R} = \arctan \dfrac{188.4}{0} = \arctan \infty = 90°.$$

Example 19. Refer to Example 11 above. Determine the voltage across L_1 and L_2 as measured by a VOM.

$$E_{L_1} = I_t \times X_{L_1},$$
$$E_{L_2} = I_t \times X_{L_2}.$$
$$I_t = \dfrac{V_s}{X_{L_t}} = \dfrac{12}{188.5} = 0.064 = 64 \text{ mA (rms)}.$$
$$E_{L_1} = I_t \times X_{L_1} = 0.064 \times 62.8 = 4 \text{ V},$$
$$E_{L_2} = I_t \times X_{L_2} = 0.064 \times 125.6 = 8 \text{ V}.$$

Note: In circuits such as illustrated in Figure 15-7, the simplest solution is to treat L_1 and L_2 as a single inductor.

Inductive Circuit Analysis

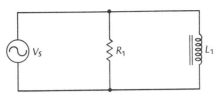

Figure 15-8

Example 20. Refer to Figure 15-8. Given that $V_s = 12$ V at 1 kHz, $R = 1000$ Ω, and $L_1 = 0.159$ H, find Z_t presented to the source.

Note that V_s, E_{R_1}, and E_{L_1} are in parallel and cannot be anything but *in phase* with each other. Resistive current, however, is in phase with V_s while inductive current *lags* V_s by 90°. Currents therefore must be added vectorially.

$$X_L = 2\pi FL = 6.28 \times 1000 \times 0.159 = 1000 \text{ Ω}.$$

To determine Z, first find branch currents.

$$I_{R_1} = \frac{V_s}{R_1} = \frac{12}{1000} = 0.012 = 12 \text{ mA}.$$

$$I_{L_1} = \frac{V_s}{X_{L_1}} = \frac{12}{1000} = 0.012 = 12 \text{ mA}.$$

To determine total current vectorially:

$$I_t = \sqrt{I_{R_1}^2 + I_{L_1}^2} = \sqrt{0.012^2 + 0.012^2}$$
$$= 0.01697 = 17 \text{ mA}.$$

Now Z can be determined.

$$Z = \frac{V_s}{I_t} = \frac{12}{0.01697} = 707 \text{ Ω}.$$

The following relationship is also valid for finding the value of Z in a simple parallel circuit.

$$Z = \sqrt{\frac{(R^2)(X_L^2)}{R^2 + X_L^2}} = \sqrt{\frac{(1000^2)(1000^2)}{(1000^2) + (1000^2)}}$$
$$= \sqrt{\frac{1 \times 10^{12}}{2 \times 10^6}} = 707 \text{ Ω}.$$

Example 21. Refer to Figure 15-9. Briefly explain the magnitude and phase relationships relative to voltage and current.

All elements are in parallel so E_{L_1} and E_{L_2} must be the same as V_s. Branch currents have a magnitude equal to V_s/X_L and each is lagging V_s by 90°. Hence I_{L_1} and I_{L_2} are *in phase with each other*, assuming that $R = 0$.

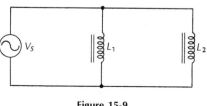

Figure 15-9

15-4 TYPES OF INDUCTORS

Strictly speaking, an inductor is any device possessing inductance. However, whenever current flows, an inductive field surrounds any conductor, so all electrical and electronic components have inductance to a greater or lesser degree. If the property of inductance predominates, and is the reason the component is used, then we call the device an inductor. In this section we shall introduce a few examples of inductors, concentrating on their physical characteristics.

Air-Core and Ferrite-Core Inductors

Several examples of small, high-frequency inductors are shown in Figure 15-10. Part a of the figure represents three slightly different rf coils wound on ceramic forms. Often the form is impregnated with silicon to reduce moisture absorption. Such inductors may or may not be adjustable by means of powdered-iron cores or ferrite cores. They are available in ranges from about 1.0 μH to perhaps 2000 μH. If adjustable, the range of minimum to maximum is about one to two. That is, an inductor might have a total range of inductance from 33 to 66 μH, for example.

Figure 15-10b illustrates a similar type, but wound with a larger coil. Typical ranges might be 40 μH to 240 μH for the smaller sizes and 180 to 750 mH for the larger sizes. Again, they may or may not be adjustable, but usually are. Often, this type is *tapped*, having in effect the same basic characteristics as an autotransformer, and they are used in transistor circuits that generate a high radio-frequency signal.

Figure 15-10c shows a nonadjustable inductor intended to be inserted directly into prepunched holes in a PC board. Part d of the figure illustrates a type of fixed inductor that has relatively large values of inductance in a small size. Such an inductor is wound on a ferrite core to increase inductance. A typical size for a 50-mH coil would be ⅞ inch in length by ⅝ inch in diameter. This type of coil is useful in high radio-frequency circuits, and is frequently designed to handle currents in the range of 50 to 200 mA.

Inductive Circuit Analysis   359

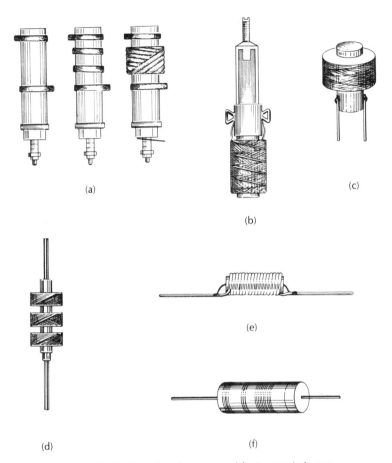

Figure 15-10 Examples of air-core and ferrite-core inductors.

Figure 15-10e illustrates a typical high-current inductor, with values from approximately 4 to 100 μH, but with resistances on the order of $1/10$ Ω. Finally, part f of the figure shows an encapsulated inductor used where environmental conditions are severe. Typically, such inductors are $1/3$ to $1/2$ inch in length, with values ranging from 0.1 μH to 1000 μH.

Iron-Core Inductors

Figure 15-11 illustrates several iron-core inductors, including both choke coils and transformers. Part a of the figure illustrates a *choke coil*, which is a *single-coil* inductor wound on an iron core. These are

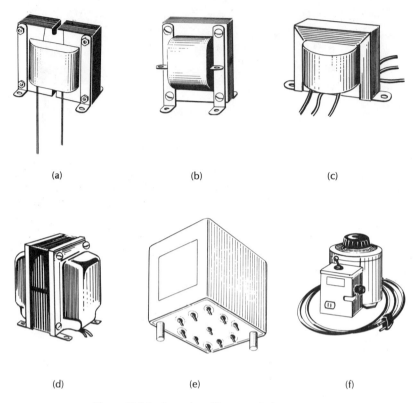

Figure 15-11 Examples of iron-core inductors.

used in power supplies to oppose a changing current. The coil must be able to handle the required current without undue temperature rise. Choke coils are often rated at currents from 50 mA to several amperes. Figure 15-11*b* illustrates a similar type, with *lugs* instead of wires for connection purposes. Figure 15-11*c* illustrates a transformer having six leads. This might mean that the unit has three separate coils, or that it has two coils, with one of them *tapped,* for several values of voltage output.

For heavier-duty use, *end bells* are often installed to protect the windings, as in part *d* of the illustration. Often, transformers are hermetically sealed, Figure 15-11*e*, to provide extreme long life and rugged performance characteristics. Finally, a variable autotransformer is shown in part *f* of the figure. Such transformers frequently can vary the 117-V line from 0 to 140 V. Autotransformers are widely used on the service bench to simulate low- and high-voltage conditions to force an equipment malfunction.

15-5 TROUBLE IN INDUCTORS

Because inductors are relatively simple devices physically, they normally exhibit few troubles. The wires from which they are wound are quite durable, so that opens ordinarily occur only at soldered connection points. This is not to say that the coil wire *never* breaks, but it is a relatively rare thing.

Shorts can be one of two kinds, turn to turn and coil to core. A turn-to-turn short occurs when the enameled insulation wears thin, perhaps from vibration. Sometimes, several adjacent turns are involved. A turn-to-turn short can have disastrous results. For one thing, such a fault can seldom be measured since the shorted turns do not appreciably change the total dc resistance of the coil. Hence, an ohmmeter cannot usually detect this kind of trouble. Then, the shorted turn or turns become a very low resistance load in which the turns ratio between it and the remaining turns is usually quite large. As a result, heavy currents flow in the shorted turns and they become very hot. This adversely affects the adjacent turns and the surrounding insulation, and usually results in the ultimate destruction of the inductor.

The second kind of short occurs when a turn of wire is shorted to the iron core. Because the core is almost always operated at ground potential, this fault can destroy not only the inductor, but any circuitry connected to it.

Generally speaking, faults in inductors are caused by devices other than the inductors themselves. That is, other devices used in conjunction with inductors can, and frequently do, short either totally or partially to ground. This causes excessive current flow, which overheats the inductor. Overheating, in turn, can cause any number of faults. The solder used for connections can melt, causing further shorting or open circuits. The paper or plastic insulation can be burned. Enamel insulation on the wires can also be burned and therefore weakened. What is worst in this kind of fault is that often the damage is concealed from view. If this hidden damage is not immediately recognized, it may cause equipment malfunction at any future time, with no apparent cause.

SUMMARY

- A complex number consists of a real part and an imaginary part.
- An imaginary number is one the magnitude of which operates at a 90° angle from the real numbers.
- Complex numbers can be written in any of three systems of notation: trigonometric, rectangular, and polar.

- Given a complex number expressed in any of the three ways, it is possible to convert the expression to either of the other two.
- To add or subtract complex numbers, use rectangular notation.
- To multiply or divide complex numbers, use polar notation.
- Zero power is dissipated in a purely inductive ac circuit.
- In an ac circuit with both R and L, power is I^2R.
- Inductors can be constructed as a single coil, or with multiple coils.
- Multiple-coil inductors are called transformers.

QUESTIONS

1. What is a complex number?
2. A circuit is described as having the following values: $45 + j60$. Describe what this expression means.
3. A circuit is described as having the following values: $75/53.1°$. Describe what this expression means.
4. A circuit is described as follows: $Z = 75\ \Omega$, $\theta = 53.1°$. Express this information in both rectangular and polar forms.
5. Refer to Figure 15-2. Describe in your own words the meaning of the values along the abscissa to the right of zero (point of origin).
6. Refer to Figure 15-2. Describe in your own words the meaning of the values along the abscissa to the left of zero (point of origin).
7. Refer to Figure 15-2. Describe in your own words the meaning of the values along the ordinate above zero (point of origin).
8. Refer to Figure 15-2. Describe in your own words the meaning of the values along the ordinate below zero (point of origin).
9. Briefly explain what $j2$ means.
10. What does $j4$ mean?
11. Briefly explain apparent power in ac circuits.
12. Briefly explain true power in ac circuits.

PROBLEMS

1. Express $25 + j15$ in polar form.
2. Express $500 + j300$ in polar form.
3. Express $87.5/53.1°$ in rectangular form.
4. Express $150/25°$ in rectangular form.
5. Refer to Figure 15-6. $R = 2500\ \Omega$ and $L = 10$ H. Express the impedance in rectangular and polar forms.
6. Refer to Figure 15-6. $R = 500\ \Omega$ and $L = 0.8$ H. Express the impedance in rectangular and polar forms.
7. Refer to Figure 15-6. $R = 333\ \Omega$ and $L = 1.33$ H. Express the impedance in rectangular and polar form. Find the value of total current.

8. Refer to Figure 15-6. $R = 3000 \, \Omega$ and $L = 8.0$ H. Express the impedance in rectangular and polar form. What is I_t?
9. Refer to Problem 5 above. Find E_R and E_L.
10. Refer to Problem 6 above. Find E_R and E_L.
11. Two quantities are expressed as $75/30°$ and $55/20°$. Find their sum.
12. Two quantities are expressed as $1250/10°$ and $1500/15°$. Find their sum.
13. Two quantities are expressed as $30 + j35$ and $40 + j25$. Find their product.
14. Two quantities are expressed as $1500 + j1500$ and $1500 + j1000$. Find their product.
15. Refer to Figure 15-6. $R_1 = 470 \, \Omega$ and $L_1 = 1.0$ H. Find the apparent power; the true power.
16. Refer to Figure 15-6. $R_1 = 560 \, \Omega$ and $L_1 = 5.0$ H. Find the apparent power; the true power.

CHAPTER 16

CAPACITANCE AND CAPACITIVE DEVICES

In this chapter, you will discover what capacitance is and how it is used to influence circuit action. You will find out why a capacitor (formerly called a *condenser*) is a necessary component in many ac and dc circuits. Basically a capacitor is a device that can act as a short circuit, or at least as a conducting path, to alternating current, but is an *open circuit to direct current*. In this respect, the capacitor is unlike any component discussed thus far. In some ways capacitance is the exact opposite of inductance, which offers large opposition to ac frequencies but acts as nearly a short circuit to dc.

The capacitor, then, has properties that in general are similar to those of an inductor, but the mathematical values are in opposition. Phasor diagrams for a simple capacitor circuit will show a 180° out-of-phase condition with respect to an inductor circuit, for example.

The major topics to be discussed in this chapter are listed below.

16-1 The Elementary Capacitor
16-2 Types of Capacitors
16-3 Capacitor Characteristics
16-4 Capacitor in Series and Parallel
16-5 Opens and Shorts in Capacitors

16-1 THE ELEMENTARY CAPACITOR

A very basic capacitor consists of two conductors having large areas, which are separated by an insulator, or dielectric. For example, two one-foot-square sheets of aluminum foil (the *plates*) are placed as shown in Figure 16-1a and b and are separated by a sheet of paper slightly larger than the foil. The two conducting surfaces must *not* touch one another. To make convenient connection to a circuit, one wire is firmly affixed to each piece of foil. Such a capacitor, then, consists of two plates separated by an insulator. Figure 16-1c shows the schematic symbols for capacitors.

In the introduction to this chapter we stated that a capacitor can pass ac but block dc. This does not mean, however, that there is no circuit

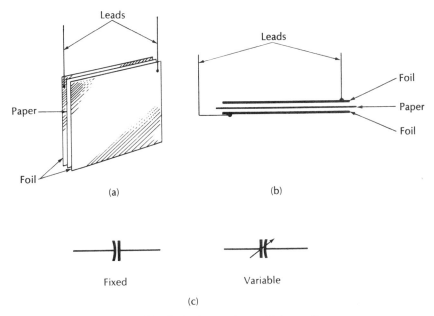

Figure 16-1 (a and b) The basic capacitor and (c) capacitor schematic symbols.

action at all when a capacitor is connected into a circuit having a dc source, as we shall see. The capacitor will become *charged* to the EMF of the dc source, and will then retain this charge for some period of time.

In an inductive circuit the property of inductance opposes a changing current. A capacitor, on the other hand, *opposes a change in voltage.* That is, if a capacitor is placed across an ac circuit, it opposes the changing voltage, and the measure of its opposition is called *capacitive reactance,* which is measured in ohms. Capacitance exists between any two conductors separated by an insulator. In a capacitor this property is, of course, desirable. Often, however, capacitance is not desirable. For example, in a long two-wire cable, the capacitance between wires can be very detrimental. Our first consideration will be to "look" inside a typical capacitor to see what makes it function.

Basic Capacitor Action

Figure 16-2 depicts the fundamental charging action of a capacitor connected to a dc source through a switch and a resistor. The left-hand drawing shows the circuit prior to the switch being closed. No current has flowed and the capacitor is in a neutral state. Note the normal

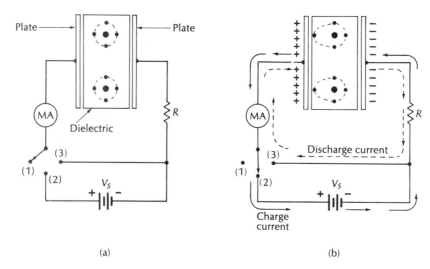

Figure 16-2 Basic capacitor action.

orbits in the atomic structure of the dielectric, where only two atoms are represented. Also, each plate is in an electrically neutral state. That is, each plate contains only the normal amount of electrons for the material of which the plates are formed.

Now, when the switch is closed (transferred to point 2), *current flows for a brief time,* and as it does, electrons pile up on the right-hand plate. As electrons become concentrated on this plate, their force *repels* electrons away from the left-hand plate on a one-for-one basis. The left-hand plate becomes more and more positive as the right-hand plate becomes more and more negative. A voltage now exists across the capacitor plates, and when its magnitude becomes equal and opposite to the applied voltage, current ceases and—as long as the circuit is not disturbed—nothing further happens. The capacitor is said to be fully charged to the supply voltage. Note that when the capacitor is in a charged condition, the orbital electrons in the dielectric operate in a *distorted orbit.* Much of the energy contained in a charged capacitor is due to the stressed orbital electrons in the dielectric.

To verify that a charged capacitor contains energy, the switch is transferred to point 3. This disconnects the battery but places a load (the resistor) in series with the capacitor. Now, with a complete circuit, the *capacitor discharges.* Again current flows, as indicated by the milliameter, until the charge across the plates has dissipated. Plate-to-plate voltage is again zero, and the capacitor is again discharged.

Moving the switch back to point 2 will again charge the capacitor until E_c, the plate-to-plate voltage, again equals V_s. If the switch contact is now moved to point 1, the capacitor will remain charged. A capacitor that is fully charged to some dc potential tends to hold its charge for a very long time. How long the full battery voltage remains across the capacitor depends on several variable factors, some of which are described below.

1. Excessive moisture in the atmosphere causes a film of water particles to appear on the surfaces; this water film tends to conduct current over the capacitor's external surfaces. For this reason, the capacitor loses its charge gradually.
2. If the dielectric is not a perfect insulator, electrons tend to migrate from the negative plate toward the positive plate, even though very slowly, which causes the capacitor to lose its charge after some period of time.
3. If the operating temperature is excessive, thermal agitation tends to cause free electrons to exist in the dielectric. This, of course, causes additional electrons to drift through the dielectric, discharging the capacitor.

Under ideal conditions, a capacitor might hold its charge for several days, while under adverse conditions it will discharge completely in a few minutes.

In summary, then, a capacitor accepts a charge from a dc source, and if allowed, returns the charge to a load. Assuming no loss in the capacitor, it returns as much energy as it received from the source. How much energy is taken by the capacitor is a function of its value of capacitance and the voltage to which it is charged.

Unit of Capacitance

Capacitance is measured in units of the *farad*, named for Michael Faraday, 1791–1867, an English physicist. Because the farad represents a large amount of capacitance, the microfarad (μF) and the picofarad (pF) are used. The μF unit corresponds to 10^{-6} F, while pF corresponds to 10^{-12} F.

A capacitor is said to have one farad of capacitance if, when charged with one coulomb, the plate-to-plate voltage is one volt. That is, if a capacitor is charged with a constant current of one ampere for a period of one second, and if the plate-to-plate voltage is then one volt, the capacitor has a capacitance of one farad.

Factors Determining Capacitance

The factors that determine the amount of capacitance in a given capacitor are:

1. The area of the plates
2. The distance between the plates
3. The substance of which the dielectric is formed

The greater the area of the plates, the greater the amount of capacitance, since the larger plates provide more area onto (or away from) which the electrons can be forced. The volume of the plates, on the other hand, has no bearing on the amount of capacitance, so the material used is often very *thin* to reduce the capacitor's physical size.

The closer together the plates are, the greater the capacitance. As plates are moved closer together, the electrostatic field that exists between them is increased, making it easier to move electrons around the circuit.

Finally, the kind of dielectric used has a large bearing on the amount of capacitance, with some materials increasing the amount of capacitance hundreds, or even thousands, of times compared to the capacitance obtained when air is used as the dielectric. That is, a given set of plates separated by a given distance have a given capacitance using air as the dielectric. Changing nothing but the dielectric to waxed paper increases the capacitance *three times*. This is due to a characteristic of insulators known as the dielectric constant, k. Table 16-1 lists several materials and gives both their dielectric constant and dielectric strength.

The dielectric constant is simply the ratio of the capacitance of a given capacitor with a specific dielectric material, to the capacitance using air as a dielectric, all else remaining the same. Dielectric strength is a measure of the insulating qualities of the dielectric, and is measured in the number of volts per mil (volts per 0.001 in.) that the dielectric is capable of withstanding without arcing (sparking) through.

The figures given in Table 16-1 are only approximate, since the purity of the material, its moisture content, and the ambient temperature can all influence the values. The values are, however, quite representative of actual cases.

TABLE 16-1   CHARACTERISTICS OF DIELECTRICS

Material	Dielectric Constant (k)	Dielectric Strength
Air	$\cong 1$	75
Waxed Paper	3	1300
Glass	7.5–10	3000
Rubber	2–35	700
Mica	3–6	5000
Ceramic	100–1000 or more	500–1500

The dielectric strength indicates the amount of voltage that the dielectric can withstand before breaking down and arcing through. Solid materials are generally non-self-healing, while air and certain oils that are used occasionally can withstand a breakdown arc and be usable again. When an arc occurs in a solid dielectric, the path taken by the discharge often becomes carbonized, which of course produces a conductive path from plate to plate. The capacitor is then said to be shorted, and must be replaced. The carbonized path may have a dc resistance value of anywhere from a few ohms to several megohms, depending upon how severe the arc was.

You can see that the choice of dielectric influences many characteristics of a capacitor. Dielectric thickness and the kind of material influence both the amount of capacitance and the *voltage rating*. Capacitors are usually rated according to the maximum voltage that can safely be placed across them. Small capacitors are designed to be used in low-voltage semiconductor circuits and may have voltage ratings as low as 5.0 V. At the other extreme, large (oil-filled) storage capacitors can safely withstand 150 kV, while special capacitors used to connect, or *couple,* telephone equipment to high-voltage, long-distance electrical power lines are designed for use with *500 kV*! Transmission systems are now in the design stage for *1,100-kV* application. Generally, ratings of more moderate values are encountered.

16-2 TYPES OF CAPACITORS

Capacitors come in a wide assortment of sizes, shapes, and types. Such capacitors fall into two broad categories, fixed and variable. Figure 16-3 illustrates several common kinds of capacitors. The construction and application of each of these is briefly described in the following paragraphs.

Paper-type (Tabular) Capacitors

Paper capacitors (Figure 16-3*a*) are probably the most common type, or at least they were for several years. They originally consisted of two long strips of foil separated by a thin strip of waxed paper slightly larger than the foil. To make as compact a package as possible they were rolled up, leads were attached, and the entire unit was encapsulated in plastic to seal out moisture and protect the capacitor. Modern counterparts are essentially similar, but usually the dielectric is a thin film of plastic, such as Teflon or Mylar. Such capacitors range in size from perhaps 50 pF to 1.0 μF, and are generally considered rugged and long-lasting. Paper-type capacitors are obviously of the fixed type.

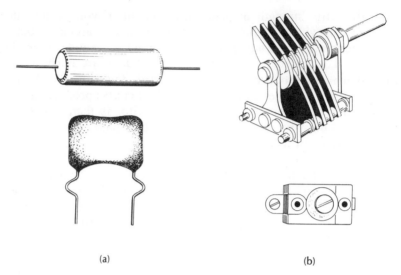

Figure 16-3 Capacitor types: (a) paper, (b) variable.

Note that one of the units illustrated has *preformed* leads for automatic insertion in a PC board.

Variable Capacitors

Variable capacitors (Figure 16-3b) are widely used in rf circuits where a continuously variable capacitance can *tune-in* one particular radio station and reject all others. In all probability, the tuning knob on your radio simply rotates a variable capacitance to tune in various stations, one at a time. The device consists of a set of *stator* (fixed) plates and a set of *rotor* (rotatable) plates. As the one set of plates is rotated, either a smaller or larger area is adjacent to the fixed plates, and so the capacitance varies. When the plates are completely unmeshed, minimum capacitance is produced, while a completely meshed condition yields maximum capacitance. One typical size of this type of capacitor has a maximum capacitance of approximately 475 pF, and is used in conventional AM (amplitude modulation) radios.

Another kind of variable capacitor, also shown in the illustration, is the *trimmer* or *padder*. This type of capacitor is used in applications that require *exact* values of capacitance. Trimmers and padders are usually adjusted when they are installed in the equipment being manufactured, and are readjusted only if the equipment is *realigned* in the field. One such application is in the *local oscillator* of a typical AM radio, which is used to *trim* or adjust, to a very fine degree, the much larger tuning capacitor. Once adjusted properly, a trimmer seldom, if ever, requires further attention.

Trimmers and padders are generally constructed with mica as the dielectric, and one popular construction method is shown. One plate is affixed to the ceramic base and covered with a thin sheet of mica. A springy top plate is compressed by a screw that pushes it closer to the bottom plate or allows it to retract to a more open position.

Electrolytic Capacitors

Electrolytic capacitors are used when large values of capacitance are required, with values extending from approximately 1 μF to something in excess of 50,000 μF. Even the larger-value units are reasonably compact in size.

The electrolytic capacitor is constructed much differently from those previously described. Its very large capacitance is due to the chemically produced dielectric, which is microscopically thin. Hence, the capacitance per unit volume is very large.

Figure 16-4 illustrates the construction of a typical electrolytic capacitor. Two long, thin strips of aluminum foil, separated by a thin paper separator, are rolled into a tubular shape. A flexible connecting lead is attached to each plate. The paper is soaked in the electrolyte, usually a salt solution. The assembly is then sealed in an aluminum can, to prevent the electrolyte from drying out. One plate is attached to the metal can, and the other is attached to a lead post mounted on the end insulator. As will soon be seen, the leads are polarized; that is, the can is negative, while the other lead is positive. For this reason *polarity must be observed* when using this type of capacitor. With other kinds of capacitors, polarity is of no consequence, but this is not true of electrolytics.

To understand why the electrolytic capacitor works as it does, we must look more closely at it. The plate that is to become the positive plate is, along with the negative one, etched in an acid bath. This greatly increases the surface area by making it irregular, rather than smooth and polished. Thus, the effective surface area is increased with no increase in bulk. Then, the positive plate (the anode) is subjected to a process that leaves a thin aluminum oxide film on the etched surface. During manufacture the capacitor is formed by subjecting the circuit to a low-value dc voltage, which causes the oxide film to develop. The outside can is made negative and the center terminal, or terminals, is made positive. To preserve the film, *this polarity must be maintained* throughout the capacitor's useful life. The film is exceedingly thin, often but a few molecules in thickness. This oxide film *is the dielectric* in

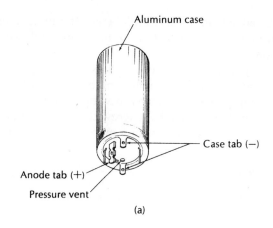
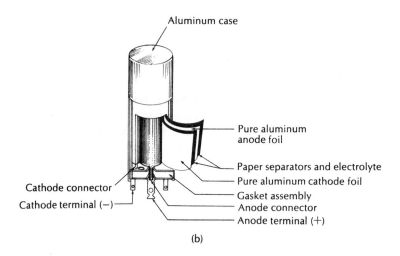

Figure 16-4 (a) A typical electrolytic capacitor and (b) a cutaway drawing of its construction.

the completed device. It is the extreme thinness of the dielectric that causes the capacitor to have large capacitance per unit volume. The fluid electrolyte makes contact with the oxide film more efficiently than by any other means. The electrolyte is therefore the true negative plate, but it, in turn, must make contact with the outside world via the other aluminum foil which is connected to the container.

Because they are polarized, electrolytics are not used in purely ac circuits, but instead are used in either true dc circuits, or those in which

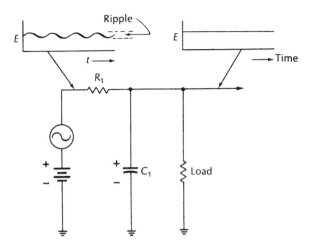

Figure 16-5 In this circuit, an electrolytic capacitor is used as a smoothing filter.

both ac and dc exist simultaneously. A few such applications are briefly described below.

A *filter* capacitor is one used to remove small amounts of ac from an otherwise dc voltage. Such an application is shown in Figure 16-5, where the electrolytic capacitor's ability to store energy at one point in time and then to return it is used to smooth out the voltage variations. As indicated on the drawing, the dc supply has a variation in the amplitude known as a *ripple voltage*. The capacitor absorbs energy when the voltage increases, and returns energy to the circuit when the source voltage decreases, thus removing the peaks and filling in the valleys. The net result is a much smoother dc voltage.

A *coupling* capacitor is used as shown in Figure 16-6. In this example, the capacitor is used to block the dc voltage from one part of the circuit, while passing the ac component. In this application, the capacitor is often called a *blocking* capacitor.

Electrolytic capacitors suffer from a major disadvantage—they exhibit a rather large leakage current, a deficiency explained later in this chapter.

Tantalum Electrolytic Capacitor

When very small size and outstanding reliability are required, a tantalum capacitor is often called for. The basic characteristics of a tantalum capacitor include low voltage, large capacitance, low leakage current, and excellent reliability. A tantalum capacitor utilizes tantalum foil as the anode, on which an oxide film is formed to act as the

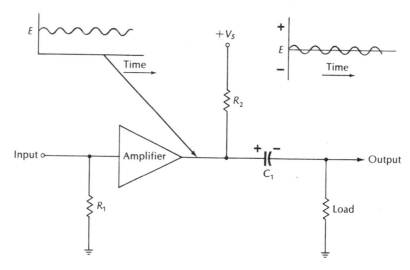

Figure 16-6 An electrolytic capacitor used as a coupling, or blocking, capacitor.

dielectric. The plates are separated by paper soaked in an acid electrolyte. The plates are often sintered to increase the effective area. In such a case, the anode is a sintered *slug* of tantalum. Capacitance values range from perhaps 0.1 to 100 μF, with voltage ranges from 5 to 50 V being typical.

Mica and Ceramic Capacitors

Mica capacitors find wide application in radio-type circuits where losses at high frequencies would be detrimental. Mica capacitors are constructed of alternate layers of mica and foil squares. The foil plates are divided into two groups, each group forming one set of plates. Mica has a relatively high dielectric constant and excellent high-voltage characteristics. Such capacitors have relatively small values of capacitance, with values from 5 pF to 500 pF typical. Voltage ratings to 1 kV are not uncommon. Typical mica capacitor construction is illustrated in Figure 16-7.

Ceramic capacitors find widespread usage in nearly all branches of electronics. They are small, reliable, and rugged, and are inexpensive to manufacture. They are often constructed with metalized foil to reduce bulk to a minimum. The ceramic material (often steatite or barium titanate) of the proper size and shape is placed in a furnace and the plates are formed by depositing the appropriate material (often silver) on the ceramic surfaces. The plates are therefore *very* thin, reducing the

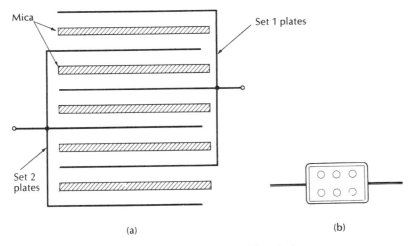

Figure 16-7 (a) The stacked construction and (b) typical appearance of a mica capacitor.

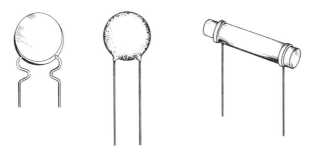

Figure 16-8 Disc and tubular ceramic capacitors.

size of the finished unit. After leads are attached, the capacitor is dipped in a protective insulating coating. Figure 16-8 shows several ceramic capacitors. These capacitors often have disk or tubular shapes.

16-3 CAPACITOR CHARACTERISTICS

In order to be able to apply capacitors in practical circuitry, we must specify their major characteristics. Capacitors are available in a wide variety of types, so to be assured of obtaining the right one, the following characteristics must be specified.

Capacitance

Obviously, the value of capacitance is the most important factor. Its value is given in farads or micro- or picofarads, as mentioned earlier.

Working Voltage

The working voltage of a capacitor is the normal maximum dc voltage that can be safely applied to the capacitor. This value is often printed on the capacitor; for example, 25 WVDC. This indicates that the maximum dc voltage to be applied is 25 V_{dc}, usually specified at the maximum allowable temperature.

Note that if an ac *ripple* voltage is riding on top of the dc, then the maximum combined excursion must not exceed the specified maximum allowable working voltage. The *surge voltage rating* (given on the capacitor specification sheet) is the absolute maximum voltage to which the capacitor can be subjected without total failure.

Tolerance

Just as resistors are rated by their tolerance ($\pm 1\%$, $\pm 5\%$, and so on), so are capacitors. Because it is difficult, and therefore expensive, to construct capacitors with tight tolerance, values of $\pm 10\%$ and $\pm 20\%$ are not uncommon. Indeed, electrolytic capacitors often have tolerances of -40%, $+100\%$, due to the many variables occurring during manufacture.

Temperature Coefficient

Because metal expands when heated, a capacitor changes value when the temperature changes. A measure of the degree of this change is the temperature coefficient, usually measured in *parts per million per degree Celsius* (ppm/°C). If the capacitance increases with a temperature increase, the temperature coefficient is *positive;* if capacitance decreases with a temperature increase, the temperature coefficient is negative.

The normal operating temperature range is sometimes specified, and is often in the range of perhaps $-50°C$ to $125°C$.

Power Factor

In many instances, the power factor has little, if any, bearing on circuit action. In the case of high-quality capacitors, power factor must be considered only if (1) very high frequencies are involved or (2) electrolytic capacitors are used in 60-Hz power applications.

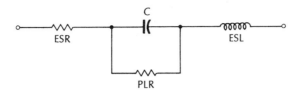

ESR = equivalent series resistance
PLR = parallel leakage resistance
ESL = equivalent series inductance

Figure 16-9 Capacitor equivalent circuit.

The equivalent circuit for a capacitor illustrates the losses that can occur, as shown in Figure 16-9. The equivalent series resistance (ESR) represents the total dc resistance of the connecting leads and plates. The parallel leakage resistance (PLR) represents the leakage current path through the dielectric and has the greatest influence in electrolytic capacitors, where leakage current is relatively large. In capacitors having a wound configuration, the coiled plates exhibit significant inductance *(ESL),* and the effects are especially pronounced at the higher frequencies.

Power factor is simply the ratio of resistance to impedance, or $\cos \theta$, or $\frac{\text{true power}}{\text{apparent power}}$:

$$PF = \frac{R}{Z} = \cos \theta = \frac{P_{\text{true}}}{P_{\text{app}}}.$$

In most simple capacitive circuits, reactive power and apparent power are the same; hence the power factor is unity if resistance is very small. In circuits having resistance, capacitance, *and* inductance, power factor is of much greater importance, because there is a significant amount of resistance in which power is dissipated. Chapter 18 deals thoroughly with such circuits.

Energy Storage

The amount of energy stored in a charged capacitor is determined as follows:

$$J = \tfrac{1}{2} C \times V^2,$$

where J = joules (watt-seconds)
 C = capacity in farads, and
 V = Voltage.

Example 1. What is the amount of energy stored in a 1.0-μF capacitor charged to 100 V?

$$J = \tfrac{1}{2} C \times V^2 = 0.5(1 \times 10^{-6})(100^2)$$
$$= 0.005 \text{ J} = 0.005 \text{ watt-seconds.}$$

That is, the capacitor will supply 1 W of power for 5 ms, or ½ W for 10ms, or 0.005 W for 1 s, or any combination of power and time whose product is 0.005 watt-seconds.

Capacitive Reactance

Capacitive reactance is the opposition to an ac current afforded by a capacitor. In many respects it is similar to inductive reactance. While an inductor opposes a change in current, a capacitor opposes a change in voltage. It is this opposition to a change in voltage that gives rise to the opposition known as capacitive reactance X_C. As is true in the case of inductance, pure capacitive reactance takes power from the source during a portion of the ac cycle, but returns it during a later time. Hence, capacitive reactance consumes no power when averaged over a number of cycles.

Capacitive reactance is measured in ohms, and its value is determined by

$$X_C = \frac{1}{2\pi FC},$$

where F = frequency in hertz,
C = capacitance in farads, and
X_C = reactance in ohms.

This suggests the *inverse* relationship between the size of the capacitor, the frequency, and the number of ohms. That is, a larger capacitor has a smaller reactance, and a given capacitor has a smaller reactance as the frequency is increased. The following examples illustrate this relationship.

Example 2. A 1.0-μF capacitor is to be connected in a 60-Hz circuit. Determine the capacitive reactance.

$$X_C = \frac{1}{2\pi FC} = \frac{1}{6.28 \times 60 \times (1 \times 10^{-6})} = \frac{1}{3.77 \times 10^{-4}} = 2.65 \text{ k}\Omega.$$

Example 3. The same 1.0-μF capacitor is now to be used in a 120-Hz circuit. Find the capacitive reactance.

$$X_C = \frac{1}{2\pi FC} = \frac{1}{6.28 \times 120 \times (1 \times 10^{-6})} = \frac{1}{7.54 \times 10^{-4}} = 1.33 \text{ k}\Omega.$$

Note that doubling the frequency *halves* the reactance.

Example 4. Refer to example 1 above. A 0.1-μF capacitor is substituted for the 1.0-μF unit. Determine the capacitive reactance.

$$X_C = \frac{1}{2\pi FC} = \frac{1}{6.28 \times 60 \times 0.1 \times 10^{-6}} = \frac{1}{3.8 \times 10^{-5}}$$
$$= 26.5 \text{ k}\Omega.$$

Note that reducing the size of the capacitor by a factor of 10 *increases* the reactance by the same factor. Obviously, the reverse is also true; increasing the size of the capacitor decreases its reactance by the same factor, given a constant frequency.

The above formula can also be used to find the value of C if F and X_C are known:

$$C = \frac{1}{2\pi F X_C}.$$

Also, if C and X_C are known, then

$$F = \frac{1}{2\pi X_C C}.$$

When a single capacitor is placed across an ac source, the circuit current is determined by V_s and X_C. For example, a capacitor having an X_C of 50 Ω, connected across 12 V_{ac}, will have an effective current of

$$I_C = \frac{12}{50} = 0.24 \text{ A}.$$

Phase Relationships

In an ac circuit containing capacitance, the current *leads* the voltage. This is illustrated in Figure 16-10a, where the current is shown leading the voltage by 60°. Alternatively, we could say that the voltage is lagging the current.

In a pure capacitive circuit (no dc resistance), the current leads the voltage by 90° as in part *b*. When resistance is introduced, depending upon the amount of resistance relative to the reactance, the amount of current lead is something less than 90°. If resistance and reactance are equal, current leads the voltage by 45°.

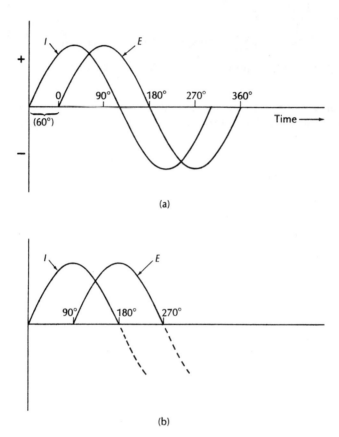

Figure 16-10 Typical phase relationships in a capacitive circuit: (a) Current leads voltage by 60° (some resistance); (b) current leads voltage by 90° (no resistance).

Capacitor Color Code

Modern practice calls for the printing of capacitor values on the body of the capacitor where possible. However, on some small mica and ceramic capacitors, a color code is still used to indicate the value in pF, as well as other characteristics.

When the color code is used on a capacitor, more information is available to the user than would be the case if values were simply printed on the body. Such items as value in pF, tolerance, working voltage, operating temperature range, drift, insulation resistance, and temperature coefficient can be specified with a color code.

Unfortunately, the capacitor color code has never been completely standardized. For example, there have been several different military standards, one for each specific type of capacitor. There is also more

Figure 16-11 Series-connected capacitors.

than one industrial standard, plus several obsolete systems that may still be encountered. All of this leads to much confusion. The reader is referred to Appendix 2 for a comprehensive digest of existing capacitor color codes.

16-4 CAPACITORS IN SERIES AND PARALLEL

When one or more capacitors are placed in an ac circuit, the total circuit response is determined by many of the concepts you have studied thus far. You will find that capacitors in series and parallel combinations require only a slightly modified method of attack compared to those used before. In this section capacitors are described in terms of circuit action for both series and parallel circuit configurations.

Capacitors in Series

When capacitors are connected in series, the total capacitance C_t is *less* than the capacitance of any one of the capacitors. This is true because this type of connection effectively increases the distance between the plates, as illustrated in Figure 16-11, where two identical capacitors connected in series have the same total, or combined, capacitance as a single capacitor with twice the distance between plates. The relationship for capacitors in series is similar to that for resistors in parallel.

$$C_t = \frac{1}{1/C_1 + 1/C_2 + 1/C_3 \ldots},$$

for as many capacitors as are in the series. If only two capacitors are involved, the following equation may be used.

$$C_t = \frac{C_1 C_2}{C_1 + C_2}.$$

Some examples follow to illustrate this determination.

Example 5. Two capacitors, each rated at 0.05 μF, are connected in series. What is the total equivalent capacitance?

$$C_t = \frac{1}{1/C_1 + 1/C_2} = \frac{1}{(1/0.05 \times 10^{-6}) + (1/0.05 \times 10^{-6})} = \frac{1}{4 \times 10^7}$$
$$= 2.5 \times 10^{-8} = 0.025 \ \mu F.$$

Note that two capacitors of equal value have a series equivalent capacitance of $C_1/2$ or $C_2/2$, or half the value of either one.

Example 6. Two capacitors having values of 750 pF and 1250 pF respectively are connected in series. What is the total equivalent capacitance?

$$C_t = \frac{1}{1/C_1 + 1/C_2} = \frac{1}{1/750 + 1/1250} = \frac{1}{2.13 \times 10^{-9}}$$
$$\cong 4.7 \times 10^{-10} = 470 \ pF = 0.00047 \ \mu F.$$

Example 7. Three capacitors have values of 10 μF, 20 μF, and 30 μF. What is the total series-connected capacitance?

$$C_t = \frac{1}{1/C_1 + 1/C_2 + 1/C_3} = \frac{1}{(1/10 \times 10^{-6}) + (1/20 \times 10^{-6}) + (1/30 \times 10^{-6})}$$
$$= \frac{1}{(1 \times 10^5) + (5 \times 10^4) + (3.333 \times 10^4)} \cong 5.45 \ \mu F.$$

As we have said, when capacitors are connected in series, the total capacitance is less than the smallest capacitor. There is, however, a distinct advantage to this connection—the total working (breakdown) voltage, or DCWV, is equal to the *sum* of the individual capacitor ratings. Referring to Example 6 above, if each capacitor is rated at 100 WVDC, the two capacitors connected in series will be rated at 200 WVDC. This technique is frequently used to obtain a working voltage in excess of the individual rated value. However, note that the voltage appearing across each capacitor in an ac circuit is a function of the *reactance* of each. Thus, if their capacitance values differ, the voltage across each capacitor will be different.

Capacitors in Parallel

When capacitors are connected in parallel, as in Figure 16-12, the effective plate area is increased, so that the combined, or total, capacitive value increases. Parallel capacitors are treated the same as series resistances:

$$C_t = C_1 + C_2 + C_3 \ldots, \text{ for as many capacitors as are parallel connected.}$$

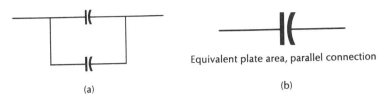

(a) (b)

Equivalent plate area, parallel connection

Figure 16-12 Parallel-connected capacitors.

Example 8. What is the total parallel capacitance of three capacitors, having values of 10 μF, 20 μF, and 30 μF, respectively?

$$C_t = C_1 + C_2 + C_3 = 10 \ \mu F + 20 \ \mu F + 30 \ \mu F = 60 \ \mu F.$$

When capacitors are connected in parallel, the same voltage appears across each, and so the unit with the lowest working-voltage rating determines the maximum safe applied voltage.

X_C in Series and Parallel

When capacitors are connected in series, the capacitive reactance of each is simply added to all others in series with it to obtain a total value. Thus,

$$X_{C_t} = X_{C_1} + X_{C_2} + X_{C_3}.$$

In Figure 16-13a, the three capacitors shown have a total capacitive reactance of 30 Ω. Because reactance is an opposition to current measured in ohms, *reactance* is treated like resistance. Note that the *values* of the capacitors in series, measured in farads, are treated by the relationship

$$C = \frac{1}{1/C_1 + 1/C_2 + 1/C_3. \ldots}$$

When capacitors are connected in parallel, the reactance of each is combined with the others by the double reciprocal formula.

$$X_{C_t} = \frac{1}{1/X_{C_1} + 1/X_{C_2} + 1/X_{C_3}}.$$

In Figure 16-13b, this is a total of 3.33 Ω for the three capacitors shown. As before, reactance is treated similarly to resistance, while the capacitors' *value*, in farads, is not.

We summarize these relationships to allow you to become familiar with them as quickly as possible.

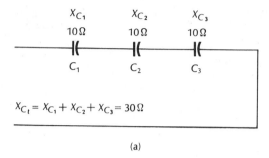

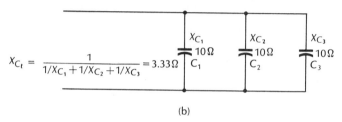

Figure 16-13 X_C in (a) series and (b) parallel.

Series: $C_t = \dfrac{1}{1/C_1 + 1/C_2 + 1/C_3 \ldots}$;

$X_{C_t} = X_{C_1} + X_{C_2} + X_{C_3} \ldots$

Parallel: $C_t = C_1 + C_2 + C_3 \ldots$;

$X_{C_t} = \dfrac{1}{1/X_{C_1} + 1/X_{C_2} + 1/X_{C_3} \ldots}$.

16-5 OPENS AND SHORTS IN CAPACITORS

Normally and ideally, a capacitor presents an open circuit to dc. However, one possible capacitor defect is a short circuit. When a short circuit does develop, the capacitor is said to have become *leaky*. This simply means that the insulation has been degraded and electrons are leaking from one plate to the other through the dielectric. Such a capacitor must be replaced if the amount of leakage current is enough to upset circuit action.

To test a capacitor other than the electrolytic types for leakage, an ohmmeter is used on its *highest* ohms scale. If the capacitor is mounted in existing circuitry, one end is disconnected to avoid *sneak parallel paths*. Then the dc resistance is measured across the capacitor. The capacitor will first charge to the battery voltage of the meter; hence the needle may *kick* upwards for a brief moment if the capacitor is of large enough value. Very quickly, however, the needle should return to infinity (∞) if the capacitor is good. Any value of resistance indicates a bad capacitor. If the capacitor is very small, no needle deflection will be noticed, since the current required to charge small values of capacitance is insufficient to overcome the inertia of the meter movement. When measuring capacitors for leakage, do not touch both leads, because body resistance will lower the reading. Note that some capacitors will not show leakage current when this technique is used, because the meter battery voltage is usually very low, and the leakage may be appreciable only when more voltage is applied. If a leaky capacitor is suspected, it is usually best to replace it with a known good capacitor.

When an electrolytic capacitor is tested with an ohmmeter, some final value of resistance is normal. A comparison with a new, identical unit will indicate whether the old one should be replaced. Because of the polarized leads on an electrolytic capacitor, make the measurement twice, reversing the meter leads, and, if the readings differ, use the *higher* resistance of the two measurements.

Capacitors can become *fully* short circuited; in such a case, an ohmmeter will read very close to zero ohms. This indicates that the plates are probably touching, and the capacitor should be immediately discarded.

An open in a capacitor is more difficult to detect. A small capacitor causes special problems, since no ohmmeter indication can be observed even if the unit is good. Direct substitution with a new capacitor is an accepted procedure. Electrolytic capacitors are especially likely to open, at least partially, because as they age the electrolyte tends to dry out. This reduces the capacitance value, and, when the electrolyte is completely dry, the capacitor is open.

When the foregoing methods fail, the use of a *capacitance checker* is indicated. This test instrument is available in a variety of sizes, shapes, and characteristics. Basically, the checker measures the *capacitive reactance* at a fixed frequency. The meter face is calibrated in μF or pF, so the value can be read directly. The capacitance checker also measures the *power factor,* which gives an idea of the internal resistance, thus indirectly giving leakage information. Instruments have recently been developed and marketed that directly measure capacitive values between 0.1 PF and 0.2 F.

PRACTICE PROBLEMS

1. Convert 0.18 F to pF.
2. Convert 140 pF to μF.

3. Convert 0.00017 F to μF.
4. Convert 0.470 μF to pF.
5. Two capacitors, each of a value of 4.7 μF, are connected in series. Find C_t.
6. Two capacitors, each of a value of 560 pF, are connected in series. Find C_t.
7. Two capacitors, each of a value of 4.7 μF, are connected in parallel. Find C_t.
8. Two capacitors, each of a value of 560 pF, are connected in parallel. Find C_t.
9. A capacitor is connected across an ac source of 12 V at 1000 Hz. If the capacitor's value is 0.0033 μF, what is X_C?
10. A capacitor is connected across an ac source of 12 V at 100 kHz. If the capacitor's value is 0.0033 μF, what is X_C?
11. A capacitor is connected across an ac source of 120 V at 60 Hz. If the capacitor's value is 4.7 μF, what is X_C?
12. Refer to Problem 9. Find the value of current.
13. Refer to Problem 10. Find the value of current.
14. Refer to Problem 11. Find the value of current.
15. Two capacitors are in series across a source of 18 V at 2.5 kHz. If the capacitors have values of 0.033 μF and 0.047 μF, what value is the current?
16. Refer to Problem 15. Find the current if the two capacitors are connected in parallel.

Answers: (1) 180,000 pF; (2) 0.00014 μF; (3) 170 μF; (4) 470,000 pF; (5) 2.35 μF; (6) 280 pF; (7) 9.4 μF; (8) 1120 pF or 0.00112 μF; (9) 48.3 kΩ; (10) 482 Ω; (11) 564 Ω; (12) 0.249 = 0.25 mA; (13) 24.88 mA; (14) 0.213 A; (15) 5.48 mA; (16) 22.6 mA.

SUMMARY

- A capacitor is a device that can pass ac to a greater or lesser degree, but does not pass (blocks) dc.
- A capacitor opposes a change in voltage.
- A charged capacitor will hold its charge for a relatively long time.
- The unit of capacitance is the farad; a capacitor has a value of one farad if it accepts a charge of one coulomb and then has a plate-to-plate voltage of one volt.
- Practical capacitors have ratings in the microfarad (μF) to picofarad (pF) range.
- One $\mu F = 10^{-6}$ F; one pF $= 10^{-12}$ F.
- The three major factors that determine the value of a capacitor are: (1) area of the plates, (2) the distance between plates, and (3) the kind of dielectric.

- Dielectric constant k is a measure of the increase in capacitance, as compared to air, for a substance used as a dielectric.
- Dielectric strength is a measure of the voltage a capacitor's dielectric can safely withstand.
- A variable capacitor has a set of stator plates and a set of rotor plates, or plates that can be squeezed together or separated by a screw adjustment.
- Electrolytic capacitors are used to provide maximum capacitance in minimum size.
- The dielectric of an electrolytic capacitor is aluminum oxide formed as a very thin film.
- Electrolytic capacitors are polarized; therefore polarity must be observed.
- Capacitors are used in many circuit applications, among which are filtering, coupling, and blocking.
- Major capacitor characteristics are (1) capacitance, (2) working voltage, (3) tolerance, (4) temperature coefficient, and (5) power factor.
- The energy stored in a charged capacitor is ½ CV^2.
- A capacitor's opposition to ac current is called capacitive reactance, X_C; $X_C = 1/(2\pi FC)$.
- For capacitors connected in series, total capacitance equals $1/(1/C_1 + 1/C_2 + 1/C_3 \ldots)$ and $X_C = X_{C_1} + X_{C_2} + X_{C_3}$.
- For capacitors connected in parallel, total capacitance equals $C_1 + C_2 + C_3 \ldots$ and $X_{C_t} = 1/(1/X_{C_1} + 1/X_{C_2} + 1/X_{C_3} \ldots)$.

QUESTIONS

1. Briefly describe the physical construction of a capacitor.
2. List three factors contributing to the value of a capacitor.
3. Two nearly identical capacitors are compared. The only difference is that A has a thicker dielectric than B. Which has the greater capacitance?
4. Two nearly identical capacitors are compared. The only difference is that A has an air dielectric and B has a ceramic dielectric. Which has the greater capacitance?
5. Two nearly identical capacitors are compared. The only difference is that A has a plate spacing of 0.01 inch and B has a plate spacing of 0.02 inch. Which will withstand greater voltage?
6. Two 0.05-μF capacitors are connected in series. What is the total capacitance?
7. Two 0.05-μF capacitors are connected in parallel. What is the total capacitance?
8. Briefly describe the voltage and current phase relationship in a capacitor circuit consisting of a single capacitor across an ac source.

9. A single resistor and capacitor are connected in series across an ac source. Briefly describe the voltage and current phase relationships in general terms.

PROBLEMS

1. Convert 2500 pF to μF.
2. Convert 68 pF to μF.
3. Convert 0.047 μF to pF.
4. Convert 0.00075 μF to pF.
5. A 0.075-μF capacitor is charged to 15 V. Find the energy stored in joules.
6. A 100-μF capacitor is charged to 250 V. Find the energy stored in joules.
7. A 0.01-μF capacitor is connected across a 12-V, 120-Hz source. What is X_C?
8. A 1.0-μF capacitor is connected across a 25-V, 1-kHz source. What is X_C?
9. Refer to Problem 7. What is I_t?
10. Refer to Problem 8. What is I_t?
11. Three 1.0-μF capacitors are connected in series. What is the total capacitance?
12. Three capacitors, with values of 5 μF, 10 μF, and 15 μF respectively, are connected in series. What is the total capacitance?
13. Three 1.0-μF capacitors are connected in parallel. Find C_t.
14. Three capacitors, 5 μF, 10 μF, and 15 μF respectively, are connected in parallel. Find C_t.
15. Two capacitors each have an X_C of 50 Ω and are connected in series. Find X_{C_t}.
16. Three series-connected capacitors have reactances of 30 Ω, 40 Ω, and 50 Ω. Find X_{C_t}.
17. Two capacitors connected in parallel have reactances of 50 Ω each. Find X_{C_t}.
18. Three parallel-connected capacitors have reactances of 30 Ω, 40 Ω, and 50 Ω. Find X_C.

CHAPTER 17

CAPACITIVE CIRCUITS

In this chapter, series and parallel circuits (ac and dc) having capacitance are investigated. Resistors and capacitors, when placed in ac circuits, influence voltage and current relationships much as do inductors and resistors. Our primary purpose, then, is to learn how to analyze such circuits for the significant values so a firm understanding of circuit action can be obtained. The topics to be covered are listed below.

17-1 Capacitance in Ac Circuits
17-2 Capacitance in Dc Circuits
17-3 Capacitive Voltage Dividers

17-1 CAPACITANCE IN AC CIRCUITS

While capacitors are used in dc circuits, their most frequent use is probably in alternating-current circuits where the frequency is something other than 60 Hz or other power-line frequencies. You remember that such ac voltages and currents are known as signals. In this section, then, we investigate the property of capacitance in all manner of ac circuitry, including signal-type circuits, where the only basic difference is that the frequency is often higher than 60 Hz.

Series RC Circuits

When a single capacitor is placed in an ac circuit, the relationship between voltage and current is such that they are not in phase. If there is no dc resistance, current leads voltage by 90°. (There is always some amount of resistance, due to the wires themselves, but this value is so low that it can ordinarily be ignored.) If the capacitive reactance is very low, or if the resistance is unusually high, then such resistance must be considered. As a rule of thumb, if R is twenty times smaller than X_C, θ is 87° or greater, and the resistance can be considered negligible (an almost purely capacitive circuit). As the resistance becomes smaller compared to reactance, θ approaches 90°. If R is one hundred times smaller than X_C, $\theta = 89.5°$, and the circuit is considered purely capacitive.)

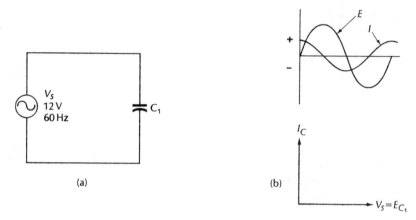

Figure 17-1 (a) A capacitive circuit and (b) the phase relationships for this circuit.

Figure 17-1 illustrates a simple series circuit configuration. If the resistance is assumed to be negligible, the voltage, current, and phase relationships are equally simple. Circuit current is determined by X_C, which is a function of the values of C and F. Assume that C_1 is 0.05 μF. Then

$$X_{C_1} = \frac{1}{2\pi FC} = \frac{1}{6.28 \times 60 \times 0.05 \times 10^{-6}} = 53{,}079 \,\Omega \cong 53 \text{ k}\Omega$$

If the total circuit resistance due to the wires is perhaps 1 Ω or less, it is negligible. Hence, $\theta = -90°$, which indicates that the current *leads* the voltage (or voltage lags current) by 90°, as shown in part *b* of the figure. The negative sign on the angle indicates that the current is capacitive (leads the voltage); positive values indicate inductive currents. Also, since $R \cong 0$, the impedance is, for all practical purposes, equal to the reactance. Circuit current is simply $I_C = V_s/Z = V_s/X_C$, and its value is

$$I_C = \frac{V_s}{X_C} = \frac{12}{53{,}000} = 2.26 \times 10^{-4} = 0.226 \text{ mA}$$

The voltage across the capacitor is, of course, in phase with the source, and so the current is leading the voltage by 90°. Note that the phasor diagram has V_s (or E_{C_1}) as the reference. Therefore, to indicate a leading current, the current phasor is drawn upward.

When appreciable resistance appears in the circuit with the capacitor, the phase angle is less than 90°. That is, the current leads the voltage by something less than 90°. How much less is a function of the ratio between X_C and R, much the same as is the case for inductive

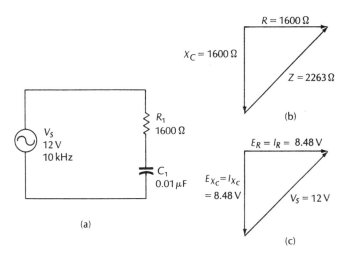

Figure 17-2 (a) A series RC circuit (b) the impedance phasor; (c) the voltage phasor.

circuits. Figure 17-2a illustrates a series circuit containing both R and C. Assume $V_s = 12$ V at 10 kHz, while $C = 0.01$ μF and $R = 1600$ Ω. To illustrate the relationships for this circuit, we shall determine the values of $X_C, Z, I_t, \theta, E_{R_1},$ and E_{C_1}. Note that, since the same current flows through all components, current becomes the reference phasor.

$$X_C = \frac{1}{2\pi FC} = \frac{1}{6.28 \times 10 \times 10^3 \times 0.01 \times 10^{-6}} = \frac{1}{6.28 \times 10^{-4}} \cong 1.6 \text{ K}\Omega$$

$$Z = \sqrt{R^2 + X_C^2} = \sqrt{1600^2 + 1600^2} = \sqrt{5{,}120{,}000} = 2263 \; \Omega.$$

$$I_t = \frac{V_s}{Z} = \frac{12}{2263} = 0.0053 = 5.3 \text{ mA}.$$

$$\theta = \arctan \frac{X_C}{R} = \arctan \frac{1600}{1600} = \arctan 1 = -45°.$$

The current leads the voltage (or the voltage lags the current) by 45°. Now, in a series circuit such as this, the current must be the same in all parts of the circuit, so it is convenient to consider current as the reference, and to relate the phase relationships of voltage drops to the current.

The magnitude of voltage drop across the resistor is a function of the current through it, and is therefore in phase with current.

$$E_{R_1} = I_t \times R_1 = 5.3 \text{ mA} \times 1600 \; \Omega = 8.48 \text{ V};$$
$$\theta = 0°.$$

The magnitude of voltage drop across the capacitor is

$$E_{X_{C_1}} = I_t \times X_{C_1} = 5.3 \text{ mA} \times 1600 \text{ } \Omega = 8.48 \text{ V}.$$

This voltage, however, lags the current by 90°. Note that the voltage across the capacitor and the voltage across the resistor are 90° displaced. To verify that the two voltage drops are in quadrature (at 90° to each other), and are of the correct amplitude, the following formula is used.

$$V_s = \sqrt{E_{R_1}^2 + E_{C_1}^2} = \sqrt{8.48^2 + 8.48^2} = \sqrt{143.8} = 11.99 \cong 12 \text{ V}.$$

The phasor diagram in Figure 17-2b illustrates the impedance vector, where R is the reference, and X_C is drawn at a 90° angle to R. The hypotenuse, then, represents the total circuit impedance. The phasor diagram of the voltage drops, drawing c, shows that they also operate at a 90° angle. Hence, since $R = X_C$, each has an 8.48V drop across it. Because of the 90° phase difference, these drops cannot simply be added, as explained above. Each voltage is displaced from the other by 90°, so they must be summed phasorially.

Parallel RC Circuits

When a capacitor and a resistor are placed in the parallel connection, as shown in Figure 17-3, the voltage phasor becomes the reference, since the same voltage appears across V_s, R, and C_1. Here, the voltage and current relationship of the resistor is simply that V_s appears across R_1, so current is in phase with both V_s and E_{R_1}. The current in the capacitor leg of the circuit, however, *leads* the applied voltage (and E_{C_1}) by 90°. The *total* current, therefore, is the phasor sum of I_{R_1} and I_{C_1}. The total current supplied by the source is at a phase angle determined by the relative magnitudes of I_{R_1} and I_{C_2}. A simple example will serve to illustrate these factors.

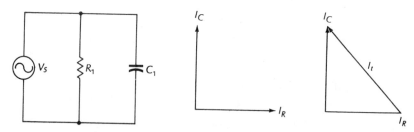

Figure 17-3 A parallel RC circuit for the discussion in text.

Example 1. In Figure 17-3, assume that V_s is 12 V at 10 kHz, while $C_1 = 0.01$ μF and $R_1 = 1.6$ kΩ. Find the values of X_C, Z, I_t, θ, I_{R_1} and I_{C_1}.

$$X_C = \frac{1}{2\pi FC} = \frac{1}{6.28 \times 10 \times 10^3 \times 0.01 \times 10^{-6}} = \frac{1}{6.28 \times 10^{-4}}$$
$$= 1592 \cong 1600 \ \Omega.$$

$$Z = \frac{RX_C}{\sqrt{R^2 + X_C^2}} = \frac{2.56 \times 10^6}{\sqrt{5.12 \times 10^6}} = 1131 \ \Omega.$$

$$I_t = \frac{V_s}{Z} = \frac{12}{1131} = 0.0106 = 10.6 \ \text{mA}.$$

$$\theta = \arctan \frac{R}{X_C} = \arctan \frac{1600}{1600} = -45°.$$

$$I_{R_1} = \frac{V_s}{R_1} = \frac{12}{1600} = 0.0075 = 7.5 \ \text{mA}.$$

$$I_{C_1} = \frac{V_s}{X_C} = \frac{12}{1600} = 0.0075 = 7.5 \ \text{mA}.$$

Note that an alternate way to find Z is to determine values for I_{R_1} and I_{C_1} first. Then, because the two currents are at a 90° phase angle,

$$I_t = \sqrt{I_{R_1}^2 + I_{C_1}^2} = \sqrt{7.5^2 + 7.5^2} = 0.0106 = 10.6 \ \text{mA},$$

and

$$Z = \frac{V_s}{I_t} = \frac{12 \ \text{V}}{10.6 \ \text{mA}} = 1132 \ \Omega.$$

In a parallel circuit such as this, the two branch currents are not in phase; instead they are 90° out of phase, so they must therefore be summed phasorially. In a *series* RC circuit, the same is true of the voltage drops across R and C, and you will recall that they, too, were summed phasorially.

Compound Parallel Circuits

A compound parallel RC circuit is shown in Figure 17-4a. In order to illustrate the full procedure for analyzing such an arrangement, we shall require nearly all of the principles studied thus far. Multiplication, division, and addition of complex numbers are required to solve all unknown factors in this circuit. To make the description easier to follow, each procedure is given a step number. The steps are organized in the simplest possible sequence, but, depending upon the calculation aids available, some of them could be combined. That is, for example, if an electronic calculator having polar and rectangular conversion is

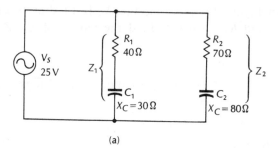

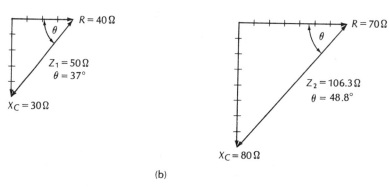

Figure 17-4 (a) A compound parallel RC circuit, and (b) the phasor diagrams for branches 1 and 2.

available, many of these steps can be combined and the entire calculation greatly simplified. If such a device is not available, the solution is still not difficult, as long as you work carefully to avoid confusion. Label all your answers clearly so when you must find an intermediate value you will not select the wrong one.

First of all, realize that the circuit values are solved for with the same basic techniques that are used in dc circuits, allowing for the phase differences in the ac circuit. Therefore, the total impedance can be found by either of the following formulas:

$$Z_t = \frac{Z_1 \times Z_2}{Z_1 + Z_2} = \frac{1}{1/Z_1 + 1/Z_2}.$$

Keep in mind that Z_t, Z_1, and Z_2 are *complex numbers*, since voltage and current values are *not in phase*. To start the calculations, then, we first determine the individual branch impedances.

Find the values of Z_1 and Z_2, including the individual phase angles.

Step 1. $Z_1 = 40 - j30 = \sqrt{40^2 + 30^2} = 50\ \Omega.$

$$\theta = \arctan \frac{X_c}{R} = \arctan \frac{30}{40} = -36.87 - 37°.$$

Step 2. $Z_2 = 70 - j80 = \sqrt{70^2 + 80^2} = 106.3 \ \Omega$.

$\theta = \arctan \dfrac{X_c}{R} = \arctan \dfrac{80}{70} = -48.8°$.

For future use, convert the rectangular form to the polar form for both values; phasor diagrams for each are given in Figure 17-4b.

Step 3. $Z_1 = 40 - j30 = 50\underline{/-37°}$;
$Z_2 = 70 - j80 = 106.3\underline{/-48.8°}$.

Once the branch impedances are known, the product-over-the-sum formula can be used to find Z_t. Because multiplication, division, and addition of complex numbers are involved, both polar and rectangular forms are used: the polar form for the numerator, and the rectangular form for the denominator.

Step 4. $Z_t = \dfrac{Z_1 \times Z_2}{Z_1 + Z_2} = \dfrac{(50\underline{/-37°}) \times (106.3\underline{/-48.8°})}{(40 - j30) + (70 - j80)}$

Next, multiply the numerator factors.

Step 5. $(50\underline{/-37°})(106\underline{/-48.8°}) = 5300\underline{/-85.8°}$.

Then add the denominator factors.

Step 6. $40 - j30$
$\underline{70 - j80}$
$110' - j110$.

Convert the rectangular sum to polar form so the division indicated in step 4 can be performed:

$110 - j110 = \sqrt{110^2 + 110^2} = 155.56 \cong 156$;

$\theta = \arctan \dfrac{110}{110} = -45°$.

Therefore, the polar form is $156\underline{/-45°}$.

Now rewrite the equation from step 4 using the product (step 5) and the sum (step 6), both expressed in polar form, and perform the indicated division.

Step 7. $Z_t = \dfrac{5300\underline{/-85.8°}}{156\underline{/-45°}} = 33.97\underline{/-40.8°} \cong 34\underline{/-40.8°}$.

Thus, the total impedance presented to the source is 34 Ω, and the phase difference between V_s and I_t is 40.8°.

Now the value of the total current can be determined.

Step 8. $I_t = \dfrac{V_s}{Z_t} = \dfrac{25}{34} = 0.735 \ \text{A}\underline{/-40.8°}$.

Note that total current could have been found in step 3, once Z_1 and Z_2 were known, by finding and combining individual branch currents, since $I_2 = \dfrac{V_s}{Z_1}$ and $I_2 = \dfrac{V_s}{Z_2}$.

Branch currents can now be found.

Step 9. $I_1 = \dfrac{V_s}{Z_1} = \dfrac{25}{50} = 0.5 \text{ A}\underline{/-37°}$ (the angle is read from step 3).

$I_2 = \dfrac{V_s}{Z_2} = \dfrac{25}{106.3} = 0.235 \text{ A}\underline{/-48.8°}$

(the angle is read from step 3).

The individual voltage drops across each component can now be found.

Step 10. $E_{R_1} = I_1 \times R_1 = 20 \text{ V}.$
$E_{C_1} = I_1 \times X_{C_1} = 15 \text{ V}.$
$E_{R_2} = I_2 \times R_2 = 16.45 \text{ V}.$
$E_{C_2} = I_2 \times X_{C_2} = 18.9 \text{ V}.$

These voltage drops can now be verified, and of course will be equal to V_s across each branch.

$V_{s\text{(branch 1)}} = \sqrt{E_{R_1}^2 + E_{C_1}^2} = \sqrt{20^2 + 15^2} = \sqrt{625} = 25 \text{ V}.$
$V_{s\text{(branch 2)}} = \sqrt{E_{R_2}^2 + E_{C_2}^2} = \sqrt{16.45^2 + 18.9^2} = \sqrt{627.8} \cong 25 \text{ V}.$

You can appreciate that the foregoing procedure is not overly difficult, but since many steps are involved, it only seems so. Work the example yourself, step by step, until all processes are clear and understandable, and until you see how the steps are organized.

Note that the same general procedure is applicable for inductive circuits. In a later chapter, *R, C,* and *L* will be combined in a single circuit, which will be analyzed in much the same manner.

Capacitive Circuit Example

To assist you in becoming familiar with capacitor applications, an example is given and step-by-step calculations are worked out. Examine each part of the problem closely, so that you will be familiar with the overall circuit action.

Figure 17-5 illustrates a widely used capacitor application that was briefly mentioned in Chapter 16. Capacitors can be effectively used to *couple* ac signals and at the same time to *block* any dc level upon which the ac signal is riding. Note V_s in the figure; it consists of an ac signal riding on a 6-V_{dc} level. The waveform of V_s is given in Figure 17-6, where this relationship is clearly seen. Such a waveform is extremely common in practical circuitry, and it is nearly always necessary to remove the

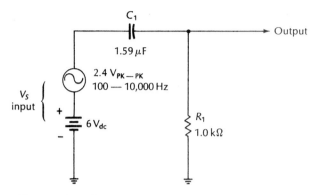

Figure 17-5 In this RC circuit, the capacitor is used to couple the input signal to the output circuit.

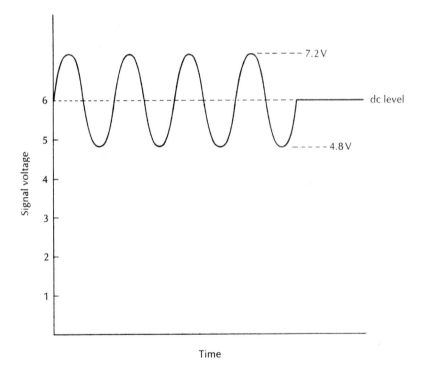

Figure 17-6 Graph of the input voltage of the circuit in Figure 17-5.

dc component from the waveform. Either a transformer or a capacitor could be used for this purpose, but the capacitor is used in this example. The voltage at the output terminal must appear as shown in Figure 17-7, where it is now a true ac voltage, operating around zero volts,

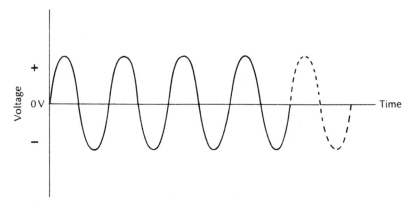

Figure 17-7 Graph of the output voltage of the circuit in Figure 17-5.

with equal excursions on either side of zero. The dc part of the waveform has been eliminated, or blocked, and the output waveform is a true alternating current.

By applying some of the foregoing principles regarding analysis of ac circuits, the action of this circuit can be described. First, note the frequency of the ac part of the input signal. It is specified as extending from a minimum of 100 Hz to a maximum of 10,000 Hz. Such a signal would be known as an audio signal, signifying that it is within the range of frequencies audible to the human ear. One of the characteristics to be determined regarding this circuit is its *frequency response*. The frequency response is a measure of how the circuit responds when signals with different frequencies are applied. For example, if the ac part of V_s is applied at 100 Hz, circuit response will be one value. When only the frequency is changed, say to 1 kHz, the response may change. And, at 10 kHz, it may be still different. It is necessary, therefore, to choose several frequencies within the *bandpass* (which extends from 100 Hz to 10 kHz in the example) and calculate the required response for each instance. Calculation points chosen are 10 Hz, 100 Hz, 1 kHz, 10 kHz, and 100 kHz. Note that 10 Hz and 100kHz are beyond the specified bandpass. These points are added to verify response beyond the specifications.

Because the ac signal voltage is given in terms of a peak-to-peak value, the first step is to find the rms value.

Step 1. $E_{\text{input (rms)}} = 0.707 \times \dfrac{E_{\text{pk-pk}}}{2} = 0.8485 \cong 0.85\ V_{\text{rms}}$.

If the output signal voltage, which is the voltage across R_1, is equal to the input signal voltage, there is no loss attributable to the capacitor, and $E_{\text{out}} = E_{\text{in}}$. This is of course an ideal situation, since if current flows through a resistance or reactance, a voltage drop occurs. If a voltage drop occurs across the capacitor, the output will then be less, since C_1 and R_1 are in series to the signal current. R and C are, in effect, a voltage divider, and any voltage drop across the capacitor will detract from the output voltage. Whether this has a significant effect on the output depends upon the relative magnitudes of R and X_C.

To determine the frequency response of this circuit, a value of reactance must be found for each frequency of interest. Once these values are known, the impedance of the circuit at each frequency can be determined. Then total current can be calculated for each condition, which will allow the individual voltage drops to be found. The voltage drop across the resistor for each applied frequency will then allow a graph to be constructed that will give a visual presentation of the overall frequency response.

Step 2. Determine X_C at the five frequencies of interest.

At 100 kHz: $X_{C_1} = \dfrac{1}{2\pi FC} = \dfrac{1}{6.28 \times 1 \times 10^5 \times 1.59 \times 10^{-6}}$

$= \dfrac{1}{0.9990264} = 1.00097 \cong 1 \, \Omega.$

At 10 kHz: $X_{C_1} = \dfrac{1}{2\pi FC} = \dfrac{1}{6.28 \times 1 \times 10^4 \times 1.59 \times 10^{-6}}$

$= \dfrac{1}{0.0999} \cong 10 \, \Omega.$

At 1 kHz: $X_{C_1} = \dfrac{1}{2\pi FC} = \dfrac{1}{6.28 \times 1 \times 10^3 \times 1.59 \times 10^{-6}} = \dfrac{1}{0.00999}$

$\cong 100 \, \Omega.$

At 100 Hz: $X_{C_1} = \dfrac{1}{2\pi FC} = \dfrac{1}{6.28 \times 1 \times 10^2 \times 1.59 \times 10^{-6}} = \dfrac{1}{0.000999}$

$\cong 1000 \, \Omega.$

At 10 Hz: $X_{C_1} = \dfrac{1}{2\pi FC} = \dfrac{1}{6.28 \times 10 \times 1.59 \times 10^{-6}} = \dfrac{1}{0.0000999}$

$\cong 10{,}000 \, \Omega.$

Tabulate all values for easy reference:

X_C at 100 kHz = 1 Ω
X_C at 10 kHz = 10 Ω
X_C at 1 kHz = 100 Ω
X_C at 100 kHz = 1000 Ω.
X_C at 10 kHz = 10,000 Ω.

Step 3. Determine the impedance and the phase angle for each of the five conditions. The phase angle is not required for the given solution, but it is interesting to see what its value is for each condition.

At 100 kHz: $Z = 1000 + j1 = \sqrt{1000^2 + 1^2} \cong 1000\ \Omega$;
$\theta = \arctan X_C/R = \arctan 1/1000 = -0.0573°$.
At 10 kHz: $Z = 1000 + j10 = \sqrt{1000^2 + 10^2} = 1000\ \Omega$;
$\theta = \arctan X_C/R = \arctan 10/1000 = -0.573°$.
At 1 kHz: $Z = 1000 + j100 = \sqrt{1000^2 + 100^2} = 1005\ \Omega$;
$\theta = \arctan X_C/R = \arctan 100/1000 = -5.71°$.
At 100 Hz: $Z = 1000 + j1000 = \sqrt{1000^2 + 1000^2} = 1414\ \Omega$;
$\theta = \arctan X_C/R = \arctan 1000/1000 = -45°$.
At 10 Hz: $Z = 1000 + j10{,}000 = \sqrt{1000^2 + 10{,}000^2} = 10{,}050\ \Omega$;
$\theta = \arctan X_C/R = \arctan 10{,}000/1000 = -84.3°$.

Convert to polar form and tabulate for easy reference:

Z at 100 kHz = $1000/\!-\!0.0573°$
Z at 10 kHz = $1000/\!-\!0.573°$
Z at 1 kHz = $1005/\!-\!5.71°$
Z at 100 Hz = $1414/\!-\!45°$
Z at 10 Hz = $10{,}050/\!-\!84°$.

Step 4. Determine the total effective current for each applied frequency.

At 100 kHz: $I_t = V_s/Z = 0.85/1000 = 0.85\ \text{mA}/\!-\!0.0573°$.
At 10 kHz: $I_t = V_s/Z = 0.85/1000 = 0.85\ \text{mA}/\!-\!0.573°$.
At 1 kHz: $I_t = V_s/Z = 0.85/1005 = 0.846\ \text{mA}/\!-\!5.71°$.
At 100 Hz: $I_t = V_s/Z = 0.85/1414 = 0.601\ \text{mA}/\!-\!45°$.
At 10 Hz: $I_t = V_s/Z = 0.85/10{,}050 = 84.6\ \mu\text{A}/\!-\!84.3°$.

Step 5. Find the voltage drops across C_1 and across R_1 for each applied frequency.

At 100 kHz: $E_{C_1} = I_t X_C = 0.85\ \text{mA} \times 1\ \Omega = 0.85\ \text{mV}$.
At 10 kHz: $E_{C_1} = I_t X_C = 0.085\ \text{mA} \times 10\ \Omega = 8.5\ \text{mV}$.
At 1 kHz: $E_{C_1} = I_t X_C = 0.846\ \text{mA} \times 100\ \Omega = 84.6\ \text{mV}$.
At 100 Hz: $E_{C_1} = I_t X_C = 0.601\ \text{mA} \times 1000\ \Omega = 0.601\ \text{V}$.
At 10 Hz: $E_{C_1} = I_t X_C = 84.6\ \mu\text{A} \times 10{,}000\ \Omega = 0.846\ \text{V}$.

At 100 kHz: $E_{R_1} = I_t R_1 = 0.85\ \text{mA} \times 1000\ \Omega = 0.85\ \text{V}$.
At 10 kHz: $E_{R_1} = I_t R_1 = 0.85\ \text{mA} \times 1000\ \Omega = 0.85\ \text{V}$.
At 1 kHz: $E_{R_1} = I_t R_1 = 0.846\ \text{mA} \times 1000\ \Omega = 0.846\ \text{V}$.
At 100 Hz: $E_{R_1} = I_t R_1 = 0.601\ \text{mA} \times 1000\ \Omega = 0.601\ \text{V}$.
At 10 Hz: $E_{R_1} = I_t R_1 = 84.6\ \mu\text{A} \times 1000\ \Omega = 0.0846\ \text{V}$.

Tabulate these values:

Frequency	E_C	E_{R_1}
100 kHz	0.85 mV	0.85 V
10 kHz	8.5 mV	0.85 V
1 kHz	84.6 mV	0.846 V
100 Hz	0.601 V	0.601 V
10 Hz	0.846 V	0.0846 V

Step 6. Calculate the percentage output in terms of input ($\% = E_{out}/E_{in} \times 100$).

Frequency	E_{in}	$E_{R_1} = E_{out}$
100 kHz	0.85 V	$0.85/0.85 \times 100 = 100\%$
10 kHz	0.85 V	$0.85/0.85 \times 100 = 100\%$
1 kHz	0.85 V	$0.846/0.85 \times 100 = 99.5\%$
100 Hz	0.85 V	$0.601/0.85 \times 100 = 70.7\%$
10 Hz	0.85 V	$0.0846/0.85 \times 100 = 9.95 \cong 10\%$

Step 7. Plot these percentage values on semilog paper as shown in Figure 17-8. The solid-line portion of the graph is a point-to-point plot of the five values calculated above. If we had determined more values

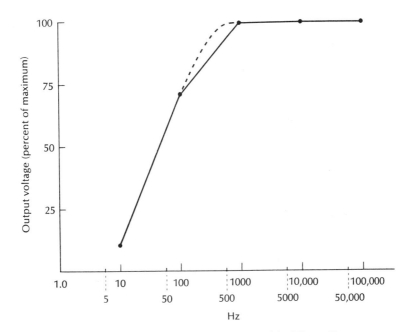

Figure 17-8 Plot of the frequency response of the RC coupling circuit.

between 100 Hz and 1 kHz, the actual response would follow the dashed line, this being a smooth, continuous curve.

Now, note how rapidly the response falls off as the frequency is reduced. This, of couse, is due to the increasing capacitive reactance as the frequency is made lower. At about 1000Hz, a point is reached where the drop across X_C is negligible as frequency is increased, and from that point on, $E_{out} = E_{in}$ because the capacitor is now essentially a short circuit to the high-frequency ac signal. To obtain more output at *lower* frequencies, either the capacitor can be made larger, the resistor made larger, or both.

17-2 CAPACITORS IN DC CIRCUITS

Capacitors are used only occasionally in purely dc circuits. If the voltage never varies, there is little point in using a capacitor. However, to illustrate the *basic action* of a capacitor in circuits using other than sinusoidal waveforms, it is convenient to use dc circuits. Futhermore, because much modern circuitry is of the pulse, or digital, type, where the waveforms are in many ways similar to dc circuits, most of this basic information is directly applicable, with only slight modification. The title of this section is therefore somewhat of a misnomer, but as you continue your studies the similarity of pulse waveforms and dc-type circuitry will become more and more apparent.

Capacitor Charge and Discharge

When a capacitor is caused to go from a state of lower charge to one of higher charge (lower to higher voltage from plate to plate), the capacitor is being (or has been) charged. When the reverse occurs (high to lower voltage, plate to plate), it is said to be discharging (or it has been discharged). A fully charged capacitor is one that has become charged to the supply voltage. If a fully charged capacitor is removed from a 6-V source (6-V, plate to plate) and placed across a 12-V battery (same polarity) it will again accept charging current until plate-to plate voltage is 12V. If the capacitor is charged to 12 V and then connected across a 6-V source (same polarity) it will *discharge* through the source until plate-to-plate voltage is 6 V. Thus, fully charged refers only to the source voltage and is only a relative term.If the source is continually increased, a capacitor will continue to charge until its dielectric breaks down.

When a capacitor charges from zero charge to full charge, the voltage from plate to plate cannot increase instantly. Figure 17-9 illustrates the way the voltage across the capacitor increases on charge and

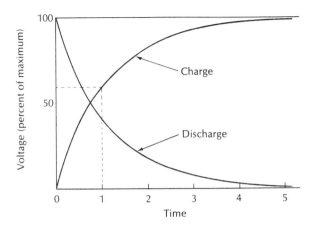

Figure 17-9 Charge-discharge curves.

decreases on discharge. The curve has the distinctive shape of an *exponential*, which is also the shape that describes the current in an inductor. The charge curve increases exponentially with respect to time, while the discharge curve decreases exponentially.

The amount of time required for a capacitor to become fully charged from a discharged state depends upon two factors, (1) the value of the capacitor and (2) the value of the resistance. For example, a large capacitor in series with a high-value resistor takes relatively longer to charge fully than a small capacitor in series with a low-value resistor.

A measure of the time required to charge a simple RC circuit is its *time constant* (t_c), which is numerically equal to the product of R and C:

$$t_c = R \times C,$$

where t_c = time constant in seconds,
R = resistance in ohms, and
C = capacitance in farads.

Note that knowing any two of these values allows you to find the third, since $C = t_c/R$ and $R = t_c/C$. For example, a 1.0-µf capacitor in series with a 1.0-MΩ resistor has a time constant of

$$t_c = R \times C = (1 \times 10^{-6})(1 \times 10^6) = 1.0 \text{ s}.$$

Now, the time constant is a period, or interval, of time during which the capacitor voltage will go from zero (discharged) to 63.21206%, or approximately 63.2%, of full charge. The abscissa of the graph is incremented in time constants. The time interval from the point of origin to 1 represents one time constant. Similarly, the interval from 1 to 2 also represents one time constant. It is apparent that it requires about 5 time constants to reach full charge. In theory, a capacitor *never* reaches absolute full charge. The reason for this is explicit in the following description.

During the interval of the first time constant, the capacitor reaches 63.2% of full charge. During the second t_c, 63.2% of the *remaining charge* is attained; during the third t_c, 63.2% of this remaining charge is attained, and so on. Thus the capacitor can never reach full charge, since only 63.2% of remaining charge is reached after each succeeding interval.

However, in practical circuitry, we consider a capacitor fully charged after five time constants. After this period, the capacitor has reached 99.3262 ≅ 99.3% of full charge. After ten time constants, a value of 99.99546 ≅ 99.99% has been attained. The percentage of full charge for one through ten time constants is listed in Table 17-1, along with discharge percentages.

During discharge, much the same action occurs. After full charge, a capacitor loses 63.2% of its charge in one t_c, and thus has left about 36.8% of the full amount of charge. After two time constants have elapsed, approximately 13.53% remains. Again, after five time constants have elapsed, the capacitor is essentially discharged, having only 0.7% of full charge left.

A circuit used to develop the charge and discharge curves is shown in Figure 17-10. The position of the switch determines the capacitor action. If the switch is in position 3, and has been for a period of time, the capacitor is discharged. If the switch is moved to position 2, nothing occurs except the series circuit is now open. Transferring the switch to position 1 connects the battery to the RC circuit; the capacitor begins to charge, and the growth of voltage e_C across its plates follows the charge curve illustrated in Figure 17-9. After five time constants, the capacitor is fully charged (99.3%), and the circuit makes no further

TABLE 17-1 PERCENT OF FULL CHARGE, 1 TO 10 t_c

t_c	Value Attained % of V_s, CHG	Value Remaining % of V_s DISCHG
1	63.2	36.79 ≅ 36.8
2	86.5	13.5
3	95.0	5.0
4	98.2	1.8
5	99.3	0.7
6	99.75	0.25
7	99.91	0.09
8	99.97	0.03
9	99.988	0.012
10	99.995	0.005

Capacitive Circuits

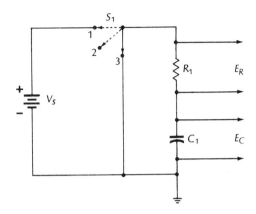

Figure 17-10 A simple RC charge-discharge circuit.

change of significance. As long as S_1 is in position 1, the capacitor voltage approximately equals V_s, and no further current flows.

Now, when the switch is transferred to position 2, the capacitor remains charged, assuming a perfect dielectric and no other discharge paths. When the switch is moved to position 3, a complete discharge path is produced, and current again flows out of the capacitor, draining it of charge. Again, the plate-to-plate voltage decreases according to the discharge curve, over a period of time, and since the circuit provides the same R and C values, the same amount of time is required to discharge as the circuit took to charge. After five time constants, the capacitor is essentially discharged, with only 0.7% of full charge remaining. The capacitor plate-to-plate voltage is essentially zero. Figure 17-11 details these occurrences, showing charge and discharge currents.

Charge and Discharge Equations

Because it is difficult to use curves such as the one given in Figure 17-9 with any degree of accuracy, the equation that describes the charge curve is used for accurate results. This equation is given below. Note its similarity to one given earlier in relation to inductive circuits

$$e_C = V_s(1 - \epsilon^{-t/RC}),$$

where e_C = instantaneous plate-to-plate capacitor voltage at the time of interest. Note that lowercase e is used for instantaneous values;
V_s = applied dc voltage;
ϵ = the base of the natural log system, $2.7182818 \cong 2.72$;
t = time of interest after charge starts; and
RC = time constant $R \times C$.

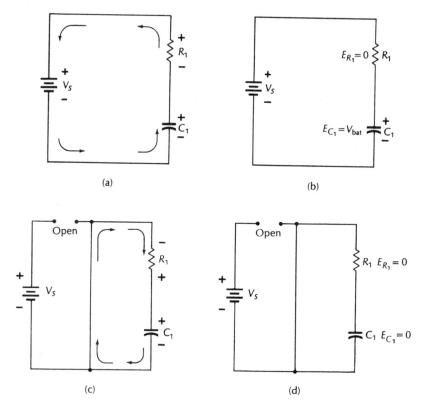

Figure 17-11 A series RC circuit showing (a) direction of current and voltage polarity during charge; (b) circuit conditions after full charge is reached; (c) direction of current and voltage polarity during discharge; and (d) circuit conditions after complete discharge.

Example 2. In Figure 17-10, assume that $V_s = 100$ V, $R_1 = 100$ kΩ, and $C = 0.1$ μF. What is the voltage across the capacitor 30 milliseconds after charge starts?

$$RC = (1.0 \times 10^5)(0.1 \times 10^{-6}) = 0.01 \text{ s} = 10 \text{ ms}.$$
$$e_C = V_s(1 - \epsilon^{-t/RC}) = 100(1 - 2.718^{-0.03/0.01})$$
$$= 100\left(1 - \frac{1}{2.718^3}\right) = 100\left(1 - \frac{1}{20.08}\right)$$
$$= 100(1 - 0.0498) = 100(0.95) = 95 \text{ V}.$$

In this example, the interval of interest is a integral number of time constants (3), which of course could be obtained from the curves of Figure 17-9 or from Table 17-1. The following example uses an interval that is a fractional part of a time constant.

Example 3. Using the same circuit, what is the capacitor voltage after a period of 6.5 ms has elapsed?

$$e_C = V_s(1 - \epsilon^{-t/RC}) = 100(1 - 2.718^{-0.0065/0.01})$$
$$= 100\left(1 - \frac{1}{2.718^{0.65}}\right) = 100\left(1 - \frac{1}{1.916}\right)$$
$$= 100(1 - 0.522) = 100(0.478) = 47.8 \text{ V}.$$

The discharge equation generates the decreasing curve of Figure 17-9. It too is an exponential equation, since the curve itself has an exponential form.

$$e_C = V_s(\epsilon^{-t/RC}),$$

where e_C = capacitor voltage at time of interest;
V_s = fully charged voltage;
ϵ = epsilon, $2.718 \cong 2.72$;
t = time of interest from the start of discharge; and
RC = time constant.

Example 4. Using the same circuit as before, assume the 0.1-μF capacitor is fully charged to 35 V. What is E_C when discharge has been occurring for 17 ms?

$$e_C = V_s(\epsilon^{-t/RC}) = 35(2.718^{-0.017/0.01})$$
$$= 35\left(\frac{1}{5.474}\right) = 35 \times 0.183$$
$$= 6.4 \text{ V}.$$

That is, after discharging from 35 V for 1.7 time constants, the capacitor has 6.4 V remaining, plate to plate.

17-3 CAPACITIVE VOLTAGE DIVIDERS

When capacitors are connected across a source of voltage in a series connection, the total voltage *divides*, with part of the total appearing across each capacitor. This is true whether the source is dc or ac, although the procedure used to calculate each capacitor's voltage is different for each case.

Figure 17-12a illustrates two series capacitors across a 100-V_{dc} source. The magnitude of voltage across each capacitor is *inversely* proportional to the size of the capacitor in farads, compared to the other. That is, the smaller-value capacitor has the higher voltage across it, while the larger-value capacitor has the smaller voltage. The reason for this is simply that a smaller unit requires fewer electrons to reach a given state of charge than does a larger one. Since in a series circuit of

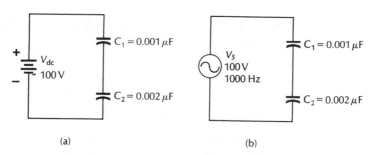

Figure 17-12 Capacitive voltage dividers: (a) dc source and (b) ac source.

any kind, current must have the same value everywhere in the circuit at a given point in time, both capacitors pass the same current. Therefore, while the capacitors are charging, and therefore current is flowing, the smaller unit accumulates its charge faster than does the larger one.

To find the voltage across each capacitor, their total series capacitance is first determined.

$$C_t = \frac{C_1 \times C_2}{C_1 + C_2} = \frac{0.001\ \mu F \times 0.002\ \mu F}{0.001\ \mu F + 0.002\ \mu F} = 0.000667\ \mu F.$$

Then, the following relationship allows us to determine individual voltages.

$$E_{C_1} = \frac{C_t}{C_1} \times V_s = \frac{0.000667\ \mu F}{0.001\ \mu F} = 66.67\ V;$$

$$E_{C_2} = \frac{C_t}{C_2} \times V_s = \frac{0.000667\ \mu F}{0.002\ \mu F} = 33.33\ V.$$

If it is desired to measure these voltages, a *very* sensitive meter must be used, a VTVM or FET-VOM, for instance, to avoid draining the capacitor of charge too quickly. The meter, of course, has an internal resistance that will upset the voltage ratio if any appreciable current is drawn.

When capacitors are placed across ac, the situation is slightly different in that current flow is continuous, rather than the brief charging current of dc. In Figure 17-12b such a circuit is shown, and the voltage across the capacitors is directly proportional to capacitive reactance. However, since X_C is *inversely* proportional to C, the net result is the same; that is, the smaller capacitor has the larger voltage across it.

To determine the individual voltages, first find X_{C_1} and X_{C_2}:

$$X_{C_1} = \frac{1}{2\pi FC} = \frac{1}{6.28 \times 1000 \times 0.001\ \mu F} = 159\ k\Omega;$$

$$X_{C_2} = \frac{1}{2\pi FC} = \frac{1}{6.28 \times 1000 \times 0.002\ \mu F} = 79.6\ k\Omega.$$

Then

$$E_{C_1} = \frac{X_{C_1}}{X_{C_1} + X_{C_2}} \times V_s = \frac{159{,}000}{238{,}600} = 66.6 \text{ V};$$

$$E_{C_2} = \frac{X_{C_2}}{X_{C_1} + X_{C_2}} \times V_s = \frac{79{,}600}{238{,}600} = 33.3 \text{ V}.$$

Note that the same relative voltages appear across the capacitors as in the dc circuit.

SUMMARY

- Where appreciable resistance and capacitance occur in an ac circuit, voltage and current are not in phase.
- In a predominately capacitive ac circuit, current leads voltage.
- In a purely capacitive ac circuit, current leads voltage by 90°.
- If $X_C = R$, the phase angle is 45°.
- In an ac series RC circuit, the voltage across the capacitor lags circuit current by 90°.
- In an ac parallel RC circuit, capacitor current leads resistive current by 90°.
- A coupling capacitor passes the alternating component while blocking the dc component.
- A filter capacitor smooths out the variations in electrical energy that are not pure dc by absorbing energy on the peaks and returning it to the circuit in the valleys, thus reducing the peaks and filling in the valleys.
- A fully charged capacitor is one that is charged to a particular source.
- One time constant t_c is equal to $R \times C$.
- The voltage attained by a capacitor in one time constant is 63.2% of the applied voltage.
- The capacitor charge equation is $e_C = V_s(1 - \epsilon^{-t/RC})$.
- The capacitor discharge equation is $e_C = V_s(\epsilon^{-t/RC})$.
- In a dc circuit, when two capacitors are connected in series, the voltage across each can be found by $E_{C_1} = (C_t/C_1)V_s$ and $E_{C_2} = (C_t/C_2)V_s$.
- In an ac circuit, when two capacitors are connected in series, the voltage across each can be found by $E_{C_1} = [X_{C_1}/(X_{C_1} + X_{C_2})] \times V_s$ and $E_{C_2} = [X_{C_2}/(X_{C_1} + X_{C_2})] \times V_s$.

QUESTIONS

1. A series RC circuit is connected across an ac source. Which is the reference phasor, voltage or current?
2. A parallel RC circuit is connected across an ac source. Which is the reference phasor, voltage or current?

3. A 0.05-μF capacitor is connected in series with a 470-Ω resistor. What is t_c?
4. A 4700-pF capacitor is connected in series with a 6.8-kΩ resistor. What is the time constant?
5. In one time constant, a capacitor will charge to what percent of final voltage?
6. In one time constant, a capacitor will discharge to what percent of total voltage?

PROBLEMS

1. Refer to Figure 17-13. V_s is 120 V_{ac} at 400 Hz, R_1 is 100 Ω, and C_1 is 2.5 μF. Find the values of X_c and θ.
2. Refer to Problem 1. Find the value of Z.
3. Refer to Problem 1. Find the value of total current.
4. Refer to Problem 1. Find the values of E_{R_1} and $E_{X_{c_1}}$.
5. Refer to Problem 1. Express the impedance in both rectangular and polar forms.
6. Refer to Problem 1. Prove that E_{R_1} and $E_{X_{c_1}}$ properly summed equal V_s.
7. Refer to Figure 17-13. V_s = 12 V_{ac} at 1.59 kHz, R_1 = 1000 Ω, and C_1 is 0.1 μF. What is the value of X_C?
8. Refer to Problem 7. Determine the phase angle θ.
9. Refer to Problem 7. Determine the impedance.
10. Refer to Problem 7. Determine the value of total current.
11. Refer to Problem 7. Determine the value of E_{R_1} and $E_{X_{c_1}}$.
12. Refer to Problem 7. Express the impedance in both rectangular and polar forms.
13. Refer to Problem 7. Prove that E_{R_1} and $E_{X_{c_1}}$ properly summed equal V_s.

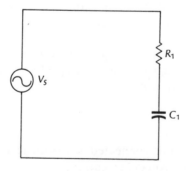

Figure 17-13

14. For the following problems, refer to Figure 17-14. Using the values given, find X_{C_1}; X_{C_2}.
15. Using the values given, find Z_1 and Z_2.
16. Using the values given, find θ_1 and θ_2.
17. Using the values given, find Z_t.
18. Using the values given, find I_1; I_2.
19. Find I_t; find the angle θ_t between V_s and I_t.
20. Refer to Problem 14. Find E_{R_1}; E_{C_1}.
21. Refer to Problem 14. Find E_{R_2}; E_{C_2}.
22. Verify that Branch 1 voltage drops equal V_s.
23. Verify that Branch 2 voltage drops equal V_s.
24. For the following problems, refer to Figure 17-15. Determine the value of C_1 that will provide a voltage drop across itself equal to the voltage drop across R_1.
25. With C_1 of the value found in the previous problem, what is the output voltage at 50 Hz?
26. With C_1 of the value calculated in Problem 24, what is the output voltage at 10 kHz?

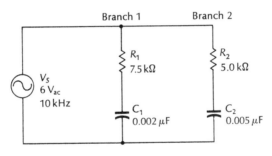

Figure 17-14

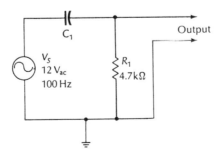

Figure 17-15

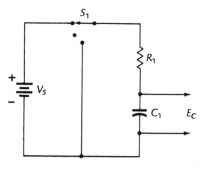

Figure 17-16

27. In Figure 17-15, given that $C_1 = 1.0 \ \mu F$, determine the output voltage and the phase angle between voltage and current.
28. For the following problems refer to Figure 17-16. $V_s = 100$ V, $R_1 = 22$ kΩ, and $C_1 = 1.0 \ \mu F$. What is t_c?
29. $V_s = 100$ V, $R_1 = 470 \ \Omega$, and $C_1 = 470$ pF. What is t_c?
30. $V_s = 100$ V, $R_1 = 1800 \ \Omega$, and $C_1 = 4700$ pF. What is e_C 12 μs after initial charging starts, if $e_C = 0$ at the start?
31. $V_s = 12$ V, $R_1 = 6800 \ \Omega$, and $C_1 = 0.01 \ \mu F$. What is e_C 6 μs after initial charging starts, if $e_C = 0$ at the start?

CHAPTER **18**

ALTERNATING-CURRENT CIRCUITS

In this chapter we examine circuit action in alternating-current *RLC* circuits. When inductance, capacitance, *and* resistance occur together in ac circuits, certain factors not yet discussed must be accounted for. Therefore, series *RLC* circuits will first be investigated, followed by parallel, and then by compound series-parallel networks. Nearly all of the concepts studied thus far will be applied and expanded. Topics to be covered are listed below.

- 18-1 Resistive Ac Circuits
- 18-2 Series *LC* Circuits
- 18-3 Parallel *LC* Circuits
- 18-4 Series *RLC* Circuits
- 18-5 Parallel *RLC* Circuits
- 18-6 Compound *RLC* Circuits
- 18-7 Conductance, Susceptance, and Admittance
- 18-8 Measuring Ac Circuits
- 18-9 Power in Ac Circuits
- 18-10 Three-phase Power
- 18-11 Rectifier Power Supplies

18-1 RESISTIVE AC CIRCUITS

When an ac circuit contains only resistance, it is analyzed by the same methods used in dc circuits. The only additional step is, of course, to select the proper value of voltage or current—rms, peak, average, or peak-to-peak. To illustrate this, a simple example is given in Figure 18-1.

Because there is no reactance in the circuit, voltage and current are *in phase* throughout the circuit. It is therefore analyzed as though it were a simple dc circuit. Usually, the rms voltage and current values are used. A brief analytical procedure is given to illustrate the similarity to dc procedures.

$$R_{2,3} = \frac{R_2 \times R_3}{R_2 + R_3} = 250 \text{ }\Omega.$$
$$R_t = R_1 + R_{2,3} = 350 \text{ }\Omega.$$

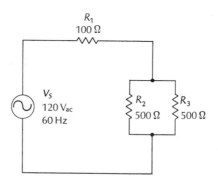

Figure 18-1 A resistive ac circuit.

$$I_t = \frac{V_s}{R_t} = \frac{120}{350} = 0.343 \text{ A}.$$
$$E_{R_1} = I_t \times R_1 = 34.3 \text{ V}.$$
$$E_{R_{2,3}} = I_t \times R_{2,3} = 85.7 \text{ V}.$$
$$V_s = E_{R_1} + E_{R_{2,3}} = 120 \text{ V}.$$

It should be noted at this point that all components have small amounts of inductance and capacitance. As we said in an earlier chapter, even a straight piece of wire has inductance, and since wires are usually used in pairs, capacitance exists between them. Such unwanted reactances are known as *stray* capacitance and inductance, and are usually significant only in very-high-frequency applications. When deliberately introduced into a circuit by physical components, the circuit is said to have *lumped* capacitance or inductance. This simply means that a capacitor or inductor has been introduced deliberately into the circuit.

18-2 SERIES *LC* CIRCUITS

When only inductance and capacitance are present in an ac circuit, current is limited only by the *net* reactance. Up to this point in your studies, inductive and capacitive reactance have been considered essentially the same. X_L and X_C have been treated as positive numbers to maintain simplicity. However, when C and L exist in the same circuit, *they have opposite effects*. You may have already noticed that capacitance causes current to lead voltage, while inductance causes current to lag—these are exactly opposite effects. As was done in Chapter 17, it is necessary to assign a positive sense to X_L and a negative sense to X_C, to account for the opposing circuit effects. The total reactance in

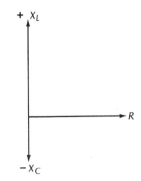

Figure 18-2 Phase relationships of R, X_L, and X_C.

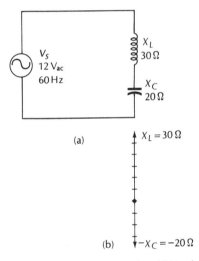

Figure 18-3 (a) A simple series LC circuit and (b) its phasor diagram.

any ac circuit, then, is the net difference between X_L and X_C. Expressed as a formula,

$$X_t = X_L - X_C.$$

Expressed as phasors, the relative values are as shown in Figure 18-2. X_L is plotted in the upward, or +, direction, while X_C is plotted in the downward, or −, direction. Resistance, if any, is of course plotted at a 90° angle to the reactance.

A simple series LC circuit is shown in Figure 18-3, along with the corresponding phasor diagram. It is analyzed as follows. Note that I_t is the reference phasor.

$$X_t = X_L - X_C = 30 - 20 = 10 \, \Omega.$$
$$I_t = \frac{V_s}{X_t} = \frac{12}{10} = 1.2 \, \text{A}.$$
$$E_{X_L} = I_t \times X_L = 1.2(30) = 36 \, \text{V} \underline{/90°}.$$
$$E_{X_C} = I_t(-X_C) = 1.2(-20) = -24 \, \text{V} \underline{/-90°}.$$
$$Z_t = X_L - X_C = 30 - 20 = 10 \, \Omega = X_T.$$

Note that the individual voltage drops across the inductor and capacitor are *greater* than V_s. This may at first seem to be impossible, but it is not. An ac voltmeter placed directly across the inductor will read 36 V, and when placed across the capacitor will read 24 V. The same meter, however, placed across *both* components, will read 12 V. This rather startling result is due, of course, to the fact that the two voltage drops are not in phase. Hence, considered singly, the excessive voltages actually exist, but considered together, they can only total to a value equal to V_s.

In this circuit, the two voltage drops are 180° out of phase with each other, but total current lags V_s by 90° since the circuit is predominately inductive. The following circuit illustrates the reverse instance.

Still referring to Figure 18-3, assume that $X_L = 20 \, \Omega$ and $X_C = 30 \, \Omega$, all else remaining the same. Note in the following calculations that the only differences are the sign of the total reactance and I_t.

Step 1. $\quad X_t = X_L - X_C = 20 - 30 = -10\Omega.$

The minus sign simply denotes that the circuit is predominately capacitive.

Step 2. $\quad I_t = \dfrac{V_s}{X_t} = \dfrac{12}{10} = 1.2 \, \text{A}.$
Step 3. $\quad E_{X_L} = I_t X_L = 1.2 \times 20 = 24 \, \text{V} \underline{/90°}.$
Step 4. $\quad E_{X_C} = I_t(-X_C) = 1.2(-30) = -36 \, \text{V} \underline{/-90°}.$
Step 5. $\quad -X_t = X_L - X_C = 20 - 30 = -10 \, \Omega = Z.$

Remember that the minus signs in steps 1, 4, and 5 simply denote that the circuit has more capacitive reactance than inductive reactance. The sign of the angle gives the phasor direction. In general, the closer X_L and X_C become in value, the lower the total impedance, and if $X_L = X_C$, $Z = 0$ if resistance is negligible.

18-3 PARALLEL *LC* CIRCUITS

In the simple *LC* circuit illustrated in Figure 18-4, *C* and *L* are in a parallel configuration. V_s is the reference phasor here, since V_s must

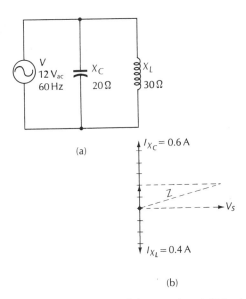

Figure 18-4 (a) A simple parallel ac circuit and (b) its phasor diagram.

appear across all components. Branch currents, therefore, are 180° displaced from each other as suggested by the phasor drawing.

Note particularly that, since I_c is *leading* the voltage, which is reference, it is shown in the leading direction on the phasor diagram. Similarly, inductive current must be shown in the lagging direction, since it is lagging the applied voltage. Total line current is the phasor sum of each branch current. Line current assumes the direction of the larger current (smaller reactance), and is at a 90° angle to V_s.

These statements indicate the difference between series and parallel ac circuits. In the series circuit, the *greatest* value of reactance determines whether the circuit is predominately capacitive or inductive. However, in parallel circuits, the *smallest* value of reactance, which of course allows the greatest current to flow, determines the overall circuit action.

For the circuit in Figure 18-4, the individual branch currents are simply the supply voltage divided by the corresponding branch reactance.

Step 1. $I_L = \dfrac{V_s}{X_L} = \dfrac{12}{30} = 0.4 \text{ A}\underline{/90°}$.

Step 2. $I_C = \dfrac{V_s}{-X_C} = \dfrac{12}{-20} = -0.6 \text{ A}\underline{/-90°}$.

Step 3. $I_t = I_L - I_C = 0.4 - 0.6 = -0.2$ A.

Step 4. $Z = \dfrac{V_s}{I_t} = \dfrac{12}{0.2} = 60\ \Omega.$

or,

$$Z = \dfrac{X_L X_C}{X_L - X_C} = \dfrac{600}{10} = 60\ \Omega.$$

Comparing Steps 1 and 2 above, note that I_C is greater than I_L. The circuit is therefore predominately capacitive. Since I_C and I_L are 180° out of phase, each tends to cancel the other, and only the larger of the two determines the total line current.

In general, the closer X_L and X_C become in value, the *higher* the total impedance; if $X_L = X_C$, then $Z = \infty$ if R is negligible. The reason for this is given in the impedance equation. As X_L and X_C approach equality, the denominator $(X_L - X_C)$ becomes smaller, so that the quotient becomes larger. If $X_L = X_C$ exactly, the denominator becomes zero and the answer is therefore infinity. This assumes no resistance in the circuit, for if any exists, the impedance will not be infinite, although it may be very large if the resistance is large.

In both the series and parallel circuits just described, a special case arises if $X_L = X_C$. As we have said, in the series case, $Z = 0$; and in the parallel case, $Z = \infty$. Such circuits are called *resonant circuits,* and as such have somewhat special attributes. Circuits that are in resonance are so widely used in electronics that Chapter 19 is entirely devoted to them.

18-4 SERIES *RLC* CIRCUITS

A series *LC* circuit as described in Section 18-2 is altered considerably when appreciable resistance is introduced. The phase angle between voltage and current decreases from 90°, with the actual number of degrees of lead or lag depending on the $X_L - X_C$ difference and on how this difference compares to R. Figure 18-5 illustrates a series *RLC* circuit, along with appropriate phasor diagrams. This circuit is treated in much the same manner as a series *RL* or *RC* circuit after the reactances are algebraically summed to determine the total, or net, reactance.

In the illustrated circuit $X_L = 30\ \Omega$ and $X_C = 60\ \Omega$. The first step is to determine which reactance is the larger and by how much.

Step 1. $X_t = X_L - X_C = 30 - 60 = -30\ \Omega.$

As before, the minus sign denotes that the circuit is predominately capacitive. The remainder of the analysis is accomplished as though there were only a 40-Ω resistance and a 30-Ω capacitive reactance.

Alternating-Current Circuits 419

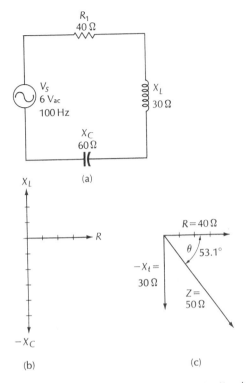

Figure 18-5 (a) A series RLC circuit; (b) the R, X_L, X_C phasor diagram for this circuit; (c) the net, or total, phasor diagram for the circuit.

Step 2. $Z = \sqrt{R^2 + X_T^2} = \sqrt{40^2 + 30^2} = 50 \ \Omega.$

$\theta = \arctan \dfrac{X_t}{R} = \arctan \dfrac{30}{40} = \arctan 0.75 = -36.9°.$

Expressed in polar form, $Z = 50\underline{/-36.9°}.$

Step 3. $I_t = \dfrac{V_s}{Z} = \dfrac{6}{50} = 0.12 \ \text{A}.$

Step 4. $E_R = I_t R_1 = 0.12(40) = 4.8 \ \text{V}\underline{/0°}.$

Step 5. $E_{X_L} = I_t X_L = 0.12(30) = 3.6 \ \text{V}\underline{/90°}.$

Step 6. $E_{X_C} = I_t(-X_C) = 0.12(-60) = -7.2 \ \text{V}\underline{/-90°}.$

Step 7. $V_s = \sqrt{E_R^2 + E_{X_t}^2} = \sqrt{4.8^2 + 3.6^2} = 6.0 \ \text{V}\underline{/-36.9°}$, (relative to circuit current),

where $E_{X_t} = E_{X_L} - E_{X_C} = 3.6 - 7.2 = -3.6 \ \text{V}.$

Again, the minus sign signifies that X_C is larger.

Note that step 7 is only for the purpose of verifying the net voltage drops around the circuit. In this example, X_C is greater than X_L. In

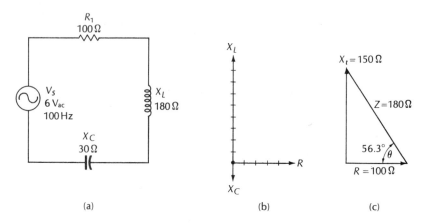

Figure 18-6 (a) A series *RLC* circuit; (b) the R, X_L, X_C phasor diagram; (c) the net, or total, current phasor.

circuits where X_L is greater, the same procedure is used, as illustrated in the next example.

Example 1. A circuit is shown in Figure 18-6, along with the attendant phasor diagrams. The analytic procedure is given below.

Step 1. $X_t = X_L - X_C = 180 - 30 = 150 \; \Omega$.
Step 2. $Z = \sqrt{R^2 + X_t^2} = \sqrt{100^2 + 150^2} = 180.3 \; \Omega$.
$\theta = \arctan \dfrac{X_t}{R} = \arctan \dfrac{150}{100} = 56.3°$.
Step 3. $I_t = \dfrac{V_s}{Z} = \dfrac{6}{180.3} = 0.0333 = 33.3 \; \text{mA}$.
Step 4. $E_{R_1} = I_t R_1 = 33.3 \; \text{mA} \times 100 = 3.33 \; \text{V}$.
Step 5. $E_{X_L} = I_t X_L = 33.3 \; \text{mA} \times 180 = 5.99 \cong 6.0 \; \text{V}$.
Step 6. $E_{X_C} = I_t X_C = 33.3 \; \text{mA} \times 30 = 0.998 \cong 1.0 \; \text{V}$.
Step 7. $V_s = \sqrt{E_{R_1}^2 + E_{X_t}^2} = \sqrt{3.33^2 + 4.99^2} = 6 \; \text{V}$.

18-5 PARALLEL *RLC* CIRCUITS

In the parallel *RLC* circuit of Figure 18-7, branch currents in the reactive branches are 180° out of phase with each other. Therefore, the total current flowing from the source is simply the difference in their magnitudes summed phasorially with the resistive current, as indicated in part *b* of the figure. The first step in analyzing for the unknown values is to determine the branch currents.

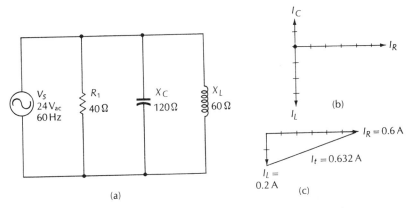

Figure 18-7 (a) A parallel *RLC* circuit; (b) the current phasor; (c) the total current phasor.

Step 1. $I_{R_1} = \dfrac{V_s}{R_1} = \dfrac{24}{40} = 0.6$ A.

$I_{X_C} = \dfrac{V_s}{X_C} = \dfrac{24}{120} = 0.2$ A.

$I_{X_L} = \dfrac{V_s}{X_L} = \dfrac{24}{60} = 0.4$ A.

$I_{X_t} = I_{X_L} - I_{X_C} = 0.4 - 0.2 = 0.2$ A

Next, the total current I_t is the phasor sum of I_{R_1} and net reactive current I_{X_t}:

Step 2. $I_t = \sqrt{I_{R_1}^2 + I_{X_t}^2} = \sqrt{0.6^2 + 0.2^2}$
$= \sqrt{0.4} = 0.632$ A.

Now, the circuit impedance can be determined.

Step 3. $Z = \dfrac{V_s}{I_t} = \dfrac{24}{0.632} = 37.975 \cong 38 \; \Omega$.

Step 4. $\theta = \arctan \dfrac{I_{X_t}}{I_{R_1}} = \arctan \dfrac{0.2}{0.6} = \arctan 0.333$
$= 18.43°$.

As an aid in visualizing circuit action it is desirable to construct a series equivalent circuit (Figure 18-8) that, from the standpoint of total values of *I*, *X*, and *Z*, would perform in the same way as the original parallel *RLC* circuit (Figure 18-7).

To find the value of R_{eq}, multiply the original circuit impedance by the cosine of the angle θ.

$R_{eq} = (\cos \theta)Z = (\cos 18.43)38 = 0.95(38) \cong 36 \; \Omega$.

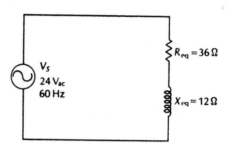

Figure 18-8 The equivalent series circuit for the parallel circuit given in Figure 18-7.

To find the equivalent value of reactance X_{eq}, multiply the original impedance by the sine of the angle θ.

$$X_{eq} = (\sin \theta)Z = (\sin 18.43)38 = 0.316 \times 38 = 12 \ \Omega.$$

Because in Step 1 above the inductive current is largest, the equivalent, or predominate, reactance must be inductive.

To verify that the series equivalent impedance and the phase angle of this equivalent circuit are the same as that produced by the original circuit, find Z and θ by conventional means.

$$Z = \sqrt{R^2 + X^2} = \sqrt{36^2 + 12^2}$$
$$= \sqrt{1296 + 144} = 37.95 \cong 38 \ \Omega;$$
$$\theta = \arctan \frac{X}{R} = \arctan \frac{12}{36} = \arctan 0.333 = 18.43°.$$

As a further check that the original circuit and the equivalent circuit yield the same results, find I_t for Figure 18-8.

$$I_t = \frac{V_s}{Z} = \frac{24}{38} = 0.63158 \cong 0.632 \ \text{A}$$

These values for Z, θ, and I_t check with the previous calculations; we have proved that the series equivalent circuit yields the same overall results as the original parallel RLC circuit.

18-6 COMPOUND RLC CIRCUITS

In the circuit shown in Figure 18-9, each branch is only partly reactive, since there is some resistance and some reactance in each branch. Therefore, total impedance is determined by the product-over-the-sum

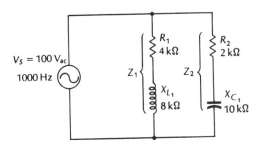

Figure 18-9 A parallel *RLC* circuit with series *RL* and *RC* branches.

formula, and we must depend on multiplication of polar forms and summation of rectangular forms, as previously explained.

To analyze such a circuit for voltage and current values as well as phasor values, we start by finding the individual branch impedances.

Step 1. Find the values of Z_1 and Z_2:

$$Z_1 = \sqrt{R_1^2 + X_L^2} = \sqrt{4000^2 + 8000^2} = \sqrt{(16 \times 10^6) + (64 \times 10^6)}$$
$$= \sqrt{80 \times 10^6} = 8944 \ \Omega \cong 8.9 \ k\Omega.$$

$$Z_2 = \sqrt{R_2^2 + X_C^2} = \sqrt{2000^2 + 10{,}000^2} = \sqrt{(4 \times 10^6) + (1 \times 10^8)}$$
$$= \sqrt{1.04 \times 10^8} = 10{,}198 \ \Omega = 10.2 \ k\Omega.$$

Step 2. Find θ for each branch: Z_1/θ_1 and Z_2/θ_2.

$$\theta_1 = \frac{X_L}{R} \text{ and arctan } \theta_1 = \arctan \frac{X_L}{R_1} = \arctan \frac{8000}{4000} = 63.4°.$$

$$\theta_2 = \frac{-X_C}{R} \text{ and arctan } \theta_2 = \arctan \frac{-X_C}{R} = \arctan \frac{-10{,}000}{2{,}000} = \arctan -5$$
$$= -78.7°.$$

Step 3. Express Z_1 and Z_2 in rectangular and polar notation:

$$Z_1 = 4000 + j8000 = 8944 \underline{/63.4°}.$$
$$Z_2 = 2000 - j10{,}000 = 10{,}200 \underline{/-78.7°}.$$

Step 4. Write the product-over-the-sum equation, using polar and rectangular forms where appropriate:

$$Z_t = \frac{Z_1 \times Z_2}{Z_1 + Z_2} = \frac{(8944\underline{/63.4°})(10{,}200\underline{/-78.7°})}{(4000 + j8000) + (2000 - j10{,}000)}$$

Step 5. Find the product:

$$8{,}944\underline{/63.4°} \times 10{,}200\underline{/-78.7°} = 91.23 \times 10^6 \underline{/-15.3°}$$

Step 6. Find the sum:

$$\begin{array}{r} 4000 + j8000 \\ 2000 - j10{,}000 \\ \hline 6000 - j2000 \end{array}$$

Step 7. Express the sum in polar form.

$$Z = \sqrt{6000^2 + 2000^2} = \sqrt{(36 \times 10^6) + (4 \times 10^6)} = \sqrt{40 \times 10^6}$$
$$= 6325;$$

$$\theta = \arctan \frac{X}{R} = \arctan \frac{2000}{6000} = -18.4°$$

and

$$Z = 6325 \underline{/-18.4°} \; \Omega.$$

Note: Because the sum in Step 6 is $6000 - j2000$, the angle is $-18.4°$. If this sum had been $6000 + j2000$, the angle would be positive.

Step 8. Write the expression for Z_t in polar form so division can be carried out:

$$Z_t = \frac{91.23 \times 10^6 \underline{/-15.3°}}{6.325 \times 10^3 \underline{/-18.4°}} = 14{,}423 \; \Omega \underline{/3.1°} \cong 14{,}400 \; \Omega \underline{/3.1°}$$

Step 9. Find the value of I_t:

$$I_t = \frac{V_s}{Z_t} = \frac{100 \underline{/0.0°}}{14{,}400 \underline{/3.1°}} = 6.9 \; \text{mA} \underline{/-3.1°}.$$

Step 10. Construct the simplified series equivalent circuit:

$$R_{eq} = \cos \theta (Z_t) = (\cos 3.1°)(14{,}400) = 14{,}378 \; \Omega \cong 14{,}380 \; \Omega.$$
$$X_{eq} = \sin \theta (Z_t) = (\sin 3.1°)(14{,}400) = /778.7 \; \Omega \cong 780 \; \Omega.$$

Hence, a resistor of $14{,}380 \; \Omega$ in series with an inductive reactance of $780 \; \Omega$ will give the same performance as the original circuit with respect to total current and phase angle.

To analyze a compound parallel circuit such as that shown in Figure 18-10a, the simplest approach is to start by combining all similar quantities. Then, a simplified equivalent circuit is constructed, using these total combined values, as illustrated in Figure 18-10b. Recall that reactances and resistances are combined in the parallel connection in the same manner.

Step 1. Combine similar quantities:

$$R_t = \frac{R_1 R_2}{R_1 + R_2} = \frac{200 \times 300}{200 + 300} = 120 \; \Omega.$$

$$X_{L_t} = \frac{X_{L_1} X_{L_2}}{X_{L_1} + X_{L_2}} = \frac{75 \times 100}{75 + 100} = 42.9 \; \Omega.$$

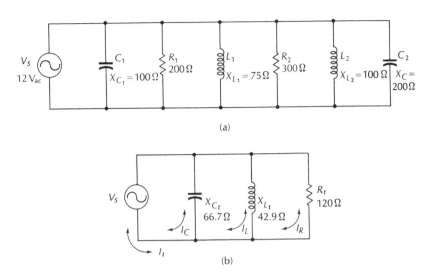

Figure 18-10 (a) A compound parallel circuit with several Rs, Cs, and Ls; and (b) the equivalent circuit.

$$X_{C_t} = \frac{X_{C_1} X_{C_2}}{X_{C_1} + X_{C_2}} = \frac{100 \times 200}{100 + 200} = 66.7 \; \Omega.$$

The remainder of the circuit is solved as before, using conventional methods. First, find the branch currents:

Step 2. Determine branch currents:

$$I_R = \frac{12}{120} = 0.1 \; \text{A}.$$

$$I_L = \frac{12}{42.9} = 0.2797 = 0.28 \; \text{A}.$$

$$I_C = \frac{12}{66.7} = 0.1799 = 0.18 \; \text{A}.$$

Step 3. Next, determine the total reactive current I_X. Because the reactive currents in the C and L branches are 180° out of phase with each other, the total current is simply the smaller subtracted from the larger.

$$I_X = I_L - I_C = 0.28 - 0.18 = 0.1 \; \text{A}.$$

Step 4. Next, determine total current I_t:

$$I_t = \sqrt{I_R^2 + I_X^2} = \sqrt{0.1^2 + 0.1^2} = 0.1414 \; \text{A}.$$

Step 5. Find the impedance.

$$Z = \frac{V_s}{I_t} = \frac{12}{0.1414} = 84.85 \cong 85 \; \Omega.$$

Step 6. Determine the circuit phase angle.

$$\theta = \arctan \frac{I_X}{I_R} = \arctan \frac{0.1}{0.1} = \arctan 1 = 45°.$$

Because I_L is greater than I_C, the total reactive current is predominately inductive. Therefore, total current lags the supply voltage by 45°.

18-7 CONDUCTANCE, SUSCEPTANCE, ADMITTANCE

It is sometimes convenient to use the reciprocals of the three major opposition functions. In an earlier chapter, mention was made of *conductance (G)* in a simple resistive dc circuit. Conductance is the reciprocal of resistance:

$$G = \frac{1}{R} \quad \text{and} \quad R = \frac{1}{G}.$$

The unit of conductance is the *mho*, the symbol of which is ℧. (In SI, conductance is measured in *siemens*.)

Similarly, reactance and impedance have reciprocal values. The reciprocal of reactance is *susceptance*, and

$$B_C = \frac{1}{X_C} \quad \text{and} \quad B_L = \frac{1}{X_L},$$

where B = susceptance, also measured in mhos or siemens.

The reciprocal of impedance is *admittance*, and

$$Y = \frac{1}{Z} \quad \text{and} \quad Z = \frac{1}{Y},$$

where Y = admittance, again measured in mhos or siemens.

To illustrate the use of these three values, a simple circuit is given in Figure 18-11. The following procedure illustrates how conductance, susceptance, and admittance can be used to find the unknown values.

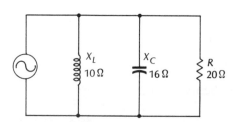

Figure 18-11 A simple parallel *RLC* circuit to demonstrate the use of reciprocal functions.

Step 1. List the known values:

$$R = 20\ \Omega.$$
$$jX_L = j10\ \Omega.$$
$$-jX_C = -j16\ \Omega.$$

Step 2. Find the conductance and the susceptance of the reactances:

$$G = \frac{1}{R} = \frac{1}{20} = 0.05\ \Omega.$$

$$-jB_L = \frac{1}{jX_L} = \frac{1}{10} = 0.1\ \Omega.$$

$$jB_C = \frac{1}{-jX_C} = \frac{1}{16} = 0.0625\ \Omega.$$

Note that when a positive reactance is divided into a real number the quotient has a negative reactance and vice versa. That is,

$$\frac{1}{jX_L} = -jB_L \quad \text{and} \quad \frac{1}{-jX_C} = jB_C.$$

Step 3. Find the net, or total, susceptance:

$$B_{X_t} = jB_C - jB_L = 0.0625 - 0.1 = -0.0375\ \Omega.$$

Step 4. Find the admittance:

$$Y = G - jB_X = 0.05\Omega - j0.0375.$$

Step 5. Find the phase angle:

$$\theta = \arctan \frac{B_X}{G} = \arctan \frac{j0.0375}{0.05} = -36.9°.$$

Step 6. Express the admittance in polar form:

$$Y = \sqrt{G^2 + B_X^2} = \sqrt{0.05^2 + 0.0375^2} = 0.0625 \underline{/-36.9°}.$$

Step 7. Determine the impedance:

$$Z = \frac{1}{Y} = \frac{1}{0.0625 \underline{/-36.9°}} = 16 \underline{/36.9°}.$$

18-8 MEASURING AC CIRCUITS

The VOM

The electrical meter movement described in Chapter 7 does not produce a deflection if used in ac circuits. The reason for this is simply that the average of alternating voltage or current is zero. The D'Arsonval movement is basically a device for measuring dc current or voltage if suitable multipliers are employed, and by itself it cannot measure ac.

However, nearly all VOMs have the ability to measure ac volts, so obviously a modification is added to the basic meter movement. This modification consists of the addition of a *rectifier*. A rectifier is a device that allows current to flow through itself *in one direction only*. One such rectifying device is the diode, as described in Chapter 8.

Reviewing very briefly, one kind of diode is a semiconductor such as germanium or silicon, arranged so that when current flows through it in one direction the diode exhibits very low resistance, and current flows readily. Thus, the diode acts essentially as a *short circuit*. When current reverses direction, the diode exhibits a very *high* resistance and appears to be an *open circuit*. Because a diode allows current to flow in one direction only, it is an efficient device for converting ac to dc. This is precisely its purpose when used in a VOM.

The ac portion of a VOM is shown schematically in Figure 18-12. Note the components labeled D_1 and D_2; these are the diodes that change ac to dc. In this figure the schematic symbol for the diode indicates that this component will allow electrons to flow readily

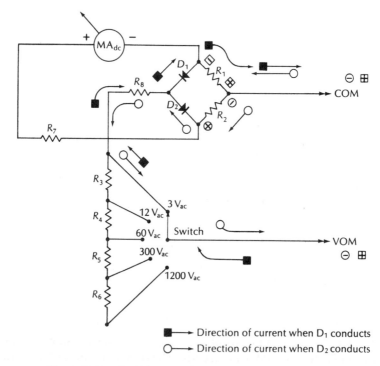

Figure 18-12 The D'Arsonval meter movement modified to read ac volts.

against the arrowhead, but not in the reverse direction. Hence, when the meter probes are connected to an ac source, the two diodes alternately conduct as the measured polarity changes. Thus, D_1 conducts when the measured polarity is such as to make the COM lead positive and the VOM lead negative. When the polarity changes, COM is negative and VOM is positive, and D_2 conducts.

Note the current-direction arrows in the illustration. These represent the current for the two possible directions. Each is identified with a symbol relating to the applied voltage symbol. The symbols verify that only one diode conducts at a given time, depending upon the polarity presented to the circuit. Not only the diodes are responsible for the ac-to-dc conversion; R_1 and R_2, plus the type of circuit connection, assist in it. To aid in visualizing circuit action, study Figure 18-13. Only the meter movement, D_1, D_2, R_1, and R_2 are shown. When D_1 conducts, D_2 is considered to be an open circuit, and when D_2 conducts, D_1 is open. It must be appreciated that the two directions of current occur at different times.

When D_1 conducts, current flows from left to right as shown, and a voltage drop occurs across R_1, with the polarity as shown. This drop has a polarity that agrees with the required meter-movement polarity, and the meter deflects accordingly. Since D_2 is open, the only current through R_2 is that required to operate the meter movement. On the next half cycle, D_2 conducts and D_1 is open, and the drop across R_2 now actuates the meter. Note that, even though current through the entire device has reversed, the drop across R_2 agrees with the meter requirement, and the meter still deflects in the required direction. As before, the only current through R_1 at this time is the required current through the meter.

It is apparent that the meter is actually measuring the voltage drops across R_1 and R_2 on alternate half cycles. Because of the mechanical inertia of the movement, the needle, and movement coil, the deflection

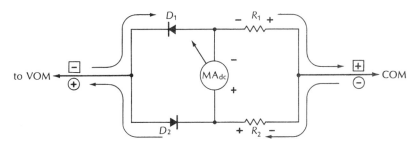

Figure 18-13 A simplified view of the meter rectifier circuit.

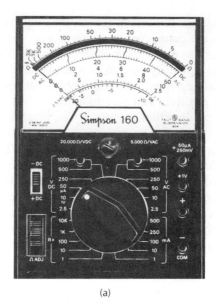

(a)

(b)

Figure 18-14 (a) A portable VOM with ac voltage scales; (b) a bench-type solid-state VOM; (c) a vacuum-tube voltmeter (VTVM). Courtesy Simpson Electric Co.

is actually proportional to the *average* of the rectified ac voltage. The meter scale is therefore calibrated (adjusted or modified) to give rms, instead of average, readings.

The other components are the range switch and range resistors (multipliers) that allow full-scale readings of 3, 12, 60, 300, and 1200 volts rms for the example shown in Figure 18-12. R_7 and R_8 are calibration resistors, which are usually hand-picked to achieve the required accuracy (often ± 2% on ac ranges).

Procedures for operation of the ac-voltage part of the VOM are very similar to those followed in using the dc voltmeter ranges. First, select the range so that deflection beyond full scale will not occur. Then simply place the probes across the device to be measured and read the correct scale. Most meters have red-colored scales for the ac ranges. Note that it is not necessary to observe polarity, since the polarity is continually changing.

Three voltmeters are illustrated in Figure 18-14, each with ac voltage scales much as described. The meter in part *a* of the figure is

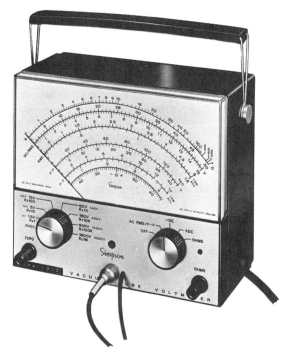

(c)

portable; it offers a wide range of scales in a small package. The larger model (*b*) is a portable solid-state VOM. This is a modern counterpart of the VTVM shown in Figure 8-14*c*. Instead of vacuum tubes, which require large amounts of power to operate, FETs, or field-effect transistors, are used. The use of an amplifying device, such as a vacuum tube or FET, greatly increases the sensitivity of the VOM.

It should be noted at this point that while VOMs usually cannot be used to measure alternating current, such use is certainly not impossible. One need only insert an accurate low-value (often 1-Ω) resistor in series with the circuit, and then measure the ac voltage drop across the resistor while the circuit is energized. Then simple Ohm's law ($I = E/R$) will indicate the rms value of current. *Be careful* when you perform this measurement in power-type circuits, since you are dealing with lethal voltages. Make certain that the circuit is deenergized before you make the connections.

Other Types of Ac Meters

Special types of ac meters are often encountered. These include thermal, electromagnetic, and electronic types. We provide a brief operating description of each in the following paragraphs. Representative meters are illustrated in Figure 18-15.

Thermal Types

The *thermocouple* meter is made from two dissimilar metals, usually in the form of wires, bonded together at one end. When the bonded end is heated, a small dc voltage proportional to the heat is produced; this voltage is easily read on a dc volt or milliameter.

A second thermal-type meter is the *hot-wire* meter. In this type of meter, a wire is heated by passing the current to be measured through it. As the wire heats, it expands, and the expansion is converted to needle deflection. Such meters are often used in radio-frequency applications.

Electromagnetic Types

In the *iron-vane* meter, the current to be measured is passed through a fixed coil, causing electromagnetic induction to exist in two small, soft-iron vanes, one of which is fixed and one of which is movable. The movable vane becomes magnetized and is repelled by the fixed vane. The movable vane is connected to the indicator, and thus provides deflection proportional to the current.

In the *Dynamometer*, the force between a large fixed coil and a small rotatable coil provides deflection. Since both coils are energized by ac, the polarities are *relatively* constant, and the smaller coil rotates to

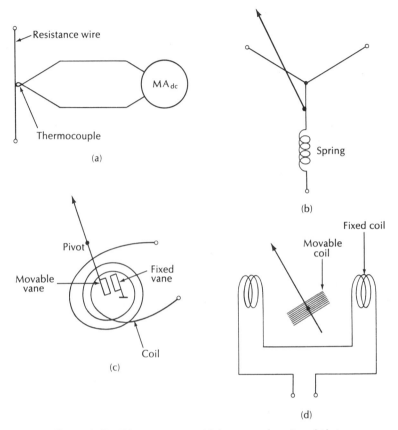

Figure 18-15 Other meter types: (a) thermocouple meter. (b) hot-wire ammeter, (c) moving-vane meter, and (d) dynamometer.

provide deflection proportional to voltage, current, or power. When a dynamometer is used to measure power, it is frequently called a *wattmeter*.

An Electronic Type

An instrument used extensively for the measurement of ac waveforms is the *oscilloscope*. Two typical oscilloscopes are shown in Figure 18-16. A full description of the oscilloscope would fill this book, but we give a brief description to acquaint you with the extreme usefulness of this remarkable instrument. The oscilloscope, or scope for short, operates in much the same manner as your TV set. It utilizes a cathode-ray tube (CRT) to trace a picture of the waveform of an ac voltage. In the

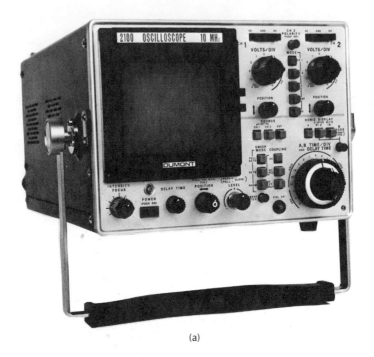

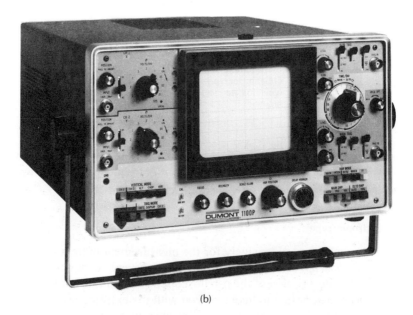

Figure 18-16 Typical oscilloscopes: (a) A 10-MHz portable model; (b) a 100-MHz portable 'scope. Courtesy DuMont Corp.

more elaborate versions, it is capable of time measurements that are as accurate as the trace can be read. Increments of time much shorter than fractions of a microsecond, as well as rapidly changing voltages ranging from millivolts to hundreds of volts can easily be measured. The exact waveform can be seen, so any distortion that exists can easily be determined. The scope can be *triggered* so as to start its trace from left to right at any specific point in time.

The indicating mechanism of the scope is a very fine beam of electrons, and because in a vacuum the beam can be caused to change direction (*deflect*) virtually instantly, the instrument responds to quantities being measured just as instantly. It is therefore used mostly in high-frequency applications. Waveform display and voltage measurement of frequencies ranging up to thousands of megahertz are not uncommon.

The oscilloscope has two major *deflection* systems, one of which moves the electron beam vertically while the other moves it horizontally. The signal voltage to be observed is connected to the vertical deflection system. The internal mechanism of the scope generates a *time-base* voltage which moves the electron beam across the cathode-ray tube in a precise interval of time. If no vertical signal exists, this yields a straight horizontal line that appears similar to the illustration in Figure 18-17.

If a sinusoidal signal is introduced into the vertical deflection system, the scope will, by the combined horizontal and vertical movement of the beam, trace out the exact waveform of the signal, thus allowing the operator to see a true graphical representation of the waveform. An

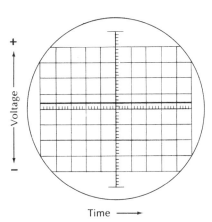

Figure 18-17 The time base with no vertical deflection, offset from zero for clarity.

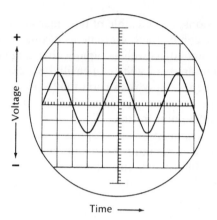

Figure 18-18 Several cycles of a sinusoid as would appear on an oscilloscope.

example is given in Figure 18-18, where several cycles of a sinusoid are drawn as they would appear on an oscilloscope screen.

The general purpose of many of the controls on the front panel of this instrument is to adjust the internal circuitry so as to make the waveform fit into the size of the cathode-ray tube face. For example, if a 60-Hz waveform is viewed and if the beam traveled from extreme left to extreme right in 0.001 second (1.0 ms), only a very small part of one cycle would be viewed. Some of these controls, then, adjust the speed of the horizontal trace so at least one complete cycle can be seen at any reasonable input frequency. Similarly, other controls adjust the vertical deflection, so that the viewed image will remain within the confines of the tube face, and will therefore be neither too small nor too large. If the image is too small, it simply cannot be seen in the proper perspective, and if too large, most of the waveform is off-screen and cannot be seen. Still other controls synchronize the horizontal deflection to the vertical signal being observed, so that a stable image is produced. An unsynchronized condition results in a jumble of meaningless traces on the viewing screen.

To assist in measuring voltages and time intervals the screen of the cathode-ray tube is covered by a transparent *graticule*, a glass cover into which accurately spaced lines are etched. Thus, one vertical space between lines etched horizontally might represent 1.0 V, or 10.0 V, or 100.0 V, depending upon the front-panel switch setting. Also, one horizontal space between lines etched vertically might represent 1.0 second, or 1.0 ms, or 1.0 μs, again depending upon the corresponding switch setting. Hence, voltage and time intervals can both be accurately measured.

18-9 POWER IN AC CIRCUITS

In previous chapters we discussed power in both inductive and capacitive circuits. We now investigate power in *RLC* circuits, where the total power delivered by the source is, in all probability, *not* equal to *EI*. As we have already said, in an ac circuit having both inductance and resistance, the current and voltage are not in phase. The product of *E* and *I* does not, therefore, give the true power being consumed by the load. The product of voltage and current is called *apparent power*, while the current and resistance determine *true power (I^2R)*.

A circuit example is given in Figure 18-19*a*, which is a series circuit consisting of L, C, and R. For this circuit, two methods can be used to find true power. (1) The total current and I^2R are used to determine the true power dissipation of the circuit. (2) The phase angle between applied voltage and total current can be determined, which allows $\cos \theta \ (V_s \times I_t)$ to be evaluated.

I^2R Method

The first step is to determine total circuit impedance; refer to Figure 18-19*b* for the equivalent circuit.

Step 1. Find the impedance:

$$Z = \sqrt{R^2 + (X_L - X_C)^2} = \sqrt{1000^2 + (7000 - 5000)^2} = 2236 \ \Omega.$$

Step 2. Find the total current:

$$I_t = V_s/Z = 12/2236 = 0.0054 \text{ A} = 5.4 \text{ mA}.$$

Step 3. Find true power:

$$P_{\text{true}} = I^2R = 5.4 \text{ mA}^2 \times 1000 \ \Omega = 0.029 \text{ W} = 29 \text{ mW}.$$

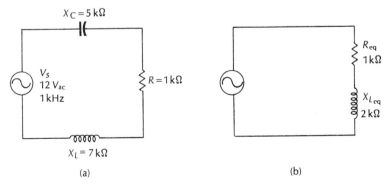

Figure 18-19 A circuit used to demonstrate apparent and real power: (a) the original circuit, and (b) the simplified equivalent circuit.

Step 4. Find apparent power:

$$P_{app} = V_s \times I = 12 \text{ V} \times 5.4 \text{ mA} = 0.0648 \text{ W} = 64.8 \text{ mW}.$$

Note that true power is much less than apparent power.

Step 5. Find power factor.

$$pF = \frac{P_{true}}{P_{app}} = 0.432$$

Cos θ ($V_s \times I_t$) Method

Step 1. Find the impedance expressed in polar form:

$$Z = \sqrt{R^2 + (X_L - X_C)^2} = 2236 \text{ }\Omega.$$
$$\theta = \arctan \frac{X_{eq}}{R_{eq}} = \arctan \frac{2000}{1000} = 63.4°.$$

Therefore, $Z = 2236 \underline{/63.4°}$.

Step 2. Find I_t:

$$I_t = \frac{V_s}{Z} = 12/2236 = 5.4 \text{ mA}.$$

Step 3. Find cos θ ($V_s \times I_t$), where cos θ = cos 63.4° = 0.432:

$$P_{true} = \cos \theta \ (V_s \times I_t) = \cos \theta \ (12 \times 5.4 \times 10^{-3})$$
$$= 0.432 \times (12 \times 5.4 \times 10^{-3}) = 2.8 \times 10^{-2} = 28 \text{ mW}.$$
$$P_{app} = V_s \times I = 12 \times 5.4 \times 10^{-3} = 64.8 \text{ mW}.$$

Step 4. Find the power factor.

$$pF = \cos \theta = \cos 63.4° = 0.448.$$

Because both reactances return power to the source, neither consumes any power. The resistance, which includes any wire resistance in the coil, *does* dissipate power in the form of heat.

An expression sometimes encountered is the VAR, or *volt-ampere reactive*. The VAR is the reactive volt-amperes, or $E \times I$ existing at a 90° angle. Just as [cos θ ($E \times I$)] yields the *resistive* power, [sin θ ($E \times I$)] gives the reactive power that is temporarily absorbed by the reactive components, but later returned to the source. For the circuit just discussed,

$$\text{VAR} = \sin \theta \ (V_s \times I_t) = \sin 63.4 \ (12 \text{ V} \times 5.4 \text{ mA})$$
$$= 0.0579 \text{ W} \cong 58 \text{ mW}.$$

That is, there is a power dissipation in true watts of 29 mW, while the reactive watts (VAR) is 58 mW. The VAR watts are not dissipated as heat, but are simply returned to the source.

The Decibel

A convenient method of comparing two or more circuits having different levels of power is the *decibel* (dB) named after Alexander Graham Bell (1847–1922). The *bel* is the fundamental unit of measure used to express the ratio between two power levels on a logarithmic scale. A more convenient unit is the decibel, which is one-tenth of a bel.

To illustrate how the decibel is used, refer to Figure 18-20. The triangle symbol represents an *amplifier* circuit, the purpose of which is to increase the power level of a signal voltage. All radios, TVs, and stereos contain a large number of amplifiers. As an example, a high-quality portable AM radio amplifies a signal generated by the antenna at a power level of, perhaps, 5 to 500 μW to a level of 0.5 W at the loudspeaker. Of course, the amplifier circuits perform other functions at the same time they are amplifying, such as station selection, and so forth.

An amplifier, then, is any circuit capable of increasing the voltage, or current, or both. As such, it has an *input*, where the power level is low, and an output, where the signal is identical except for the amplification.

Refering to Figure 18-20, note that the signal level at the input has a power level of 0.02 W, or 20 mW, while the level at the output is 2.50 W. The simple ratio of output to input, P_{out}/P_{in}, is, in this instance, 125. That is, the power level has been increased 125 times. However, this number is misleading, especially in reference to audio signals. One might expect the amplified signal to sound 125 times louder than the input, but this is not true. Because the human ear responds to sound loudness *logarithmically*, the above ratio must be modified so it can be expressed in decibels.

$$dB = 10 \log_{10}\left(\frac{P_2}{P_1}\right),$$

where P_2 is the output power and
P_1 is the input power.
In the example,

$$dB = 10 \log_{10}\left(\frac{2.50}{0.020}\right) = 10 \log_{10} 125$$
$$= 10(2.09691) = 20.97 \cong 21.$$

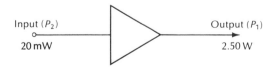

Figure 18-20 A circuit to illustrate the use of the decibel.

The output level, then, does not sound 125 times as loud as the input but only 21 times as loud, and the circuit is said to have a power *gain*, or amplification, of 21 dB.

The decibel is a relative unit of measure. That is, it is derived from the ratio of two levels of power, neither of which is necessarily related to the other, and either of which can be of any value. Therefore, the decibel is useful for indicating the increase or decrease of power, which is known as *gain*. The decibel cannot be used to provide a measure of a fixed amount of power. To indicate a specific amount of power, (for example, 10 W) the dBm is used. Zero dBm is a specific amount of power at a specific impedance level. Zero dBm is equal to 0.001 W, or 1 mW, being dissipated in an impedance of 600 Ω. Table 18-1 lists values of dBm versus watts for a range of powers.

To use the table to convert watts to dBm, find the correct value of watts and read the value of dBm at the left. For example, if a 10-W audio amplifier is to be specified in terms of dBm, its output is said to have a maximum level of +40 dBm. The voltage and current values required to provide the level in dBm are also given in the table.

The device being measured must have an impedance of 600 Ω, or any reading (measurement) taken will be incorrect using a standard ac voltmeter. Many instruments used in the communications field are calibrated to give correct values when used in 600-Ω circuits. Using a test instrument that is so calibrated across any other impedance requires a correction factor. By applying the formula given below, the corrected reading for *any* impedance can be found:

$$dBm_{(cor)} = dBm_{(meas)} + 10 \log_{10}\left(\frac{600}{Z_{(meas)}}\right)$$

The corrected value is equal to the measured value plus ten times the log to the base 10 of the ratio of 600 Ω to the impedance of the circuit being measured. As an example, assume that a meter cali-

TABLE 18-1 dBm AND ASSOCIATED VALUES

dBm	Watts	Volts	mA
+50	100 W	245	0.408
+40	10 W	77.5	0.129
+30	1 W	24.5	0.041
+20	0.1 W	7.75	0.0129
+10	0.01 W	2.45	0.0041
0.0	0.001 (1.0 mW)	0.775	0.00129
−10	0.1 mW	0.245	0.41 mA
−20	0.01 mW	0.0775	0.129 mA
−30	0.001 mW (1.0 μW)	0.0245	0.041 mA

brated for 600 Ω is used to measure across a 75-Ω circuit and the meter reads −25 dBm. Then

$$dBm = -25 + 10 \log_{10}\left(\frac{600}{75}\right) = -25 + 10 \log_{10} 8$$
$$= -25 + 9.0309 \cong -16 \text{ dBm}.$$

18-10 THREE-PHASE POWER

Whenever large amounts of electrical power must be transported or consumed, it is not practical to use single-phase. Instead, electrical power is generated in a *three-phase system*, and the resultant three-phase power is transported in this form. Large industrial complexes that use tremendous amounts of energy use this energy directly in three-phase form, often at 440 or 880 volts.

Figure 18-21 illustrates a simplified three-phase generator. Instead of the single coil of a single-phase generator, there are three coils, physically oriented at 120° intervals around the axis. Each coil can be considered as a single-phase generator by itself. However, when all three are considered together, *the voltage generated by each is displaced 120° from the other voltages*, as shown on the waveforms. The three phases are commonly designated A, B, and C. Such a generator, or alternator, can supply power to either single- or three-phase loads.

The three coils can be connected in one of two basic configurations, depending upon the end application. One of these configurations is the *delta* connection, so called because of its resemblance to the Greek letter delta, Δ. This is shown in Figure 18-22, where the three coils are connected end to end. If each coil generates 110 V$_{ac}$(rms), then total

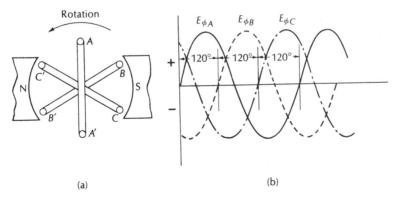

Figure 18-21 Generating three-phase power: (a) the coil structure, and (b) the resulting waveforms.

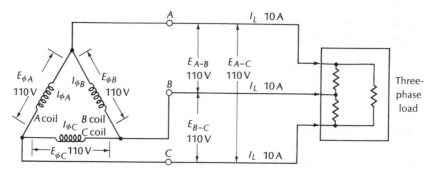

Figure 18-22 A three-phase delta circuit.

voltage for each coil will be distributed to the three-phase load as shown. The voltage from any one wire to either of the others is 110 V$_{ac}$, of course, since any two wires are connected directly across one of the coils. Hence, phase *A*-to-*B*, *B*-to-*C*, and *A*-to-*C* voltage is 110 V$_{ac}$ per phase.

Now assume that the load draws 10 A per leg. That is, in each of the three distribution wires a 10-A current is flowing. If the current in *each coil* were to be measured, it would be found to be only 5.77 A. The reason for this is simply that each distribution wire is fed by more than one phase! However, these currents are *not* in phase and so cannot be directly added. A specific relationship determines the coil current versus line current in a delta-connected circuit: line current is 1.732 times greater than coil current, or conversely, coil current is smaller than line current by the factor 0.5773. For the circuit illustrated in Figure 18-22:

$$I_{line} = I_{coil} \times 1.732 = 5.77 \text{ A} \times 1.732 = 9.99 \text{ A} = 10 \text{ A}.$$
$$I_{coil} = I_{line} \times 0.5773 = 10 \text{ A} \times 0.5773 = 5.77 \text{ A}.$$
$$E_{line} = E_{coil} = 110 \text{ V}.$$

An alternate type of connection is shown in Figure 18-23*a*. This is called the *wye* connection, since it resembles the letter Y. Note in this instance that while the line-to-line voltage is 110 V, the individual coil voltage is only 63.5 V. In this circuit configuration, each line-to-line set of terminals, *A* to *B*, for example, is *across two coils*. As before, the two voltages cannot be simply added, since each is displaced 120° from the other. However, the factors used to determine current in a delta network are applicable to wye networks as well.

$$E_{line} = E_{coil} \times 1.732 = 63.5 \text{ V} \times 1.732 = 109.98 \text{ V} \approx 110 \text{ V}.$$
$$E_{coil} = E_{line} \times 0.5773 = 110 \text{ V} \times 0.5773 = 63.5 \text{ V}.$$
$$I_{line} = I_{coil} = 10 \text{ A}.$$

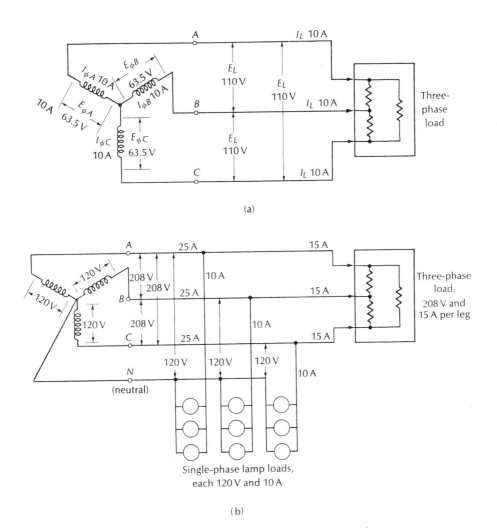

Figure 18-23 (a) A three-phase wye circuit; (b) a three-phase four-wire circuit.

Finally, Figure 18-23b shows the method of power delivery to residential users. The addition of the tap (N) from the center of the wye connection allows either three separate single-phase distribution branches, or a three-phase load, or both. Since each coil produces 120 V, the tap and one outside line provide 120 V, single phase, to the appropriate loads. At the same time, a three-phase load can be driven from the three phase wires. The three-phase line voltage is

$$E_{line} = E_{coil} \times 1.732 = 120 \text{ V} \times 1.732 = 207.8 \text{ V} \cong 208 \text{ V}.$$

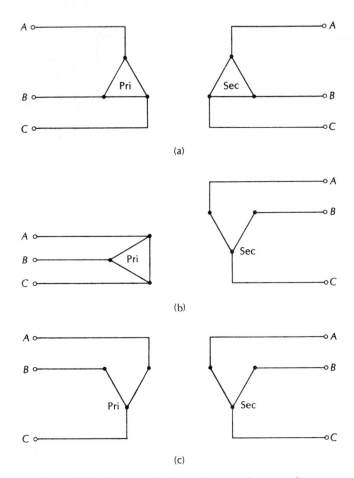

Figure 18-24 Three possible three-phase transformer configurations: (a) delta-delta; (b) delta-wye; (c) wye-wye.

Note that the coils shown connected in either wye or delta configuration represent either the armature coils on a generator or the coils that are the secondary windings on three-phase transformers. When considering three-phase transformers, the primary and secondaries can assume the configurations shown in Figure 18-24. To simplify such drawings, the coils themselves are often omitted, as illustrated. Figure 18-25 illustrates a typical three-phase transformer.

18-11 RECTIFIER POWER SUPPLIES

As has been previously mentioned, it is very often necessary to convert ac to dc, because most electronic circuitry operates on dc, while ac is

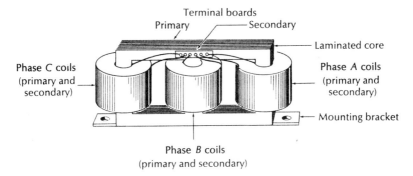

Figure 18-25 A typical three-phase transformer.

the most commonly available supply. Ac to dc conversion is accomplished by the *rectifier* circuit, of which there are many kinds. The first element of a rectifier circuit is the diode, which, as we explained in Section 18-8, passes current only in one direction. The second element is a *filter*, which consists of various values of R, C, and L to smooth the rectified voltage to resemble pure dc. The entire circuit arrangement is called the *rectifier power supply*.

A block diagram of such a power supply is given in Figure 18-26a. The 120-V, 60-Hz input feeds the power transformer, which is used to provide any required amount of voltage. The power transformer is connected to any of a number of possible diode arrangements that perform the ac-to-dc conversion. The rectifier output is by no means pure, smooth dc, but the filter smooths out the variations. In some cases, a voltage divider follows the filter to provide other, smaller values of the main filter output. The voltage divider is optional and in many cases is not used. However, a resistor, usually of large value, is often used across the filter, to provide a discharge path for the capacitors. Such a resistor is known as a *bleeder* resistor, since its purpose is to bleed off any charge on the capacitors and thus to minimize shock hazards.

Figure 18-26b illustrates the schematic diagram for a typical rectifier power supply. The ac line voltage is stepped up or down, depending on the load circuit requirement. Then, as the correct value of ac voltage is presented to the diodes, each diode conducts on alternate half cycles. The waveforms shown illustrate the output of the diodes before filtering. Note that this *is* dc, although it is not yet pure. This dc rectified voltage is then filtered by C_1, L_1, and C_2. Both capacitors are large in value (tens to thousands of microfarads) and the iron-core inductor (or choke coil) usually has a value of 1 to 50 henrys.

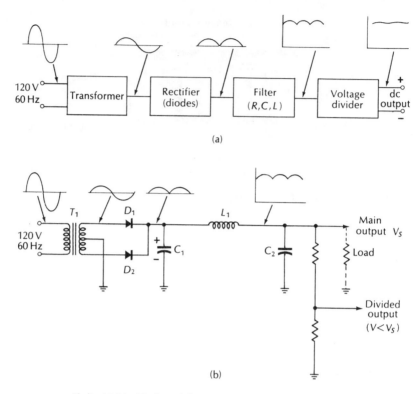

Figure 18-26 The basic full-wave rectifier circuit: (a) block drawing and (b) schematic.

To understand how this circuit works, consider Figure 18-27a. Two half cycles are shown, and the resultant currents are identified by solid lines and dashed lines. When the upper end of the transformer is positive, then the lower end is negative. Diode D_2 is reverse biased, but since the transformer center tap is also negative, a complete path exists for forward-biased D_1. Current (dashed lines) flows through D_1, down to the center tap, and up through the load, yielding the polarity shown. For the full half cycle, this polarity exists, and the waveform follows the ac input. If the input is a sine wave, so is the output.

Now, when the ac input changes polarity, the *lower* end of the transformer is most positive, so D_2 becomes forward biased and D_1 becomes reverse biased. Now D_2 conducts (solid lines) and current flows through D_2, through one half of the transformer secondary, and again up through the load, yielding a voltage across the load in the *same polarity* as the preceding half cycle. This, of course, is direct current, but since it is not a smooth dc, it is called *pulsating dc*.

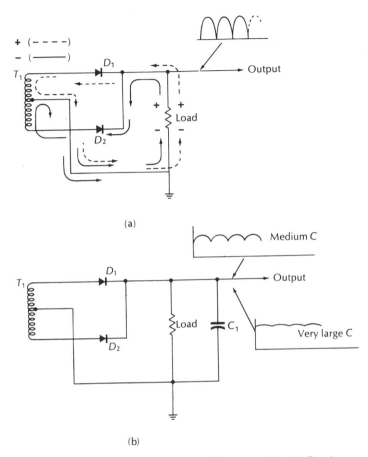

Figure 18-27 Circuit details for the full-wave rectifier: (a) Circuit action; (b) the effect of adding a capacitor that serves as a filter.

Pulsating dc cannot be used to supply the power needs of most electronic equipment. The rectified output must therefore be filtered, to reduce the peaks and to fill in the valleys of the waveform. Figure 18-27b illustrates the simplest possible filter that, depending on the value of C_1, will produce better-quality dc. The capacitor stores energy during the maximum voltage time, and returns this energy to the load during minimum voltage times, thereby smoothing out the pulsating waveform. As shown, the larger the value of C, the smoother the dc waveform.

To keep the size of capacitors used to reasonable values, two capacitors and an inductor are usually used, as was shown in Figure 18-26b. Because the inductor also stores energy, this arrangement is the

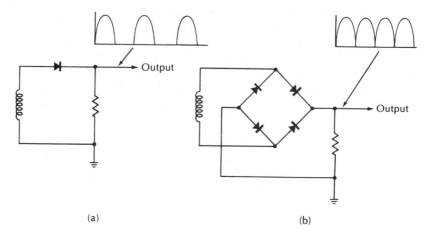

Figure 18-28 (a) A half-wave rectifier; (b) a bridge rectifier.

most economical for achieving a reasonably variation-free dc voltage. Any variation left in the dc output is called "ripple," which represents the amount of ac left in the filtered output.

Passive filters contain only L and C. *Active* devices (such as transistors) can eliminate this remaining ripple voltage almost entirely. While a detailed discussion of active filters is beyond the scope of this book, it is interesting to note that electronically regulated power supplies are, generally, far superior to those using only passive filters. Most modern equipment uses dc power from an electronically regulated power supply.

Two other variations of diode rectifier configurations are shown in Figure 18-28. The *half-wave rectifier* (part *a* of the figure) is rarely used, since it delivers power to the load only on alternate half cycles. A further drawback of half-wave rectification is that it requires very large values of L and C to produce smooth dc, so that it is usually not an economical rectifier for single-phase circuits. However, halfwave rectification does find application in three-phase circuits, where power is delivered to the load at all times.

The *full-wave bridge rectifier* (Figure 18-28b) has a number of advantages. First, it delivers power continuously (both half cycles) as does the full-wave circuit discussed previously. Second, unlike the full-wave circuit, it uses the *full secondary voltage*. Third, the transformer requires no center tap; hence costs are reduced. (The full-wave rectifier shown earlier must have a secondary winding *twice* as large as the bridge circuit for a given output voltage, since only half of the secondary is used at one time.) The bridge rectifier has the same requirements for filtering that the full-wave circuit has.

SUMMARY

- In a series LC circuit, total current is reference, and the inductive reactance operates at 180° in relation to the capacitive reactance.
- In a series LC circuit, X_L and X_C cancel each other's effects, and the numerically larger value predominates.
- In a series LC circuit, E_L leads by 90° and E_C lags by 90°.
- In a series LC circuit, the impedance is low, and, if $X_L = X_C$, $Z = 0$.
- In a parallel LC circuit, V_s is reference, and inductive current operates at 180° in relation to capacitive current.
- In a parallel LC circuit, I_L and I_C cancel each other's effects, and the numerically larger value predominates.
- If, in a parallel LC circuit, X_L and X_C are equal, the total impedance is very high; in theory $Z = \infty$.
- In a series RLC circuit, total current is reference, and the angle of lead or lag between voltage and current is greater than 0° but less than 90°, and is equal to arctan X_T/R.
- In a series RLC circuit, E_L and E_C operate at an angle of 180° relative to each other, and E_R is in phase with total current.
- In a parallel RLC circuit, V_s is the reference, and reactive branch currents are 180° out of phase. Resistive current is in phase with the source.
- Conductance (G) equals $1/R$.
- Susceptance (B) equals $1/X$.
- Admittance (Y) equals $1/Z$.
- An ac voltmeter uses diodes to change ac to dc.
- An oscilloscope is an electronic device that measures ac voltages and displays the waveform on a CRT.
- Power in an ac circuit may be found by I^2R.
- True power is $\cos \theta \, (V_s \times I_t)$.
- VAR $= \sin \theta \, (E \times I)$.
- The decibel (dB) is one-tenth of a bel, and is useful in expressing power ratios.

QUESTIONS

1. Comparing ac and dc circuits containing resistance only, why can the ac circuit be analyzed the same as a dc circuit?
2. Name at least three types of ac meters.
3. Briefly explain why the phasor for inductance is opposite to that for capacitance.
4. A reactance phasor is always at an angle of 90° to the resistance phasor. Briefly explain why an impedance phasor is usually at an angle of less than 90°.

5. Name the three reciprocal current-opposition functions and identify their counterparts.
6. Briefly explain how a D'Arsonval meter movement can be modified to measure ac voltage.
7. Why is the bel, or decibel, useful in comparing two power levels?
8. Briefly describe the VAR.

PROBLEMS

1. Refer to Figure 18-29. $V_s = 5.0$ V$_{ac}$ at 60 Hz, $C_1 = 66.3$ μF, and $L_1 = 0.0796$ H. Find the values of X_C, X_L, X_t, Z, and θ.
2. Refer to Figure 18-29. $V_s = 120$ V$_{ac}$ at 60 Hz, $C_1 = 1.0$ μF, and $L_1 = 2.0$ H. Find the values of X_C, X_L, X_t, Z, and θ.
3. Refer to Problem 1 above. Find the values of E_{X_L} at $\underline{/90°}$; E_{X_C} at $\underline{/-90°}$.
4. Refer to Problem 2 above. Find the values of E_{X_L} at $\underline{/90°}$; E_{X_C} at $\underline{/-90°}$.
5. Refer to Figure 18-30. $V_s = 12$ V$_{ac}$ at 100 Hz, $C_1 = 1.0$ μF, and $L_1 = 1.0$ H. Find I_L, I_C, I_t, and Z.

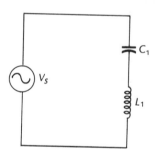

Figure 18-29

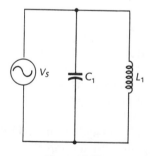

Figure 18-30

6. Refer to Figure 18-30. $V_s = 12$ V_{ac} at 1 kHz, $C_1 = 1.6$ μF, and $L_1 = 0.032$ H. Find I_L, I_C, I_t, Z.
7. Refer to Figure 18-31. $V_s = 9$ V_{ac} at 150 Hz, $R_1 = 60$ Ω, $X_{C_1} = 45$ Ω, and $X_{L_1} = 90$ Ω. Find the values of $X_t, Z, \theta, I_t, E_R, E_L$, and E_C. Also verify that the total voltage drops equal V_s.
8. Refer to Figure 18-31. $V_s = 24$ V_{ac} at 10 kHz, $R_1 = 1400$ Ω, $X_{C_1} = 1000$ Ω, and $X_{L_1} = 1500$ Ω. Find $X_t, Z, \theta, I_t, E_R, E_{L_1}$, and E_{C_1}. Show that the total voltage drops equal V_s.
9. Refer to Figure 18-32. $V_s = 16$ V_{ac} at 1 kHz, $R_1 = 27$ Ω, $X_{C_1} = 8$ Ω, and $X_{L_1} = 40$ Ω. Find the values of $I_R, I_{X_C}, I_{X_L}, I_{X}, I_t, Z$, and θ.
10. Refer to Figure 18-32. $V_s = 48$ V_{ac} at 1 kHz, $R_1 = 81$ Ω, $X_{C_1} = 100$ Ω, and $X_{L_1} = 120$ Ω. Find the values of $I_R, I_{X_C}, I_{X_L}, I_{X}, I_t, Z$, and θ.
11. Refer to Problem 9 above. Find the values of C; of L.
12. Refer to Problem 20 above. Find the values of C; of L.
13. Refer to Figure 18-33. $V_s = 12$ $V_{ac}, R_1 = 5$ k$\Omega, R_2 = 3$ k$\Omega, X_{L_1} = 10$ kΩ, and $X_{C_1} = 12$ kΩ. Find total circuit impedance (Z) and the phase angle θ between V_s and total current.

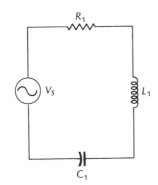

Figure 18-31

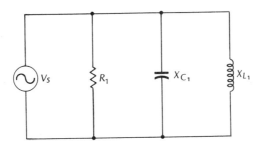

Figure 18-32

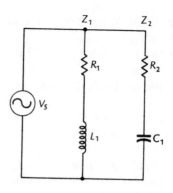

Figure 18-33

14. Refer to Figure 18-33. $V_s = 24$ V_{ac}, $R_1 = 1.5$ kΩ, $R_2 = 2$ kΩ, $X_{L_1} = 2$ kΩ, and $X_{C_1} = 1.5$ kΩ. Find total circuit impedance (Z) and the phase angle θ between V_s and total current.
15. An amplifier circuit has an input-signal power level of 1 mW and an output-signal level of 1 W. What is the power gain in decibels?
16. An amplifier circuit has an input signal of 5 W and an output signal of 10 W. What is the power gain in decibels?
17. A three-phase delta-connected secondary has a coil current of 14 A. What is line current?
18. A three-phase wye-connected secondary has a coil voltage of 254 V. What is line-to-line voltage?

CHAPTER **19**

RESONANCE AND RESONANT CIRCUITS

We have already mentioned that the special case where $X_L = X_C$ is called *resonance*, and that a circuit where $X_L = X_C$ is a *resonant circuit*. Because X_L and X_C tend to cancel each other out, circuit action in such a case is different than in any circuit previously discussed.

Resonant circuits find wide application in electronics. For instance, every radio and TV set contains several resonant circuits. In fact, it is the resonant circuit that allows the equipment to accept a certain broadcast station and reject all others. Just how this is accomplished is explain briefly in this chapter. We also examine the action of various resonant circuit configurations and examine other important applications of resonance.

The topics to be covered in this chapter are listed below.

19-1 Series-Resonant Circuits
19-2 Parallel-Resonant Circuits
19-3 Applications of Resonant Circuits
19-4 Bandpass of Resonant Circuits
19-5 Filter Circuits
19-6 Filter Traps
19-7 The Varactor Diode

19-1 SERIES-RESONANT CIRCUITS

A simple series-resonant circuit is given in Figure 19-1a. Note that it does not appear any different than many circuits already covered; indeed, it is not! The one thing that makes the resonant circuit unique among all circuits is that the source is providing energy *at the resonant frequency*. If the frequency of the source is changed significantly, then the circuit action changes; it is then *not* a resonant circuit but simply another example of whatever configuration it has, excluding resonance. In that case, either X_L or X_C is larger than the other, and the circuit is treated in the same manner as an inductive or capacitive circuit.

However, when operated at the one frequency that causes X_L to equal X_C, the two reactances cancel ($X_L - X_C = 0$), giving rise to the

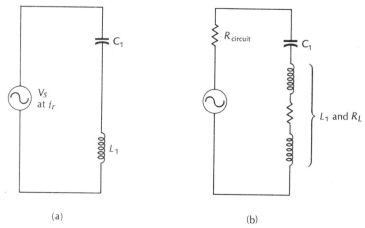

Figure 19-1 A series-resonant circuit (a); The equivalent circuit for the circuit given in Figure 19-1a; this equivalent circuit shows minimum inherent resistance R_{circuit} and R_L.

unique characteristics of the resonant circuit. Figure 19-2 illustrates how this one single frequency comes about. Recall that as frequency is increased, X_L increases, while X_C decreases. This is shown as two slanting lines, one for X_C and one for X_L. The abscissa of the graph is incremented in units of ascending frequency (logarithmic plot), while the ordinate is in units of reactance (also logarithmic). As frequency is increased (left to right) X_C decreases while X_L increases. At one frequency, and one frequency only, the two plots cross, and this is the frequency where $X_L = X_C$.

Referring again to the diagram in Figure 19-1a, note that no physical resistance is shown. In theory, if there were no resistance at all, current would be infinitely large whenever voltage was applied. In a real circuit, of course, resistance exists in several places. All voltage sources have internal resistance to a greater or lesser degree; all wiring has some resistance; and the coil itself must, of course, have some appreciable resistance, since it consists of wire. This is shown as a resistance in Figure 19-1a as the internal resistance of the coil; note that this circuit is redrawn from Figure 19-1a.

Because all circuits capable of operating in the resonant mode have resistance, it is meaningless to analyze such a circuit without some value of resistance, no matter how small its value. The resistor shown as part of the coil must therefore be considered, as well as the other resistances, shown as R_{circuit} on the drawing. This represents all other resistances in the circuit, these being primarily the wiring resistances and the internal source resistance.

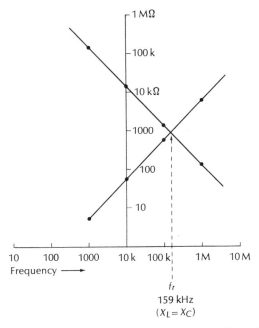

Figure 19-2 A graph of X_L and X_C for the circuit in Figure 19-1.

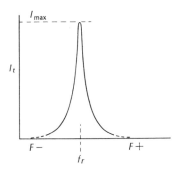

Figure 19-3 A plot of I_t versus F for the series RLC circuit.

Briefly recapping, then, a series-resonant circuit is simply a series RLC circuit that operates in the resonant mode when supplied with a source operating at the resonant frequency f_r. As mentioned, if any other frequency is provided, the circuit is not operating in resonance, and it functions simply as a series RLC circuit.

Series-Resonant Circuit Analysis

In analyzing such a circuit, it is necessary to find the one frequency f_r that will set X_L equal to X_C. Such a relationship is easily derived from the fact that at resonance, $X_L = X_C$. Recall that:

$$X_L = 2\pi FL$$

and

$$X_C = \frac{1}{2\pi FC}.$$

These values can be substituted in the expression $X_L = X_C$, so that

$$2\pi FL = \frac{1}{2\pi FC}.$$

This expression can be solved for F, to provide an equation that will allow the resonant frequency to be found. First, multiply both sides of the equation by F to consolidate like terms:

$$(F)2\pi FL = \frac{1}{2\pi \cancel{F} C}(\cancel{F}),$$

which results in

$$2\pi F^2 L = \frac{1}{2\pi C}.$$

Next, multiply both sides by $\frac{1}{2\pi}$:

$$\left(\frac{1}{2\pi}\right)\cancel{2\pi} F^2 L = \frac{1}{2\pi C}\left(\frac{1}{2\pi}\right),$$

which gives

$$F^2 L = \frac{1}{2^2 \pi^2 C}.$$

Now, multiply both sides by $\frac{1}{L}$:

$$\left(\frac{1}{\cancel{L}}\right) F^2 \cancel{L} = \frac{1}{2^2 \pi^2 C}\left(\frac{1}{L}\right),$$

which leaves

$$F^2 = \frac{1}{2^2 \pi^2 CL}.$$

Because we wish to find F, not F^2, we take the square root of both sides:

$$\sqrt{F^2} = \sqrt{\frac{1}{2^2\pi^2 LC}},$$

and

$$F = \frac{1}{2\pi\sqrt{LC}}.$$

In the original expressions for X_L and X_C, F was a general frequency, so we abbreviate resonant frequency f_r, and

$$f_r = \frac{1}{2\pi\sqrt{LC}},$$

where f_r = resonant frequency in hertz
L = inductance in henrys, and
C = capacitance in farads.

This equation allows determination of the series-resonant frequency for any combination of L and C. Note that resistance has no bearing on resonant frequency; it does, however, have a bearing on other characteristics of the resonant circuit, which we discuss later in this chapter. We can modify the expression of f_r to express it in terms of L and C:

$$L = \frac{1}{4\pi^2 f_r^2 C},$$

and

$$C = \frac{1}{4\pi^2 f_r^2 L}.$$

Example 1. Assume that, in Figure 19-1, $C_1 = 0.001$ μF and $L_1 = 0.001$ H. Find the resonant frequency.

$$f_r = \frac{1}{2\pi\sqrt{LC}} = \frac{1}{2\pi\sqrt{0.001 \times 10^{-6} \times 1 \times 10^{-3}}}$$
$$= 159.2 \text{ kHz (actually, 159.15494 kHz)}.$$

Note that Figure 19-2 shows a graph of the variation of X_L and X_C versus frequency. At the point where the two lines cross, $X_L = X_C = f_r$.

Now that the resonant frequency can be easily determined, we will investigate the characteristics of such a circuit when operated at resonance. This of course assumes that the source is a generator in which the frequency can be varied, so that it can be adjusted to the exact resonant condition.

Because $X_L = X_C$ at resonance, there is no current limitation other than dc resistance. Hence, current is at a *very* high value in a series-resonant circuit. If the frequency of the source is varied above and below resonance, current decreases from its maximum value, since X_L or X_C will predominate, causing further current limitation. At resonance, circuit current is simply $I_{f_r} = V_s/R$.

As an example, let us assign values for the circuit in Figure 19-1. $V_s = 12\ V_{ac}$ at 159.2 kHz, and the *total* series resistance is 12 Ω. Then,

$$I_{f_r} = \frac{V_s}{R} = \frac{12}{12} = 1.0\ A.$$

If the source frequency is varied above f_r, current decreases, and if it is varied below f_r, current also decreases. Hence, in a series *RLC* circuit, current is maximum when the applied voltage is at the resonant frequency. This is shown graphically in Figure 19-3.

It is interesting to determine the voltage across the capacitor and inductor when the source is operating at the resonant frequency. In a series circuit such as the one shown in Figure 19-4, the voltage existing across each component is simply the product of I_t (or I_{f_r}) and the reactance or resistance. First, find X_L and X_C.

$$X_L = 2\pi FL = 6.28 \times 159.2 \times 10^3 \times 0.001 = 1000\ \Omega.$$
$$X_C = \frac{1}{2\pi FC} = \frac{1}{6.28 \times 159.2 \times 10^3 \times 0.001 \times 10^{-6}} = 1000\ \Omega.$$

Note that within reasonable accuracy limits, X_L does indeed equal X_C. Now, I_t can be combined with these to find E_C and E_L (see Figure 19-4).

$$E_C = I_t \times X_C = 1.0 \times 1000 = 1000\ V.$$
$$E_L = I_t \times X_L = 1.0 \times 1000 = 1000\ V.$$

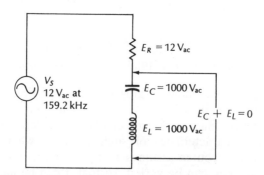

Figure 19-4 Voltage drops around a series-resonant circuit.

Note how much larger E_C and E_L are than the 12 V supplied by the source. This is typical of series-resonant circuits, where a distinct *gain* in voltage is usually obtained.

Also of interest is the drop across the resistance:

$$E_R = I_t \times R = 1.0 \times 12 = 12 \text{ V}.$$

In this kind of circuit, the voltages as calculated *would actually be measured* across each respective component. However, if a reading were to be taken across *all three* components, the meter would read 12 V_{ac}. This, of course, is true because E_C and E_L are 180° out of phase, and they therefore cancel each other.

Another factor influencing circuit action is the total circuit impedance. If $X_L = X_C$ at resonance and if each cancels the other's effects, then (at resonance only) total Z equals R, and the circuit phase angle is zero (E in phase with I). In the example,

$$Z = \sqrt{R^2 + (X_L - X_C)^2} = \sqrt{R^2} = R = 12 \text{ }\Omega \underline{/0.0°}.$$

Q of the Series-Resonant Circuit

The *Q (quality)* of a series-resonant circuit is a figure of merit. It is the ratio of X_L to R, and it determines how broad or narrow the shape of the response curve will be, as well as its height. In Figure 19-5, three curves are shown with accompanying values of Q. Note the sharp, highly peaked curve resulting when Q is on the order of 100. As Q decreases, the curve broadens out and its maximum amplitude is much

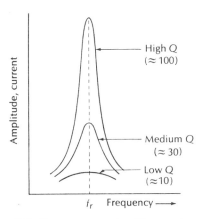

Figure 19-5 Response curves for different values of Q.

less. Ideally, then, Q should be as high as possible for most applications. Thus, the Q of the circuit in Figure 19-4 is

$$Q = \frac{X_L}{R} = \frac{1000}{12} = 83.3.$$

Such a circuit would have a response curve not unlike the curve labeled "high Q" in Figure 19-5.

19-2 PARALLEL-RESONANT CIRCUITS

When the inductance and capacitance are operated in parallel across the source (Figure 19-6), circuit action is much different. Because I_L and I_C are 180° out of phase with each other, they cancel (if R is very small). Theoretically, if $R = 0$, and $X_L = X_C$ exactly, line current or total current would be zero, suggesting that total circuit impedance would be infinitely high. However, as we have pointed out, the coil must have some resistance, and even the circuit wiring, no matter how small, is finite. Thus, the impedance in the inductive branch will differ from the impedance in the capacitive branch. Since parallel resonance is most accurately defined as the frequency where $Z_1 = Z_2$, the amount of resistance in the inductive leg has an effect on the resonant frequency.

Parallel-Resonant Frequency with $Q > 10$

If the *tank* (parallel-resonant-circuit) Q is 10 or higher, the same formula used for series circuits can be used to find f_r with reasonable accuracy. Since obtaining a Q of 10 is not difficult, this will suffice for most purposes.

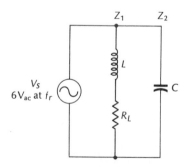

Figure 19-6 A parallel-resonant ("tank") circuit.

Example 2. In Figure 19-6 assume that $R = 10\ \Omega$, $L = 0.01592$ H, and $C = 0.01592\ \mu\text{F}$. Determine the resonant frequency. Using the formula presented in the previous section, we find

$$f_r = \frac{1}{2\pi\sqrt{LC}} = \frac{1}{6.28\sqrt{0.01592 \times 0.01592 \times 10^{-12}}}$$

$$= \frac{1}{6.28\sqrt{2.535 \times 10^{-10}}} = \frac{1}{6.28 \times 1.592 \times 10^{-5}}$$

$$= \frac{1}{1 \times 10^{-4}} = 10{,}000\ \text{Hz} = 10\ \text{kHz}.$$

As a check, determine if $Q \geq 10$:

$$X_L = 2\pi FC = 1000\ \Omega,$$

and

$$Q = \frac{X_L}{R} = \frac{1000}{10} = 100.$$

Since Q is greater than 10 by a rather large margin, the calculated frequency is proved accurate.

Parallel-Resonant Frequency with $Q < 10$

As briefly mentioned, when appreciable resistance appears in the inductive branch, current does not lag by 90°. In the capacitive branch, however, current *does* lead by 90°. Therefore I_L and I_C do not cancel, and the frequency at which I_L and I_C do cancel is not exactly $1/(2\pi\sqrt{LC})$. Thus, in a circuit where Q is less than 10, a correction factor must be applied to assure accuracy.

To illustrate the application of the low-Q correction factor, refer to Figure 19-7. To find the true resonant frequency, first determine the circuit Q, assuming that the coil resistance is the only significant

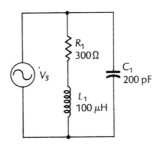

Figure 19-7 A parallel-resonant circuit having low Q.

resistance. Note that X_L cannot be found because, at this point, frequency is not known. However, X_L at the *approximate* frequency (the same as series resonance) will give an adequate answer.

$$f_{r(series)} = \frac{1}{2\pi\sqrt{LC}} = \frac{1}{6.28\sqrt{100 \times 10^{-6} \times 200 \times 10^{-12}}}$$

$$= \frac{1}{6.28\sqrt{0.02 \times 10^{-12}}} = \frac{1}{6.28 \times 0.1414 \times 10^{-6}} = \frac{1}{8.88 \times 10^{-7}}$$

$$= 1.1259 \times 10^6 \cong 1.13 \text{ MHz}.$$

Now, X_L at this frequency can be determined:

$$X_L = 2\pi FL = 6.28 \times 1.13 \times 10^6 \times 100 \times 10^{-6} = 709.64 \cong 710 \text{ }\Omega.$$

Solving for Q at this frequency gives

$$Q = \frac{X_L}{R} = \frac{710}{300} = 2.37.$$

Knowing that Q is much less than 10 at a frequency close to the true resonant frequency indicates a need to apply the correction factor:

$$f_r = \frac{1}{2\pi\sqrt{LC}} \left(\sqrt{\frac{Q^2}{1+Q^2}}\right)$$

Note that the frequency of resonance depends on the Q of the circuit, and that if Q is equal to or greater than 10, the second term is nearly 1; hence no correction is required. Substituting in the parallel-resonant-circuit formula now allows the true resonance to be determined:

$$f_r = \frac{1}{6.28\sqrt{100 \times 10^{-6} \times 200 \times 10^{-12}}} \left(\sqrt{\frac{2.36^2}{1+2.36^2}}\right)$$

$$= 1.13 \text{ MHz} \left(\sqrt{\frac{5.57}{6.57}}\right) = 1.13 \text{ MHz} \times 0.921 = 1.04 \text{ MHz}.$$

Note that the true resonant frequency is significantly less than the apparent series-resonant frequency calculated from the same components. Table 19-1 illustrates the correction factor for several values of Q.

As mentioned, the parallel-resonant circuit draws minimum line current. However, the *tank* current, which flows back and forth between the inductor and the capacitor, is normally quite large. The line current, then, is only large enough to supply the losses in the circuit. In the simple circuits illustrated, the losses are only the I^2R losses around the resonant tank.

Comparing series- and parallel-resonant circuits, we find that the total circuit impedance of the series circuit is low, while that of the parallel circuit is quite high. The total current of the series circuit is therefore high, while that of the parallel circuit is low (although tank current is

TABLE 19-1 CORRECTION FACTOR FOR Q BETWEEN 1 AND 10 FOR THE SERIES-RESONANT CIRCUIT

$Q = \dfrac{X_L}{R}$	$\sqrt{\dfrac{Q^2}{1+Q^2}}$	Approximate Correction (%)
10/1 = 10	0.995	0.5
10/2 = 5	0.98	2
10/3 = 3.33	0.958 ≅ 0.96	4
10/4 = 2.5	0.928 ≅ 0.93	7
10/5 = 2.0	0.89	11
10/6 = 1.67	0.857 ≅ 0.86	14
10/7 = 1.43	0.819 ≅ 0.82	18
10/8 = 1.25	0.78	22
10/9 = 1.11	0.74	26
10/10 = 1.0	0.707 ≅ 0.71	29

high). If the Q of the circuit is high (≥ 10) both circuits resonate at the frequency where $X_L = X_C$. If the Q of the circuit is low (< 10) the series circuit resonates at the frequency where $X_L = X_C$, but the parallel circuit resonates at the frequency where the impedance of the inductive branch equals that of the capacitive branch.

19-3 APPLICATIONS OF RESONANT CIRCUITS

In nearly all communications circuits, including both radio and TV, resonant circuits are used extensively. As we said in the introduction to this chapter, one such use is to *select,* or *tune,* one broadcast station, while rejecting all others. Because the resonant circuit is highly frequency selective, this is an excellent application. Both series and parallel combinations are used, and it is often difficult to determine whether a particular circuit is series or parallel. Often, a circuit configuration is drawn in such a manner that it appears to be parallel, when it is actually a series circuit.

Figure 19-8 illustrates two such applications. A parallel-resonant (or parallel-tuned) circuit is shown in part *a* of the figure, where it is placed in the antenna circuit of a radio receiver. In a practical circuit, either the capacitor or the inductor is variable, so that the desired broadcast can be selected. If the resonant circuit is tuned exactly to the frequency at which the signal is being broadcast, the tank currents are large at this frequency (and small at all other frequencies), so that the circuit rejects all others and readily passes the desired frequency.

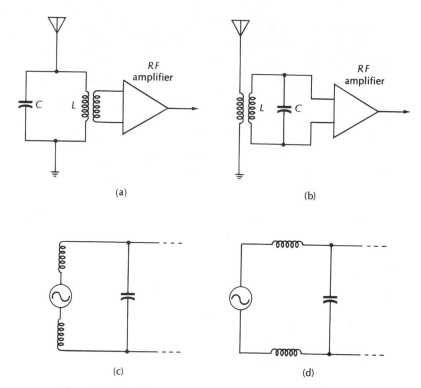

Figure 19-8 Applications of resonant tank circuits: (a) parallel, (b) series, (c) source (V_S) in series, and (d) part c redrawn.

Rather than being placed in the primary, the tank circuit is often placed in the secondary. In such an instance (part *b* of the figure), rather than being a parallel circuit, it is a series-tuned circuit. Note Figure 19-8*c*, which indicates that the source is actually in the winding of the secondary coil. This can be redrawn as in *d* to emphasize the true series connection. The overall function, however, is the same—to select one frequency, but to reject all others.

In addition to selecting a desired frequency and rejecting other frequencies, it is often necessary to perform just the opposite function. A *wave trap* is another application of a resonant circuit, in which an *undesirable* frequency is eliminated. By properly placing and tuning the resonant circuit, an undesirable frequency can be trapped, and thus prevented from continuing through the receiver to produce an unwanted output. Figure 19-9 illustrates this circuit action. The first drawing shows a series-resonant circuit connected across the signal line to ground. Because the impedance of a series circuit is very low at reso-

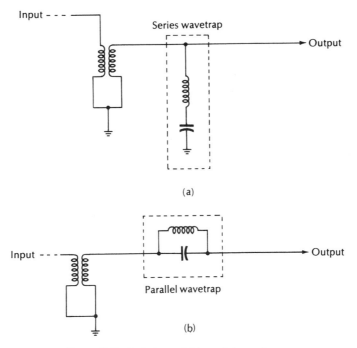

Figure 19-9 (a) Series and (b) parallel wavetraps.

nance, it provides essentially a short-circuit path to ground for the undesirable frequency. As was the case with the tuning circuit, either C or L is usually adjustable in the wave trap to allow precise adjustment.

Figure 19-9b shows a parallel-tuned circuit connected in series with the signal line. Now, this resonant tank has maximum impedance at its resonant frequency, so it acts as very nearly an open circuit to the undesirable frequency. If the desirable frequency is higher than the unwanted one, it readily passes through the capacitor; if lower, it passes through the inductor. Only at resonance does the circuit act as a high impedance.

19-4 PASSBAND OF RESONANT CIRCUITS

It was earlier illustrated (Figures 19-4 and 19-5) that the response curves for resonant circuits consist of smoothly shaped curves, the values of which depend on the circuit Q. Inspecting such a curve makes it easy to understand that the resonant circuit responds to frequencies that are close to f_r. That is, we have considered up to now that f_r

consisted of only one frequency, usually $1/(2\pi\sqrt{LC}$, but obviously a frequency within five or ten cycles of resonance will pass through almost as easily as f_r.

It is therefore necessary to determine a point on the resonant curve that is a point of division. Amplitudes greater than this point are considered to be *within the passband* of the resonant circuit, while amplitudes below are considered to be outside the passband. In practice, an amplitude of 70.7% is considered the dividing line, as shown in Figure 19-10. This value was chosen because it is lower in amplitude by 3 dB than the maximum amplitude, and is often called the *half-power point*. The lower and upper limits of the passband are called f_1 and f_2, respectively.

Example 3. Let us examine the numerical relationships between the quantities in Figure 19-10. Assume that the circuit in question is a series-resonant circuit, that f_r is 1 MHz, and that circuit Q is 100. From this information, the bandwith Δf_{BW} as well as f_1 and f_2 can be derived. Total bandwidth (passband) Δf_{BW} is determined as follows.

$$\Delta f_{BW} = \frac{f_r}{Q} = \frac{1 \times 10^6}{100} = 10{,}000 \text{ Hz}.$$

That is, the total passband between f_1 and f_2 is 10 kHz. Note that if Q were higher in value, then the passband would be narrower, and if

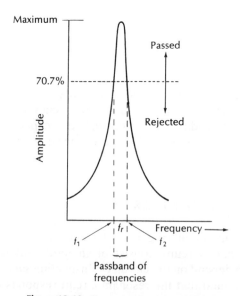

Figure 19-10 Resonant circuit passband.

Q were lower, the passband would be wider. Now, f_1 and f_2 can be found:

$$f_1 = f_r - \frac{\Delta f_{BW}}{2} = (1 \times 10^6) - \frac{10,000}{2} = 0.995 \text{ MHz};$$

$$f_2 = f_r + \frac{\Delta f_{BW}}{2} = (1 \times 10^6) + \frac{10,000}{2} = 1.005 \text{ MHz}.$$

The passband therefore extends from 0.995 MHz to 1.005 MHz at an amplitude equal to or greater than 70.7% of maximum. Since this is a series-resonant circuit, the ordinate of the graph represents current.

19-5 FILTER CIRCUITS

A *filter* circuit consists of some combination of R, L, and C that passes certain frequencies while blocking, or eliminating, others. A filter containing inductance or capacitance has a certain varying response to changing frequency. Such circuits may or may not be resonant, and so are included in this chapter.

It was shown in Figure 19-9 that a series- or parallel-resonant circuit can be used to eliminate certain frequencies, and when so used is called a wave trap. A wave trap is one example of a filter circuit using resonance to obtain the desired results. There are many other filter circuits, some of which must *not* be resonant to obtain the required filtering. One excellent example of a nonresonant filter circuit is the filter circuit used in the output of a rectifier power supply. This power supply uses the ac line voltage to produce a dc voltage and current, usually at a different voltage, by means of a step-up or step-down transformer.

Figure 19-11 shows the circuit diagram for a rectifier power supply. The input to the circuit is the 120-V, 60-Hz line. This is stepped up or down, depending upon the application, by transformer T_1. The necessary value of voltage supplied by T_1 is then *rectified,* in a manner similar to that described for the rectifiers for the ac voltmeter in Chapter 18. The diode rectifiers provide a pulsating dc voltage that, while dc in character (it never changes direction), is anything but smooth, unvarying dc. The filter, then, consists of C_1, L_1, and C_2. The capacitors oppose any change in voltage, while the inductor opposes any change in current through itself, and the net result is that the output voltage across C_2 is very much smoother and can be made to approach the ideal unvarying voltage, as shown.

Generally, the filter capacitors are quite large in value, with typical values of 10 to 100 μF. The filter inductance is also large in value. Note that it is an iron-core choke coil with typical values of 1 to 50 H. To assure nonresonance, we can assign values to the filter components and

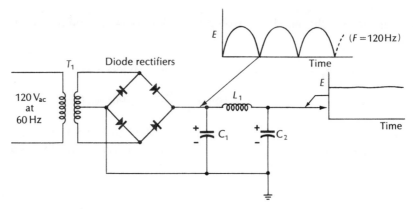

Figure 19-11 Example of the application of a filter in a rectifier power supply.

determine whether or not a condition close to resonance exists. Assume that C_1 and C_2 are 20 μF in value, while L_1 is 20 H. C_1 and C_2 are in series as far as any oscillating tank currents are concerned. Hence, C_t is 10 μF.

$$f_r = \frac{1}{2\pi \sqrt{LC}} = \frac{1}{6.28 \sqrt{20 \times 10 \times 10^{-6}}} = 11.25 \text{ Hz}.$$

Since the frequency across the filter circuit is 120 Hz, the circuit is nonresonant.

The filter just described is known as a *low-pass* filter. That is, it readily passes low-frequency voltages, but it does not allow high-frequency voltages to pass. There are three other basic kinds of filters: high pass, band pass, and bad reject.

Figure 19-12 illustrates the basic characteristics of high- and low-pass filters, as well as the basic circuit configurations. The circuit must have at least one reactive component to be frequency selective. The response curve for the low-pass filter shows generally that all frequencies *below* the 70.7% point are passed, while those above this point are rejected. Note that the output of the filter still contains some of these frequencies above the cutoff point (70.7%), but they are attenuated much more than those frequencies within the passband.

The high-pass curve is a mirror image of the low-pass curve, and all frequencies *above* the cutoff point are considered to be passed (that is, they pass with zero or minimum attenuation), while those below cutoff are considered to be rejected. Low-pass and high-pass circuits are also shown in the figure.

Resonance and Resonant Circuits   469

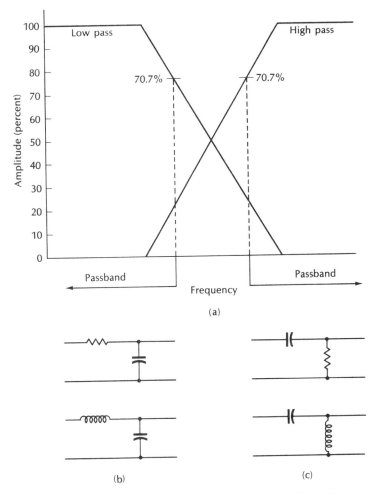

Figure 19-12 (a) Response curves for high- and low-pass filters; (b) low-pass circuits; (c) high-pass circuits.

Figure 19-13 shows the response curves for the bandpass and band-reject filters, as well as the circuits themselves. These circuits must usually pass or reject a relatively narrow band of frequencies, and hence must be resonant. The width of the passed or rejected band is primarily a function of the Q of the circuit: higher Q means a narrower band, while lower Q means a wider band. The passed or rejected band is, as always, measured between the 70.7% points. The bandpass circuits are arranged so that the resonant elements aid the passage of a

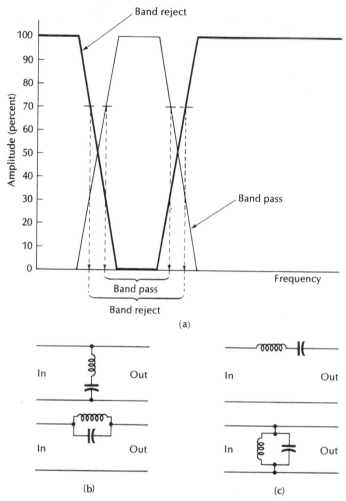

Figure 19-13 (a) Response curves for band-pass and band-reject filters; (b) band-reject circuits; (c) band-pass circuits.

narrow band of frequencies from input to output. In the series configuration, the current is maximum at resonance, so that the greatest output current occurs at or near resonance. In the parallel circuit, the impedance is maximum, at or near resonance, and so the largest voltage output occurs at this point.

Note that other circuits than those illustrated are often used. In practice, multiple-section filters are common, one of which is shown in

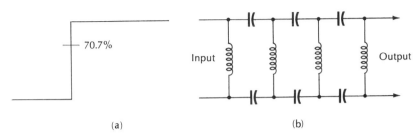

Figure 19-14 (a) The ideal high-pass response curve; (b) a multiple-section filter approximating the ideal response.

Figure 19-14. Increasing the number of sections makes the response curve steeper, to approximate more closely an ideal response, which would have completely vertical sides.

Closely related to *RLC* filter circuits are the *waveshaping circuits* covered in the next chapter. One major difference is that a waveshaping circuit is usually not a resonant circuit; another difference is that waveshaping circuits are not usually used with sinusoidal waveforms.

19-6 FILTER TRAPS

An application of both series- and parallel-tuned circuits used as *traps* that is taken from actual operating equipment is shown in Figure 19-15. As mentioned in Chapter 1, in telephone communications it is often necessary to transmit signaling information over the same wires or channel as the voice frequency (VF). Signaling information includes on-hook/off-hook information as well as dial-pulse (*address*, or phone-number) information. This is particularly true if the transmission is by way of a radio-frequency carrier system.

One method of accomplishing this is to use *in-band* signaling. That is, the band of frequencies used for the VF extends from 300 to 3400 Hz. A single, pure tone at 2600 Hz (within the voice *band*) is used to transmit signaling data, superimposed on the voice channel. The circuit shown in Figure 19-16 is used, along with some automatic switching circuitry, which removes the 2600-Hz tone that is transmitted together with conversation. If the switching circuitry was not used, an annoying high-pitched tone would be heard.

The VF plus any signaling tones present arrive at T_1, which simply isolates the line from the equipment. The filter trap passes all frequencies from about 300 to 3500 Hz. If there is no signaling tone, automatic

Figure 19-15 Telecommunications signaling unit. Courtesy Lynch Communications Systems, Inc.

switching circuits (not shown) choose the path through T_1 to R_1 and R_2 to output A and disconnect the other two output paths.

During dialing, *only* signaling tone bursts at 2600 Hz appear at T_1, and now the path is from T_1 through T_2, whose primary is sharply tuned to 2600 Hz, to output B. Again, automatic switches choose this output, excluding the others, and the dial pulses in the form of 2600-Hz tone bursts are transmitted.

During *free-call* conditions (a subscriber and an operator conversing), *both* VF and signaling tones appear at T_1. To avoid 2600-Hz tone interference, the path is switched again. VF signaling goes from T_1 to the series trap L_1 and C_1. This trap is sharply tuned to 2600 Hz, and removes only a narrow band of frequencies near 2600 Hz, leaving all other voice frequencies unattenuated. To ensure that no 2600-Hz tone passes through the circuit, a second trap, L_2 and C_2, is used. This is simply a parallel-tuned trap, and it reduces still further any residual 2600-Hz tone. Thus, conversation goes on unimpaired, since only a small part of the energy in a normal human voice is in the 2600-Hz region.

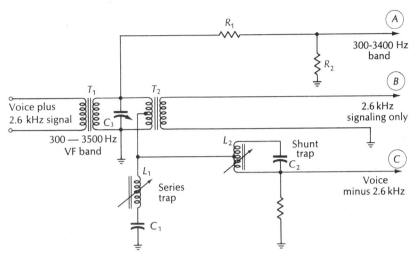

Figure 19-16 Resonant circuits used to channel selected frequencies onto one of three possible paths.

In the photograph (Figure 19-15), the cluster of four cylindrical devices is composed of T_1, T_2, L_1, and L_2. Each component is surrounded by a metallic shield, and L_1 and L_2 have adjustable ferrite cores so that the resonant frequency can be adjusted to 2600 Hz exactly.

19-7 THE VARACTOR DIODE

A varactor diode, or varactor, is a voltage-variable capacitor. That is, the value of capacitance can be made to change with applied dc voltage. The varactor therefore finds wide usage in circuits where automatic frequency control is desirable.

The internal mechanism and action of a varactor is shown in Figure 19-17. The upper drawing shows a typical silicon diode with no external bias applied. Recall from Chapter 8 that the region in the vicinity of the junction consists of a volume of material containing essentially *no* current carriers and is called the barrier region. The *n*-type material has a relatively large number of free electrons as a result of doping, while the *p*-type material contains holes. Both electrons and holes contribute to conduction, so that the device is actually a pair of conductors separated by an insulating medium (a dielectric). This, of course, is the definition of a capacitor. Any diode, then, if it is not forward biased (conducting), exhibits appreciable capacitance, with the *n*- and *p*-sides performing as the plates, while the barrier region is the dielectric.

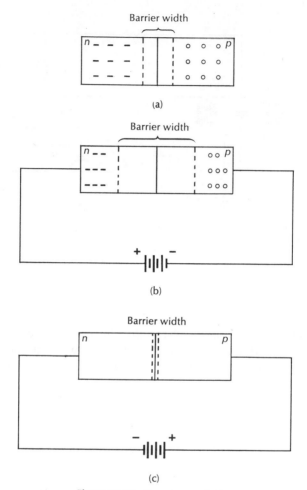

Figure 19-17 The varactor diode.

If an external dc voltage is applied to the diode in such a way as to cause reverse bias (Figure 19-17*b*), the current carriers (electrons and holes) are attracted *away* from the junction, and the capacitance *decreases*. The greater this voltage, the smaller the value of capacitance. The limit in this direction, of course, is the maximum voltage that can safely be applied before the diode is destroyed. Somewhere between the maximum safe voltage and the point where the diode begins to conduct (Figure 19-17*c*), is the normal operating range for the varactor. A conducting diode, of course, is equivalent to a shorted capacitor; hence operation in this mode is forbidden.

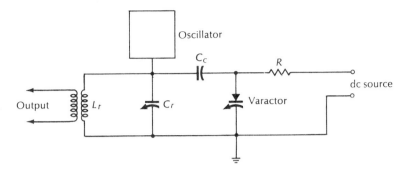

Figure 19-18 A varactor diode application.

Figure 19-18 illustrates a typical circuit arrangement utilizing a varactor to keep L_r and C_r in a resonant circuit close to the desired resonant frequency. Much of the required circuitry is not shown, since this involves such devices as transistor amplifiers, oscillators, and so on, which you have not yet studied. For example, the device that causes L_r and C_r to produce a continuous sine wave is called an oscillator, but it is shown only as a block, and, until you study oscillators, you must assume that L and C work in conjunction with the oscillator to produce a continuous sine wave at a frequency of

$$F = \frac{1}{2\pi \sqrt{LC}}.$$

The balance of the circuit is easily explained. The varactor is part of the resonant circuit since, if the coupling capacitor C_c has essentially zero reactance, then the varactor is in parallel with C_r and thus influences the resonant frequency.

Now, somewhere else in this equipment is a device that senses the frequency of the output of this circuit, and converts any error to a dc voltage that is proportional to the error. This dc voltage is then applied to the varactor in just the amount to correct the error by changing its capacitance and therefore changing the resonant frequency. If the output frequency of the resonant circuit drifts to a higher frequency, the dc voltage applied to the varactor *decreases,* to increase the total capacitance and therefore reduce the frequency of L_r and C_r. If the output frequency decreases, the dc voltage applied to the varactor *increases,* thereby reducing the total capacitance and forcing the resonant frequency higher.

Varactor diodes have relatively small values of capacitance, typically in the tens to hundreds of picofarads, since the area of the "plates" is small. They must therefore be used in high-frequency circuits, where small changes in capacitance will yield desired results.

SUMMARY

- A resonant circuit is an *LC* or *RLC* circuit operating in the unique condition where $X_L = X_C$.
- For a series-resonant circuit, $f_r = 1/(2\pi \sqrt{LC})$.
- For a parallel-resonant circuit, if $Q \geq 10$ then $f_r = 1/(2\pi \sqrt{LC})$.
- For a parallel-resonant circuit, if $Q < 10$,

$$f_r = \left(\frac{1}{2\pi \sqrt{LC}}\right)\left(\frac{Q^2}{1+Q^2}\right)$$

- Total impedance in a series-resonant circuit is equal to circuit R.
- Total impedance in a parallel-resonant circuit is very high, depending somewhat on circuit Q.
- Total current in a series-resonant circuit is high.
- Current supplied by V_s to a parallel-resonant circuit is low.
- The bandwidth of a tuned circuit is f_r/Q, where $Q = X_L/R$.
- Bandwidth is measured at the 3-dB, or 70.7%, points on the response curve.

QUESTIONS

1. State the prime requisite for resonance.
2. How many frequencies will cause resonance for a given value of L and C?
3. In a series-resonant circuit, why is current never infinitely large?
4. State the fundamental relationship from which the formula $f_r = 1/(2\pi \sqrt{LC})$ is derived.
5. In a series circuit containing L and C, how do X_L and X_C vary as the frequency is decreased?
6. In a series circuit containing L and C, how do X_L and X_C vary as the frequency is increased?
7. In a series-resonant circuit, the voltage across L may be larger than V_s. Briefly describe how this is possible.
8. In a series-resonant circuit, the voltage across C may be larger than V_s. Briefly describe how this is possible.
9. How does a parallel-resonant circuit differ from a series-resonant circuit in regard to Z_t and I_t?
10. Under what condition can the resonant frequency of a parallel *RLC* circuit be considered to be equal to $1/(2\pi \sqrt{LC})$?
11. In a parallel-resonant circuit, why does f_r not exactly equal $1/(2\pi \sqrt{LC})$ if $X_L/R < 10$?
12. Briefly describe how f_r varies in a parallel-resonant circuit if Q is made to be lower in value.
13. What is meant by *passband*?
14. Briefly describe a low-pass filter; a high-pass filter.

PROBLEMS

For the following problems, refer to Figure 19-19.

1. $V_s = 12$ V_{ac}, $L = 100$ μH, $C = 0.001$ μF, and $R = 50$ Ω. Determine f_r.
2. $V_s = 6$ V_{ac}, $L = 0.1$ mH, $C = 1000$ pF, and $R = 27$ Ω. Determine f_r.
3. $V_s = 24$ V_{ac}, $L = 0.001$ H, $C = 0.1$ μF, and $R = 10$ Ω. Determine f_r.
4. $V_s = 5$ V_{ac}, $L = 1$ mH, $C = 0.0159$ μF, and $R = 20$ Ω. Determine f_r.
5. $V_s = 20$ V_{ac}, $L = 100$ μH, $C = 470$ pF, and $R = 30$ Ω. Determine f_r and I_t.
6. $V_s = 18$ V_{ac}, $L = 250$ μH, $C = 1500$ pF, and $R = 108$ Ω. Find f_r and I_t.
7. Refer to Problem 1. Determine the values of I_{f_r}, Z, Q, and E_L.
8. Refer to problem 2. Determine the values of I_f, Z, Q, E_L.

For the following problems, refer to Figure 19-20.

9. Determine the value of f_r and Q if $R = 5$ Ω, $L = 100$ μH, $C = 300$ pF, and $V_s = 6$ V_{ac}.

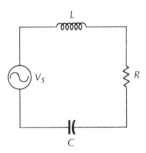

Figure 19-19

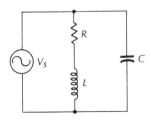

Figure 19-20

10. Determine the value of f_r and Q if $R = 7.5 \,\Omega$, $L = 150 \,\mu\text{H}$, $C = 450$ pF, and $V_s = 6 \,\text{V}_{\text{ac}}$.
11. $V_s = 6 \,\text{V}_{\text{ac}}$, $L = 100 \,\mu\text{H}$, $C = 300$ pF, and $R = 289 \,\Omega$. Find f_r.
12. $V_s = 6 \,\text{V}_{\text{ac}}$, $L = 150 \,\mu\text{H}$, $C = 450$ pF, and $R = 190 \,\Omega$. Determine the value of f_r.
13. A series-resonant circuit is resonant at 1.0 MHz and has a Q of 10. Determine the total bandwidth Δf_{BW} and the f_1 and f_2 cutoff frequencies.
14. A series-resonant circuit is resonant at 1.0 MHz and has a Q of 100. Determine the total bandwidth Δf_{BW} and the f_1 and f_2 cutoff frequencies.

CHAPTER **20**

NONSINUSOIDAL CIRCUITS

Electronic equipment that uses waveforms other than sinusoidal is becoming more and more commonplace. Applications such as the modern digital computer, your pocket calculator (some of which can do much more than the first computers), telemetry data transmission, the digital volt-ohm-milliammeter, to name but a few, are increasingly in evidence. Their operation depends upon *digital* signals, or pulses, and they do not use sinusoidal waveforms at all, except possibly to draw power from an ac line. In such a case, the ac is immediately converted (rectified) to dc.

It is therefore necessary to investigate such waveforms, and to learn certain fundamentals regarding their application. You cannot directly study such equipment at this time, because that study presupposes a thorough knowledge of semiconductors, particularly integrated circuits. We can, however, introduce the fundamentals of the waveforms used, and we can also discover how they can be changed by the use of resistors, capacitors, and inductors. The major topics to be covered are listed below.

20-1 Nonsinusoidal Waveforms
20-2 *RC* Networks
20-3 *RL* Networks
20-4 Pulse Transformers

20-1 NONSINUSOIDAL WAVEFORMS

The most common nonsinusoidal waveform, the square wave, is illustrated in Figure 20-1*a*. Note the dissimilarity of this waveform compared to a sinewave; the abrupt changes in level, from one value to another, are one distinguishing feature. The square wave is similar to a sinewave, however, in that it can be described in terms of cycles, each cycle being one complete chain of events. In the illustration, two complete cycles are shown. Other waveforms are shown, to illustrate the wide variety of nonsinusoidal waveforms encountered in digital circuitry.

Waveform *a* is a symmetrical square wave with equal excursions around the base line. Waveform *b* is similar, but it does not go negative

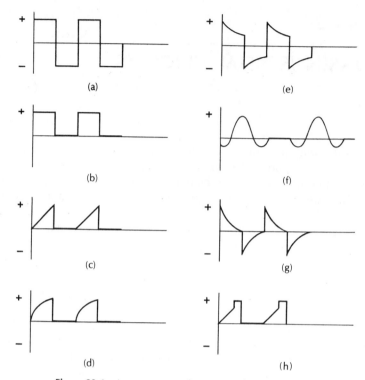

Figure 20-1 An assortment of nonsinusoidal waveforms.

at any time. Waveform *c* is a *sawtooth*, or *ramp*, voltage with positive, but no negative, excursions. Waveform *d* is a symmetrical waveform having rounded corners, sometimes called an *integrated* square wave. Waveforms *e* and *g* are called *differentiated* square waves. Waveform *f* is simply a pulse with no sharp edges. Finally, waveform *h* is a combined waveform consisting of a ramp, followed by a square excursion in the positive direction.

Most of these waveforms are produced by rather simple *RC* or *RL* networks. The basic pulse is the square wave, which is, when necessary, altered to produce one of the other forms. We must, therefore, first investigate the basic square wave. As in the case of the sinusoid, the square wave has its own unique characteristics.

A perfect square wave pulse is shown in Figure 20-2*a*, where all corners are absolutely square. Such a pulse is not possible to attain, although with the aid of modern technology, we can come very close. The typical pulse, when observed on an oscilloscope, will appear more as in Figure 20-2*b* and *c*. The rounding of the corners is due to the

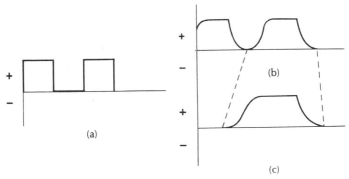

Figure 20-2 (a) A perfect square wave; (b) a practical square wave; (c) an expanded view of the practical squarewave.

inevitable stray capacitance and inductance that occur even in very short lengths of wire or other connections.

Certain names are given to various parts of the pulse to identify the units of measurements used. Referring to Figure 20-3, an important measurement of the pulse is its maximum amplitude, in the example +10 V. This is the maximum excursion above (or below if the pulse is negative) the base or zero-volt line. The *rise time* (t_r) of a pulse is a measure of how rapidly the amplitude increases from minimum to maximum on the leading edge of the pulse. It is measured between the 10% and the 90% values, in units of seconds (ms, μs, and so on). The reason for measuring between the 10% and 90% points is that between 0% and 10%, and between 90% and 100%, the waveform changes very slowly, and it is not possible to specify 0% and 100% as accurately as the 10% and 90% points. On the trailing edge of the pulse, the *fall time* (t_f) is analogous to the rise time, and is also measured between the 90% and 10% values.

As will be shown later, the rise time (and fall time) are indicative of the frequencies making up the pulse. We have already demonstrated that a square wave actually consists of an infinite number of sine waves, in proper combination of frequency, amplitude, and phase.

Pulse duration (t_d) is the length of time the pulse exists measured between either the 10% points or the 50% points. This measurement has not been standardized as yet. Because it appears to be used more often, we shall consider pulse duration as measured between the 50% points.

Pulse period (t_p) is analogous to the cycle in ac waveforms. It is a measure of the time required for one complete cycle of events, measured from any point on the waveform to the same point on the suc-

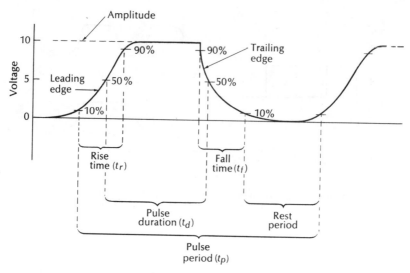

Figure 20-3 Pulse values.

ceeding waveform. As shown, pulse period is measured from the 10% point on one pulse to the 10% point on the next pulse, but the 50% or 90% points could be used as well.

The *rest period* is the time between pulses when no pulse exists. Rest period plus duration time equals the pulse period.

Not shown on the waveform of Figure 20-3 are the repetition rate and duty period or duty cycle of a pulse. The *repetition rate* of a succession of recurrent pulses is simply the number of pulses per second. Repetition rate is similar to the frequency of an ac waveform, and is measured in hertz. The *duty period* or duty cycle of a train of pulses is the ratio of "on" time to total time, or duration time divided by the period.

Finally, the *frequency*, or pulse repetition rate, is simply the number of cycles or periods per second, just as in ac circuits:

$$F = \frac{1}{P} \quad \text{and} \quad P = \frac{1}{F}.$$

The Square Wave

A square wave is often thought of as being one whose sides are of equal length. As shown in Figure 20-4, *any* rectangular wave is a square wave, depending on the time selected for the base line. Hence, as shown in drawing *a*, the time base is incremented in 1-ms units, which makes the waveform appear square. It has a duration time of 1 ms and a rest time of 1ms, while its duty period is 0.5, or 50%.

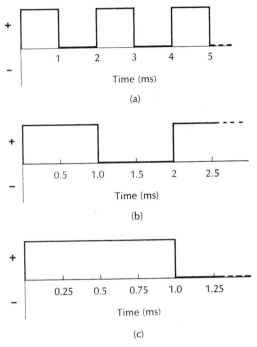

Figure 20-4 Square waves may not appear square.

In drawing *b* the same pulse is shown. It has the same amplitude, but the time base is incremented in 0.5-ms intervals, instead of 1-ms intervals as in the first part of the figure. The identical pulse no longer appears to be square, simply because it is plotted on a different time base. When the time base is spread out further (shorter time intervals) the pulse is even more elongated, as indicated in drawing *c*. If the time-base intervals were increased instead of decreased, the pulse would appear narrower, rather than wider. Therefore, the notion of "square" is a function of the time-base intervals for any rectangular pulse, and all such pulses can be thought of as square waves.

We have already said that a square wave is really composed of sine waves. Figures 20-5 and 20-6 illustrate the explanation of this fact. First, refer to Figure 20-5*a*; this shows a halfcycle of a sine wave at some frequency. Because this is the lowest frequency of all frequencies of interest, it is called the *fundamental* frequency. The next waveform shown (part *b* of the illustration) is the *third harmonic*. That is, it is exactly three times the fundamental—if the fundamental frequency is 1.0 kHz, the third harmonic is 3.0 kHz; if the fundamental is 25 kHz,

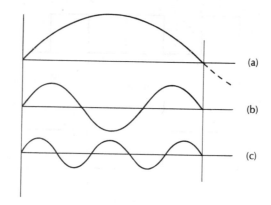

Figure 20-5 (a) A fundamental wave; (b) that wave's third harmonic; (c) the fifth harmonic.

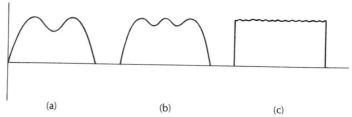

Figure 20-6 (a) Summation of first and third harmonics; (b) summation of first, third, and fifth harmonics; (c) summation of several additional odd-order harmonics.

the third harmonic is 75 kHz. Note also two additional facts: the third harmonic is *in phase* with the fundamental, and it is smaller in amplitude. Figure 20-5c illustrates the fifth harmonic (five times the fundamental frequency) which is also in phase and has a still smaller amplitude.

When these three waveforms are combined (summed) into a single waveform, the result begins to approximate a square wave. Refer to Figure 20-6a; note the result of adding only the first and third harmonics. While this is far from a true square wave, it nevertheless begins to assume a crude, square shape. Adding the fifth harmonic, in drawing *b*, makes an even closer resemblance to a square wave. If more odd-order harmonics (the seventh, the ninth, the eleventh, and so on) are added as indicated in drawing *c*, the familiar square-wave shape is clearly evident. The added harmonics must be in phase and be succeedingly smaller in amplitude.

To illustrate how the foregoing information is applied, refer to Figure 20-7. The purpose of the amplifier shown is to increase the amplitude of

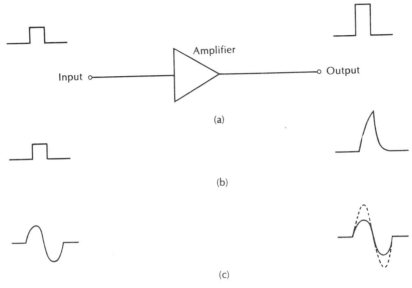

Figure 20-7 (a) A perfect square wave is amplified by an amplifier with good high-frequency response; (b) a square wave is distorted by an amplifier with poor high-frequency response; (c) when a sine wave is passed through an amplifier with poor high-frequency response, its shape is unchanged, but its amplitude is increased only slightly.

the input signal without changing or distorting it. Three sets of waveforms are illustrated. The upper ones represent the ideal case: a small, perfect square wave is amplified and emerges from the amplifier unchanged except in amplitude. The amplifier is said to have good high-frequency response. That is, it amplifies all frequencies equally well. The other two sets of waveforms illustrate the case of an amplifier with poor high-frequency response. Note how badly the center output is distorted; the input square wave has been severely changed, which is very undesirable. The bottom set of waveforms shows how a sine wave is affected by poor high-frequency response. This is the *only* waveform that is *not* altered in shape, but note that the amplitude is much less than it would be if the amplifier had better response (or if the input frequency were made lower).

Dc Component

It is quite often the case that a train of square waves has a dc component. That is, if a dc voltmeter is used to measure the waveform, a very definite dc reading would be obtained. Figure 20-8 illustrates five waveforms, four of which possess a dc component.

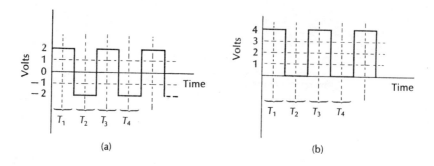

(a)

(b)

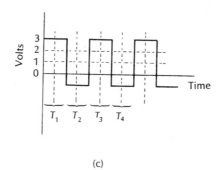

(c)

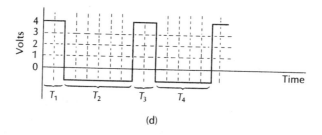

(d)

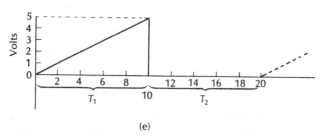

(e)

Figure 20-8 Several square waves to illustrate average dc level.

Figure 20-8a shows a symmetrical square wave operating around the zero-volt base line. The *average* level for one complete cycle is zero, since the waveform excursion above the base line equals the excursion below the base line, both in amplitude and time. The area under the curve for each half cycle consists of four squares, and $4 - 4 = 0$. Hence, a dc voltmeter used to measure this waveform would read zero.

Figure 20-8b shows essentially the same waveform, but extending *only* in the positive direction. In this instance, there *is* a dc level associated with the waveform. During the period T_1, a voltage of 4 V exists, but during the equal period of T_2, no voltage exists. One can intuitively appreciate that the average dc voltage must therefore be 2 V. This relationship is conveniently expressed as

$$E_{av} = \frac{T_1}{T_p} \times E_{max},$$

where T_1 = the period for which the pulse exists and
T_p = the period for one cycle ($T_1 + T_2$, or $T_3 + T_4$, and so forth).
In the example of Figure 20-8b,

$$E_{av} = \frac{T_1}{T_p} \times E_{max} = \frac{2}{4} \times 4 = 2 \text{ V}.$$

Note that the time base might represent any time unit (seconds, milliseconds, microseconds, and so on).

In Figure 20-8c, a slightly different condition exists. Here, the positive excursion during T_1 is 3 V, while the negative excursion during T_2 is 1 V. We must now account for the fact that *both* positive and negative voltage peaks exist.

$$E_{av} = \frac{T_1}{T_p}(E^+) + \frac{T_2}{T_p}(E^-),$$

where E^+ is the positive voltage excursion and
E^- is the negative voltage excursion.
For the waveform in Figure 20-8c, this is calculated as follows:

$$E_{av} = \frac{T_1}{T_p}(E^+) + \left(\frac{T_2}{T_p}(E^-)\right) = (0.5 \times 3) + (0.5 \times -0.5)$$
$$= -1.0 \text{ V}.$$

Hence, a dc voltmeter used to measure this waveform will read -1 V.

As frequently happens, a waveform can be nonsymmetrical, an example of which is given in Figure 20-8d. The average dc level, as would be read by a meter, is

$$E_{av} = \frac{T_1}{T_p}(E^+) + \frac{T_2}{T_p}(E^-) = \frac{2}{8}(4) + \frac{6}{8}(-1)$$
$$= 1 + (-0.75) = 0.25 \text{ V}.$$

This formula is valid only if the sides of the waveform are reasonably straight and square; any significant curvature requires analysis procedures beyond the scope of this book.

A final example is given in Figure 20-8e. Here, a sawtooth, or ramp, waveform is shown, and the analytic procedure is slightly different in a case such as this. To determine the average dc level, assume that the triangular portion of the waveform is a full rectangle, as suggested by the dashed line. The area of the triangular portion is exactly half of the area of the rectangle. This allows the average value of the positive portion to be found, which is then used in the standard relationship for average dc level. First, find the average value as if the rectangle represented the waveform, then simply divide by two to obtain the true dc value.

$$E_{av} = \frac{(10/20) \times 5}{2} = \frac{0.5 \times 5}{2} = \frac{2.5}{2}$$
$$= 1.25 \text{ V}.$$

Again, this relationship is valid only if the waveform has reasonably straight sides.

20-2 RC NETWORKS

Resistor-capacitor networks are widely used in circuits utilizing nonsinusoidal waveforms. Such networks are used as couplers and in wave-shaping applications such as *differentiators* and *integrators*, to name the most important. It is suggested that you review Section 17-2, to refresh your memory in the fundamentals of capacitors. Capacitor charge and discharge characteristics, as well as time constants, are necessary prerequisites to the following discussions.

Couplers are used in pulse applications for much the same reasons that they are used in sinusoidal ac circuits: They block the dc component while allowing the varying component to pass essentially unaltered. Figure 20-9 illustrates an *RC* coupling circuit and the attendant input and output waveforms. The input waveform is a square wave riding on a dc level in such a way that the voltage never goes negative or falls to zero. The capacitor blocks the dc part of the input, and, if the values of *R* and *C* are properly chosen, the output consists of the

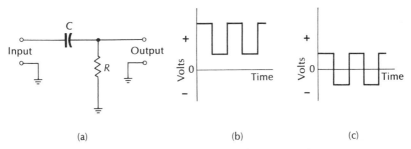

Figure 20-9 (a) An RC coupling circuit; (b) input waveform; (c) output waveform.

varying component only, operating around the zero-volt base line with equal positive and negative excursions. Note that the varying output is identical to the varying part of the input. No change in waveshape has occurred, the only change being that the dc has been blocked.

The correct values for R and C depend on the pulse period, if no waveshape distortion is to occur. As an example, assume that the input square wave has a frequency of 1.0 kHz; pulse period t_p is then

$$t_p = \frac{1}{F} = \frac{1}{1000} = 0.001 \text{ s.}$$

The period for a complete cycle is therefore 1.0 millisecond. In order to assure no pulse distortion, the time constant t_c of R and C must be equal to or greater than 100 times the pulse width. For a symmetrical waveform as shown, pulse width, or duration time, is one-half the period, or $0.001/2 = 500$ μs. Then $R \times C$ must be equal to 500 μs $\times$ 100 $= 0.05$ s $= 50$ ms. That is, $t_c = R \times C = 50$ ms. Any combination of R and C whose product is equal to or greater than 50 ms will pass this pulse with less than 1% distortion. To choose proper values, simply assign a value to either R or C, and then calculate the other necessary value. In other words, since

$$t_c = R \times C,$$

then

$$R = \frac{t_c}{C} \quad \text{or} \quad C = \frac{t_c}{R}.$$

Example 1. Assume that C is a 0.001-μF capacitor. Find the minimum value of R to properly pass the given pulse.

$$R = \frac{t_c}{C} = \frac{50 \times 10^{-3}}{0.001 \times 10^{-6}} = 50 \times 10^6 = 50 \text{ M}\Omega.$$

Since a 50-MΩ resistor is not a practical value (22 MΩ is the largest value in the ±5% resistor table), choose a larger value of capacitor, say 0.1 μF. Now

$$R = \frac{t_c}{C} = \frac{50 \times 10^{-3}}{0.1 \times 10^{-6}} = 500{,}000 \ \Omega,$$

which is a much more practical result. Therefore, if $R = 500$ kΩ and $C = 0.1$ μF, the 1.0-kHz pulse will easily pass with practically no distortion.

To verify that the voltage across the resistor at the end of a 50-ms period is essentially the same as the input excursion, the voltage attained by the capacitor after 50 ms of charging can be determined. Assume a 10-V peak-to-peak excursion.

$$E_c = V_s(1 - \epsilon^{-t/RC}) = 10(1 - 2.72^{-500 \times 10^{-6}/50 \times 10^{-3}})$$
$$= 0.0995 \cong 0.1 \ \text{V}.$$

That is, the capacitor will attain 0.1 V during the 500-μs pulse width, leaving 9.9 V across the resistor, only a 1.0% change. Larger values of R and C will result in an even smaller change. In general, then, the time constant RC must be very long to couple a pulse without distortion.

Short Time Constant: Differentiating

If, in the case of the coupler shown in Figure 20-9, the product of R and C is significantly smaller than 100 times the pulse width, severe waveform distortion will occur. The amount of distortion depends on the ratio of time constant to the pulse period, t_c/t_p. The waveforms shown in Figure 20-10 illustrate the fact that as the ratio t_c/t_p becomes smaller, the waveform loses its flat top and becomes a sharp-pointed spike, as in part c of the figure.

The reason for this is simply that, as t_c/t_p decreases, the capacitor accepts more charge, and since R and C are in series any voltage attained by C detracts from the voltage across R, which *is* the output voltage. When t_c/t_p is smaller than 1 by a significant amount, the capacitor attains full charge and leaves no voltage across R, hence the lower waveform. Stated another way, if t_c/t_p is equal to 0.2 or less, the capacitor has five time constants or more to accept a charge during the pulse duration, and, of course, becomes fully charged. Once fully charged, no further current flows, so that no drop occurs across R.

The waveform shown in Figure 20-10c is known as a *differentiated* waveform, and the RC circuit that produces this waveform is called a *differentiator*. Any similar RC configuration is capable of differentiating if $t_c/t_p = 0.2$, and will produce voltage spikes similar to those shown.

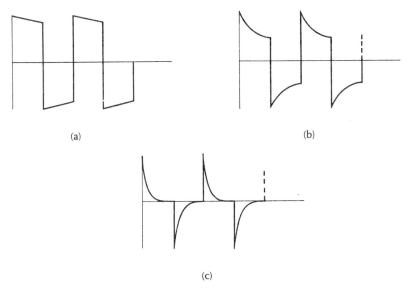

Figure 20-10 Illustrating the effects of changes in the ratio tc/tp on waveforms.

Summarizing, then, the time constant RC must be quite short relative to the pulse width, in order to produce differentiation. The circuit diagram is identical to that of a coupler, the only difference between the two being the t_c/t_p ratio. Indeed, a circuit that acts as a coupler for a very short pulse will differentiate a much longer pulse.

Long Time Constant: Integrating

When the positions of the resistor and capacitor are exchanged, circuit action is reversed. The voltage across the capacitor becomes the output voltage, and the dc level at the input is not blocked. One widely used application of the integrating network is the detection of pulses having a different width than others in the pulse train. For example, horizontal synchronizing pulses are transmitted with the TV picture information to insure that the TV receiver and the TV camera are exactly synchronized at the beginning of each horizontal line. Horizontal pulses are approximately 10 μs in width. Vertical synchronizing pulses are also part of the broadcast signal, to insure that the receiver is ready to start at the top of the picture at the same instant that the TV camera starts to scan at the top.

It is obviously necessary for the TV receiver to distinguish between horizontal and vertical pulses, both of which have equal amplitudes. In order to accomplish this, the vertical pulses are made to be much wider, approximately 25 to 30 μs. Such a pulse train is shown in Figure 20-11a. The output of the integrator is shown in part b, and the integrator circuit configuration appears in Figure 20-11c. As long as the input to this circuit consists of very short-duration pulses, the output consists only of low-amplitude pulses. But when the wider pulses appear, the capacitor has a longer time in which to charge, and the output rises to a much larger amplitude, thus signaling the arrival of the wide pulses to synchronize the vertical field.

Base-Line Shift

When a square wave, or other similar signal, is passed through an *RC* coupler, the signal is reproduced more or less exactly. However, there will be no dc component on the output side of the coupler because the

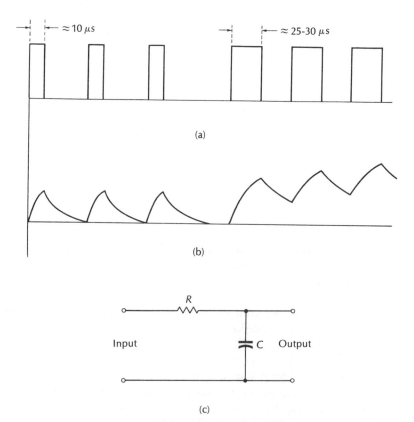

Figure 20-11 (a) Integrator network input and (b) output; and (c) an integrator circuit configuration.

capacitor will not pass the dc component. The change in the base line from input to output is called the *base-line shift,* where the base line is defined as the average input voltage.

To illustrate the effect of base-line shift, examine Figure 20-12. Part *a* of the figure shows the input signal as a unidirectional symmetrical square wave extending only in the positive direction. When passed

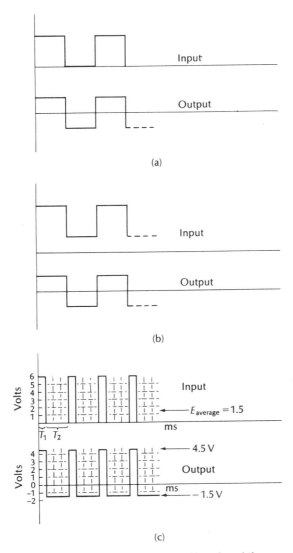

Figure 20-12 The effects of base-line shift.

through an *RC* coupler, the waveform centers itself around the zero-volt line so that equal excursions exist in the positive and negative directions. If the area above the zero-volt base line equals the area below the zero-volt base line for the period of one complete cycle, the average dc level is zero, as it should be.

Figures 20-12*b* and *c* illustrate two other possibilities: the symmetrical square wave with a large positive dc level and a very nonsymmetrical waveform. In order to determine the midpoint for the output waveform, the average dc level of the signal excursion on the input side must be determined. As a numerical example, consider Figure 20-12*c*. The input waveform consists of a 6-V rectangular wave that exists for 1 ms and that is at zero volts for 3 ms. The output waveform is also shown. To find the position of the zero-volt base line, the average dc level of the input waveform must be determined.

$$E_{av} = \left(\frac{t_d}{T_1 + T_2}\right) E_{max} = \left(\frac{t_d}{t_p}\right) E_{max} = \frac{1}{4} \times 6 = 1.5 \text{ V}.$$

The average dc level on the input waveform becomes the output zero-volt base line. This is shown on the output waveform, where the positive excursion is +4.5 V, and the negative excursion is −1.5 V. To verify that this waveform is accurate, recall that the area under the curve above the base line must equal the area under the curve below the base line for one complete cycle. The number of squares in the positive direction is 4½, and the number of squares in the negative direction is also 4½; hence the output waveform has *no dc component*.

Compensated Voltage Dividers

In simple dc circuits, a resistive voltage divider consists only of two or more resistors. Such a divider will divide the applied voltage according to the ratio of the resistors. If, however, the resistive voltage divider is supplied a square-wave pulse, the divided output may be a badly distorted waveform. To avoid this distortion in pulse circuits, the voltage divider can be *frequency compensated*.

Figure 20-13 illustrates these principles. In part *a*, a voltage divider is shown that will divide the dc input by 10; with the 100-V_{dc} input shown, the output is 10 V. Because the input is a steady dc voltage, the output is also a steady voltage of a value one-tenth that of the input.

Part *b* of the illustration shows the same circuit with a 100-V square-wave pulse applied at the input. Note that the output pulse is badly distorted; its maximum amplitude may well be less than the required 10 V. The reason for such severe distortion is the inevitable stray capacitance, both in the external circuitry and directly across the resistors. The higher the value of resistors used in the voltage divider,

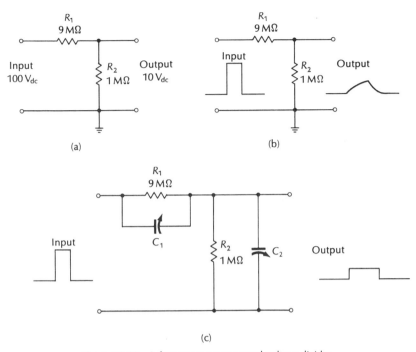

Figure 20-13 A frequency-compensated voltage divider.

the more pronounced the distortion. To correct for the distortion, two additional capacitors are used, C_1 and C_2, shown in part c of the figure. Note that these capacitors are variable, because their values must be exact to eliminate distortion.

To understand how these additional capacitors can eliminate pulse distortion, refer to Figure 20-14, where the circuit shown in Figure 20-13c is redrawn slightly to resemble a *bridge* circuit. Electrically, the two circuits (20-13c and 20-14a) are identical. In a bridge circuit such as this, if the voltage drops across R_1 and C_1 are the same (circuit open at X) as the drops across R_2 and C_2, the circuit is *not* frequency sensitive. That is, the output will depend *only* on the voltage division produced by the components at *any* frequency. The added capacitors must be larger than the stray capacitance that occurs naturally in the circuit. The added capacitors, then, dictate circuit response, and the stray capacitance has little if any effect.

To determine the values of capacitance, recall that when capacitors are connected in series, the value of the voltage drop occurring across each capacitor is inversely proportional to the size of the capacitor. Hence, the smaller capacitor will have the larger voltage drop. We

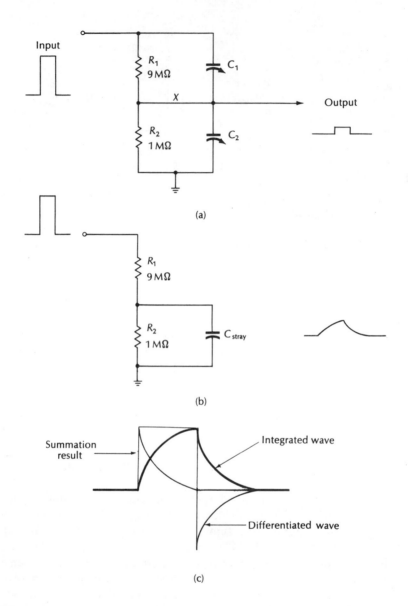

(a)

(b)

(c)

therefore wish E_{R_1} and E_{C_1}, as well as E_{R_2} and E_{C_2}, to be equal. In a proportion, this yields: R_1 is to C_2 as R_2 is to C_1, or, in formula format:

$$\frac{R_1}{C_2} : \frac{R_2}{C_1}$$

Cross multiplying:

$$R_1 C_1 = R_2 C_2$$

so

$$C_1 = \frac{R_2 C_2}{R_1}.$$

Now, choosing a value for C_2 that is larger than the stray capacitance will allow a value of C_1 to be found. As an example, assume the total stray capacitance across R_2 is estimated to be 10 pF. We shall use the rule of thumb that C_2 must be five times larger; hence $C_2 = 50$ pF. ($C_{2(\text{total})}$ is $C_{\text{stray}} + C_2 = 60$ pF.) The correct value of C_1 is therefore determined as follows:

$$C_1 = \frac{R_2 C_{2(\text{total})}}{R_1} = \frac{(1 \times 10^6)(60 \times 10^{-12})}{9 \times 10^6} = 6.67 \times 10^{-12}$$
$$= 6.67 \text{ pF}$$

To achieve exact values, variable trimmer capacitors having the calculated values in their midrange are used. In practice, the capacitors are physically installed (only one need be variable), and the circuit is adjusted while viewing the output waveform on an oscilloscope. Adjusting the capacitors to exactly the correct value will give an output that has no more distortion than the input waveform. If the input is a relatively perfect square wave, then the output will be a relatively perfect, although attenuated, square wave. Note that if the attenuator consists of very low-value resistors, the need for frequency compensation is much less, and may not be required at all if R_1 and R_2 are of the order of hundreds of ohms.

Figure 20-14c illustrates how the added capacitors can remove all distortion from the output. The heavy line shows the result of no compensation: the output waveform is badly distorted, since the 9-MΩ resistor must supply charge and discharge current to the stray capacitance. A resistor of so large a value *must* produce long time constants even with capacitance of very small values. Such a waveform can be described as a *partially integrated* waveform, in which the rounding of the edges is due to the loss of high frequencies. Recall that the reverse of integration is differentiation; hence, if we can inject a differentiated waveform as shown, into the circuit, the adverse effects can be eliminated. This is the purpose of the added capacitor. Note that while R_1 and C_2 form an integrator (output across the capacitor) C_1 and R_2 form a differentiator (output across the resistor). With the previously calculated values for the four components, the differentiator puts back into the waveform *exactly* the high-frequency energy that the integrator removes. The net result is a return of the perfect waveform shown, labeled *summation result*.

An alternate way of analyzing such a circuit is to consider it to be a bridge circuit. Then R_1 and R_2 form a *resistive* voltage divider, while C_1 and C_2 form a *capacitive* voltage divider. If the voltage drops across R_1 and C_1 and across R_2 and C_2 are equal, the bridge is balanced. With a balanced bridge, there is no current in the center arm, since the voltage is exactly the same at the junctions of resistors and capacitors. If there is no current in the center leg, then the capacitors do need not charge and discharge through the resistors, and so the waveform is not altered due to the charge-discharge action.

The preceding example is actually that of the 10-to-1 *attenuator probe* on an oscilloscope. Such a probe is illustrated in Figure 20-15. The 9-MΩ resistor is mounted in the probe handle and as shown, the variable capacitor is mounted in a small box attached to the UHF connector. To adjust the probe, the probe tip is connected to a high-quality square-wave source. Then C_1 is adjusted for zero rounding or overshoot. The probe is now adjusted so that it will not alter the pulse in any way *except* to attenuate it.

20-3 *RL* NETWORKS

Inductors are used in nonsinusoidal circuits relatively infrequently. They are, however, encountered often enough that you should be familiar with their operation and with the associated circuit action.

The induced voltage generated by a coil is given by the expression

$$e_{\text{ind}} = L \frac{di}{dt}$$

where L = inductance of the coil in henrys;
  di = change in current, minimum to maximum or maximum to minimum; and
  dt = time required for current to make one full change.

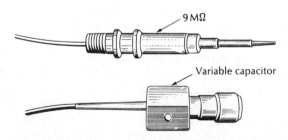

Figure 20-15. An oscilloscope probe assembly.

As a numerical example, assume a 0.01-H coil has a direct current of 10 mA flowing in it. The circuit is suddenly opened, and the current falls to zero in 50 μs. What is the induced voltage across the coil?

$$e_{ind} = L\frac{di}{dt} = 0.01\left(\frac{0.01}{0.00005}\right) = 2 \text{ V}.$$

A circuit having both inductance and resistance can be described in terms of its *time constant* t_c. This constant allows the period of time to be determined for current to go from zero to maximum, or from maximum to zero:

$$t_c = \frac{L}{R}$$

As an example, Figure 20-16 illustrates a circuit having a 1.0-H coil in series with a 500-Ω resistor. With the switch as shown in *a*, a dc current is flowing, limited by V_s and R_1:

$$I_{max} = \frac{V_s}{R_1} = \frac{12}{500} = 0.024 = 24 \text{ mA}.$$

When the switch is transferred to the alternate position, *b*, the energy stored in the magnetic field is returned to the circuit. At the first instant of switch closure, current is of a value equal to 24 mA, but starts decreasing at once. It will decrease exponentially over a period of time, according to the ratio *L/R*, eventually reaching zero. In one time constant, it will decrease by 63.2% and in five time constants it can be considered to be zero. For this circuit, one time constant is

$$t_c = \frac{L}{R} = \frac{1}{500} = 0.002 \text{ s} = 2 \text{ ms},$$

so current will decrease to zero in 5 times 2 ms, or 10 ms.

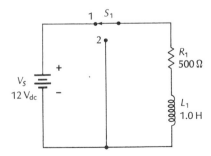

Figure 20-16. Illustrating inductance in a basic pulse circuit.

To determine the value of current at *any* point in time, we use the following relationship:

$$i = I(1 - \epsilon^{-Rt/L}) \quad \text{for increasing current}$$

$$i = I(\epsilon^{-Rt/L}) \quad \text{for decreasing current,}$$

where i = instantaneous current at the time of interest,

$I = I_{max} = \dfrac{V_s}{R}$,

$\epsilon = 2.718 \cong 2.72$,
R = resistance in ohms,
t = time interval in seconds, and
L = inductance in henrys.

As a numerical example of this expression, assume a circuit similar to Figure 20-16, where $V_s = 9$, $R = 100\ \Omega$, and $L = 1$ H. Find the value that current has fallen to 15 ms after the switch is transferred from position 1 to 2. First, determine I_{max}:

$$I_{max} = \frac{V_s}{R} = \frac{9}{100} = 0.09 = 90\ \text{mA}.$$

Then,

$$i = I(\epsilon^{-Rt/L}) = 0.09(2.72^{-100 \times 0.015/1})$$
$$= 0.02\ \text{A} = 20\ \text{mA}.$$

That is, 15 ms after the switch transfers, current has fallen from 90 mA to 20 mA.

Conversely, assume it is necessary to find the value of current 15 ms after the switch is transferred from position 2 to 1 if the switch has been in position 2 for a long time ($i = 0$). Then

$$i = I(1 - \epsilon^{-Rt/L}) = 0.09(1 - 2.72^{-100 \times 0.015/1})$$
$$= 0.0699 = 0.07\ \text{A} = 70\ \text{mA}.$$

That is, current will rise from zero to 70 mA 15 ms after the switch is transferred.

20-4 PULSE TRANSFORMERS

It is sometimes necessary to couple a pulse from one circuit to another by means of a transformer. One reason for doing this is that the dc level at the input (primary side) is blocked, just as in *RC* coupling. Furthermore, the voltage (or current) can be stepped up or down as required, and impedances can be matched. However, probably the most important advantage of pulse-transformer coupling is that the pulse length can be made a function of the transformer itself.

The pulse transformer is not constructed as is a typical 60-Hz transformer, because square-wave pulses contain very high frequencies that an ordinary transformer would simply not pass. Hence, a pulse transformer contains no laminated core and is usually constructed with a core of *ferrite* (or powdered iron) or simply uses an air core. One popular construction uses a ferrite pot core as illustrated in Figure 20-17. The pot core itself has a recess into which two separate coils are placed, one being the primary and the other the secondary. Any reasonable turns ratio can be used, providing only that the coils will fit into the recess. When the unit is assembled, it becomes a high-efficiency transformer for high frequencies.

To illustrate how the pulse transformer works, study Figure 20-18. A pulse-type transformer is connected in series with a current-limiting resistor and a switch. In the first drawing, the switch is closed and a steady current is flowing. There is *no* voltage at the secondary, since the magnetic field surrounding the primary is static. At time 1, the switch is opened (transferred to position 2). Current begins to decrease, and while it decreases a voltage *will* exist at the secondary for a period of approximately five time constants.

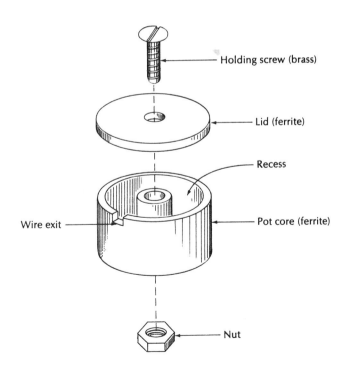

Figure 20-17 The pot core used with a pulse transformer.

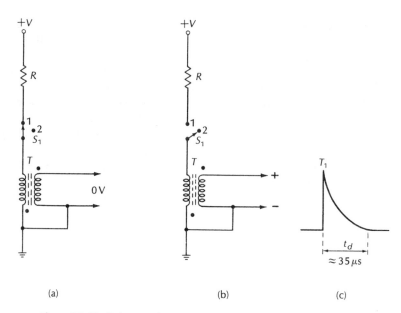

Figure 20-18 Pulse transformer operation.

Note the solid dots below the primary and above the secondary. These indicate points of similar polarity. Thus, if the dotted end of the primary coil is positive at any instant, the dotted end of the secondary is also positive. To determine the polarity of the pulse appearing at point A during the time the current is decreasing, observe that current (electrons) is flowing upward toward $+V_s$ when the switch is closed. At the instant the switch is transferred, the magnetic field starts to collapse toward the coil, inducing a voltage across the primary (and across the secondary, too). Now, the coil is the source, and it tries to keep current flowing in the same direction as before. Hence, the lower end (dot) of the primary must be *positive* (current flows from positive to negative within a *source*) and so the dotted end of the secondary becomes positive.

This is shown in Figure 20-18c, where the positive excursion is evident. The pulse continues for about five time constants if the switch remains in position 2. To determine the actual pulse length, find the time constant if $L = 70$ μH and $R = 10$ Ω:

$$t_c = \frac{L}{R} = \frac{70 \times 10^{-6}}{10} = 7.0 \, \mu s,$$

and the pulse width t_d is 5 times this:

$$t_d = 5 \, t_c = 5 \times 7.0 = 35.0 \, \mu s.$$

SUMMARY

- The rise time t_r of a pulse is that part of a pulse that rises from zero to maximum value, and is measured between the 10% and 90% points.
- The fall time t_f of a pulse is analogous to the rise time but is the decreasing edge; fall time is also measured between 90% and 10% points.
- The pulse duration or width t_d is the duration time of the pulse and is most often measured between the 50% points.
- The pulse period t_p is analogous to the ac cycle and is the time for one complete cycle of events.
- The rest period is that part of the period that the pulse does not exist; rest period plus duration equals pulse period.
- Frequency is the number of periods per second.
- A square wave is composed of the sum of a fundamental sine wave plus all odd harmonics in phase and in decreasing amplitude.
- A simple *RC* network can produce either coupling, integration, or differentiation.
- A long *RC* time constant yields either coupling (output across resistor) or integration (output across capacitor).
- A short *RC* time constant yields differentiation (output across resistor).
- The base line is defined as the average dc level of a waveform.
- A resistive voltage divider, when used with pulses having fast rise and fall times, must be *compensated* by the addition of exact amounts of capacitance to form a bridge circuit.

QUESTIONS

1. Define the rise time of a pulse.
2. Define the fall time of a pulse.
3. Define the pulse period.
4. What is the name given to the pulse time *plus* the rest time?
5. Briefly define the pulse repetition rate.
6. What is the pulse duty period?
7. List the requirements for adding harmonics to form a square wave.
8. What instrument would you use to measure a square wave to determine if it has a dc component?
9. List the most important characteristic of an *RC* coupler for pulses.
10. An *RC* circuit appears to be a coupler, but its time constant is very short for the pulse width used. What is the actual function?

PROBLEMS

1. A repetitive square-wave pulse train has a pulse width or duration of 50 μs and a rest period of 70 μs. Determine the frequency (repetition rate) and pulse period.
2. A repetitive square-wave pulse train has a pulse width or duration of 0.5 μs and a rest period of 0.85 μs. Determine the frequency (repetition rate) and pulse period.
3. A symmetrical, repetitive square wave has a frequency (repetition rate) of 1.0 kHz. Find t_p.
4. A symmetrical, repetitive square wave has a repetition rate of 150 kHz. Find the pulse period.
5. Refer to Figure 20-19a. Find the frequency, period, duration, and average dc level.
6. Refer to Figure 20-19a. The time base is to be changed from milliseconds to microseconds. Find the frequency, period, duration, and average dc level.
7. Refer to Figure 20-19b. Find the frequency, period, duration, and average dc level.
8. Refer to Figure 20-19b. The time base is to be changed from milliseconds to microseconds. Find the frequency, period, duration, and average dc level.

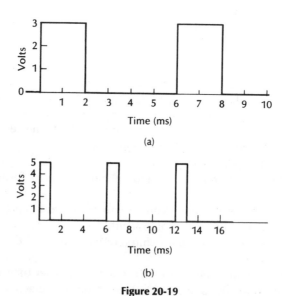

Figure 20-19

9. An RC coupling network is to be used with square-wave pulses. The pulse width (duration) is 50 μs. What is the minimum value of time constant t_c?
10. An RC coupling network is to be used with square-wave pulses. The pulse width is 0.58 μs. What is the minimum value of t_c?
11. An RC coupling network is to be used with square-wave pulses. The pulse width is 33 μs and $C = 0.001$ μF. What minimum value of R is required?
12. An RC coupling network is to be used with square-wave pulses. The pulse width is 0.047 μs, and $R = 10$ kΩ. What minimum value of C is required?
13. A series resistor string (R_1 and R_2) is to be used as a voltage divider for square-wave pulses. The total shunt capacitance across R_2 is 30 pF, and $R_1 = 4$ MΩ and $R_2 = 1$ MΩ. What value of C_1 is necessary, shunted around R_1, to assure that the output is not distorted?
14. A series resistor string (R_1 and R_2) is to be used as a voltage divider for square-wave pulses. The total shunt capacitance across R_2 is 130 pF, and $R_1 = 10$ MΩ and $R_2 = 20$ MΩ. What value of C_1 is necessary, shunted around R_1, to assure that the output is not distorted?

CHAPTER 21

CIRCUIT ANALYSIS TECHNIQUES

Having a firm foundation in Ohm's law and associated techniques allows one to solve a great many practical problems in electricity. However, there are also many kinds of circuits that we have not yet discussed, because to analyze them using only Ohm's law is either very difficult and tedious, or impossible.

There are a number of techniques available that either simplify more complex analytical procedures, or simply provide a systematic approach that eases the drudgery and minimizes the chance of error. In this chapter, we introduce certain of these procedures and provide a number of examples to illustrate how they are used. The subject headings are listed below.

- 21-1 Current and Voltage Dividers
- 21-2 Kirchhoff's Laws
- 21-3 Thevenin's Theorem
- 21-4 Norton's Theorem
- 21-5 Superposition
- 21-6 Milliman's Theorem
- 21-7 Delta-Wye Transformation

21-1 CURRENT AND VOLTAGE DIVIDERS

Current and voltage dividers are extremely important to the technician. They appear in numerous types of fundamental circuits, and solid-state equipment, both analog and digital, uses dividers to provide proper voltage and current values to such devices as diodes, transistors, and integrated circuits. This section, then, is at least as important as any other in this book. You should work toward a thorough understanding of this discussion and of related information in Chapters 4, 5, and 6.

While current and voltage dividers have already been covered, our present interest is to provide a better-organized method of attack and to show how ratio and proportion can be used to simplify analysis. A voltage divider is shown in Figure 21-1; it consists of three resistors and a voltage source, all connected in a series configuration.

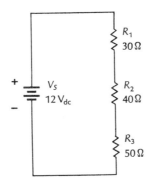

Figure 21-1 A series voltage divider.

Example 1. From your earlier studies, recall that the total voltage drop across the resistors must equal the applied voltage. Also, because the same current flows through each resistor, the voltage drop across each is determined by the resistance value and current.

In the material presented earlier, you determined the voltage drops as follows:

$$I_t = \frac{V_s}{R_1 + R_2 + R_3} = \frac{12}{120} = 0.1 \text{ A}.$$
$$E_{R_1} = I_t \times R_1 = 0.1 \times 30 = 3.0 \text{ V};$$
$$E_{R_2} = I_t \times R_2 = 0.1 \times 40 = 4.0 \text{ V};$$
$$E_{R_3} = I_t \times R_3 = 0.1 \times 50 = 5.0 \text{ V}.$$

It is more efficient to do the calculations all in one step:

$$E_{R_1} = \frac{V_s R_1}{R_t} = \frac{12 \times 30}{120} = 3.0 \text{ V};$$
$$E_{R_2} = \frac{V_s R_2}{R_t} = \frac{12 \times 40}{120} = 4.0 \text{ V};$$
$$E_{R_3} = \frac{V_s R_3}{R_t} = \frac{12 \times 50}{120} = 5.0 \text{ V}.$$

Note that this is actually no different from the previous calculation, since V_s/R_t is the total circuit current. However, this relationship has the advantage of emphasizing the fact that the voltage across an individual resistor is a function of the ratio of that resistor *to the total resistance*. That is, E_{R_1} is determined by $(R_1/R_t) \times V_s$, and E_{R_2} is determined by $(R_2/R_t) \times V_s$. Hence, one can often determine the drop across a particular resistor by simply figuring the ratio mentally and multiplying by V_s. For instance, in the case of R_1 above, $30/120 = 3/12 = 1/4$, and $1/4 \times 12 = 3.0$ V.

Example 2. Assume that, in Figure 21-1, $R_1 = 33{,}000\ \Omega$, $R_2 = 47{,}000\ \Omega$, and $R_3 = 68{,}000\ \Omega$; V_s remains 12 V_{dc}. Find the drop across each resistor.

$$E_{R_1} = \frac{V_s R_1}{R_t} = \frac{12 \times 33 \times 10^3}{148 \times 10^3} = 2.68 \text{ V};$$

$$E_{R_2} = \frac{V_s R_2}{R_t} = \frac{12 \times 47 \times 10^3}{148 \times 10^3} = 3.81 \text{ V};$$

$$E_{R_3} = \frac{V_s R_3}{R_t} = \frac{12 \times 68 \times 10^3}{148 \times 10^3} = 5.51 \text{ V}.$$

This procedure is particularly useful if you have an electronic calculator with addressable memory. If so, you can store V_s/R_t, then recall and multiply it by R_1, R_2, and R_3 in turn to obtain each voltage drop very quickly and easily.

Example 3. In Figure 21-2 a similar circuit is shown, except that connection points are given for each value of voltage with respect to ground. The preceding voltage-divider procedure can easily be used to find the drop across two resistors by simply including *both* values in the numerator:

$$E_B = \frac{V_s(R_3 + R_2)}{R_t} = \frac{12(1000 + 800)}{2400} = 9.0 \text{ V};$$

$$E_C = \frac{V_s R_3}{R_t} = \frac{12(1000)}{2400} = 5.0 \text{ V}.$$

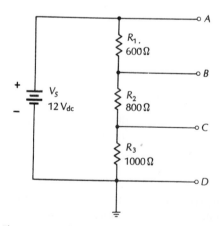

Figure 21-2 The voltage divider for Example 3.

This relationship is valid only if one end of the voltage divider string is at ground (zero volts). However, by redefining the quantity to the left of the equal sign, the relationship *is* valid for finding the drop across the resistors placed in the numerator. For example, if E_B is replaced with $R_{R2,3}$ the formula will determine the drop across $R2$ and $R3$, regardless of ground reference.

Ratio and Proportion

It is often desirable to be able to compute the values of a voltage divider given only a few specifications. The following example, based on Figure 21-3, illustrates how this can be done.

Example 4. A voltage divider is to be between $+25$ V_{dc} and ground to provide two voltages of $+15$ V_{dc} and $+10$ V_{dc}. Total current is to be 10 mA; find the required resistor values.

First find the total resistance:

$$R_t = \frac{V_s}{I_t} = \frac{25}{0.01} = 2500 \; \Omega.$$

Hence $R_1 + R_2 + R_3 = 2500 \; \Omega$.

Now, the voltage across each resistor is a function of its ratio to the total resistance. We can therefore solve for the individual resistor values by the use of proportions. To find the value of R_3, for instance, we can say, "The ratio of the voltage across R_3 to the value of R_3

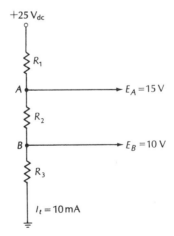

Figure 21-3 Ratio and proportion can be used to find the unknowns in this voltage divider.

must stand in the same proportion as the ratio of total voltage to total resistance." Thus

$$\frac{E_{R_3}}{R_3} : \frac{V_s}{R_t}$$

Cross multiplying yields

$$R_3 V_s = E_{R_3} R_t.$$

To isolate and solve for R_3, divide both terms by V_s:

$$R_3 = \frac{E_{R_3} R_t}{V_s} = \frac{10 \times 2500}{25} = 1000 \ \Omega.$$

Because there is to be 5 V across R_2, its value can be found in the same manner:

$$\frac{E_{R_2}}{R_2} : \frac{V_s}{R_t}$$
$$R_2 V_s = E_{R_2} R_t,$$
$$V_s = \frac{E_{R_2} \times R_t}{V_s} = \frac{5 \times 2500}{25} = 500 \ \Omega.$$

An alternate way to find the value of R_2 is:

$$\frac{(R_3 + R_2)}{(E_{R_3} + E_{R_2})} : \frac{R_t}{V_s},$$
$$(R_3 + R_2) = \frac{(E_{R_2} + E_{R_3})R_t}{V_s} = \frac{15 \times 2500}{25} = 1500 \ \Omega,$$

and,

$$R_2 = (R_3 + R_2) - R_3 = 1500 - 1000 = 500 \ \Omega.$$

Now only R_1 is unknown. The same procedure can be used to find its value, but it is simpler to subtract the known values of R_3 and R_2 from R_t. Since $R_t = R_1 + R_2 + R_3$,

$$R_1 = R_t - R_3 - R_2 = 2500 - 1000 - 500 = 1000 \ \Omega.$$

The voltage divider therefore consists of three resistors: $R_1 = 1000 \ \Omega$, $R_2 = 500 \ \Omega$, and $R_3 = 1000 \ \Omega$. The fact that these values *will* provide the proper voltages can be verified by the methods discussed earlier in this book.

Example 5. A similar problem, although somewhat more involved, is given in Figure 21-4. Here a bare minimum of known values is given, and finding the value of R_3 requires some deductive reasoning. A method of solving such a problem is first to consider what *is* known, and to locate a set of values that will allow an unknown to be found.

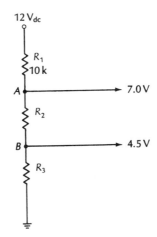

Figure 21-4 The voltage divider for Example 5.

Such a set of values is given in the voltage at point A, which allows the voltage across R_1 to be found, and hence the current through R_1.

$$E_{R_1} = V_s - E_A = 12 - 7 = 5 \text{ V};$$
$$I_{R_1} = I_t = \frac{E_{R_1}}{R_1} = \frac{5}{10,000} = 0.5 \text{ mA}.$$

I_{R_1} is the same as I_t, so the current through R_2 and R_3 is now known. At this point, the calculations may be made using any of a number of methods, all of which arrive at the same values. One of these methods is to determine the total resistance:

$$R_t = \frac{V_s}{I_t} = \frac{12}{0.5 \times 10^{-3}} = 24,000 \text{ }\Omega.$$

Then

$$R_2 + R_3 = R_t - R_1 = 24,000 - 10,000 = 14,000 \text{ }\Omega.$$

R_2 can now be given a value, since the voltage across it and the current through it are known.

$$R_2 = \frac{E_{R_2}}{I_t} = \frac{7 - 4.5}{0.5 \times 10^{-3}} = 5000 \text{ }\Omega.$$

An alternative approach is to write a proportion:

$$\frac{E_{R_2}}{R_2} : \frac{V_s}{R_t},$$
$$R_2 V_s = R_t E_{R_2},$$
$$R_2 = \frac{R_t E_{R_2}}{V_s} = \frac{24,000 \times 2.5}{12} = 5000 \text{ }\Omega.$$

Finally, a value for R_3 can be found:

$$R_3 = \frac{E_{R_3}}{I_t} = \frac{4.5}{0.5 \times 10^{-3}} = 9000 \: \Omega.$$

Note that we could have determined R_3 as soon as I_t had been given a value, since the drop across R_3 was given. Also, once total resistance was known, the values of R_2 and R_3 could have been found by proportion. As mentioned earlier, there are several ways to solve for the unknowns in such a circuit, and whichever seems the easiest and most logical to you is the method to use.

Current Dividers

Just as a string of series-connected resistors divides the available voltage according to the values of the resistors, so do parallel resistors divide the current.

Example 6. The circuit given in Figure 21-5 will be analyzed by conventional means.

Find total equivalent resistance:

$$R_t = \frac{R_1 \times R_2}{R_1 + R_2} = \frac{10 \: k\Omega \times 15 \: k\Omega}{25 \: k\Omega} = 6000 \: \Omega.$$

Now find total current:

$$I_t = \frac{V_s}{R_t} = \frac{12}{6000} = 2 \: \text{mA}.$$

Finally, the individual values of branch current can be determined.

$$I_{R_1} = \frac{V_s}{R_1} = \frac{12}{10,000} = 1.2 \: \text{mA};$$

$$I_{R_2} = \frac{V_s}{R_2} = \frac{12}{15,000} = 0.8 \: \text{mA}.$$

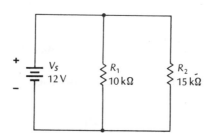

Figure 21-5 Finding unknowns in a current divider.

It is often desirable to be able to compute the current through a parallel-connected resistor when applied voltage is not known but total current is. Such a circuit is shown in Figure 21-6, and the following relationships allow branch currents to be found:

$$I_{R_1} = \frac{I_t R_2}{R_1 + R_2} = \frac{0.02 \times 30,000}{20,000 + 30,000} = 12 \text{ mA};$$

$$I_{R_2} = \frac{I_t R_1}{R_1 + R_2} = \frac{0.02 \times 20,000}{20,000 + 30,000} = 8 \text{ mA}.$$

Note carefully that, to find I_{R_1}, the value of R_2 is placed in the numerator; and to find I_{R_2}, R_1 is placed in the numerator. This indicates the inverse relationship between resistance and branch current. That is, the larger resistance draws the smaller current, and vice versa. Also, this relationship is valid only for two resistors.

In such a circuit, the value of V_s can easily be found. First, find R_t.

$$R_t = \frac{R_1 R_2}{R_1 + R_2} = \frac{20,000 \times 30,000}{50,000} = 12,000 \text{ }\Omega.$$

Then

$$V_s = I_t \times R_t = 0.02 \times 12,000 = 240 \text{ V}.$$

Occasionally very few values are known, such as is the case with the circuit shown in Figure 21-7. Here, only I_t, I_{R_1}, and R_2 are known. This, however, is sufficient to solve for all unknown values.

Probably the best starting point is to find the current through R_2:

$$I_{R_2} = I_t - I_{R_1} = 0.01 - 0.006 = 4 \text{ mA}.$$

Now the voltage across R_2 ($E_{R_2} = V_s$) can be determined.

$$V_s = I_{R_2} \times R_2 = 0.004 \times 15,000 = 60 \text{ V}.$$

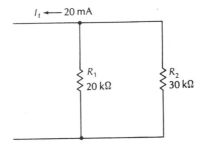

Figure 21-6 Determining branch currents when total current is known.

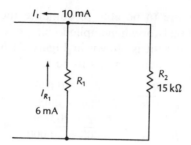

Figure 21-7 Unknowns can be found even when very few values are given.

Finally, the value of R_1 can be found.

$$R_1 = \frac{V_s}{I_{R_1}} = \frac{60}{0.006} = 10{,}000 \; \Omega.$$

Voltage and Current Divider Examples

Several examples of voltage and current dividers are given below. The problems are worked out step by step, but little is given in the way of explanation, since you should be able to follow the calculations on your own. The first circuit is given in Figure 21-8, which is a series voltage divider.

Example 7. $V_s = 9.0$ V, $R_1 = 20{,}000 \; \Omega$, $R_2 = 30{,}000 \; \Omega$, $R_3 = 40{,}000 \; \Omega$. Find E_A (the value of voltage at point A), and E_B (the value of voltage at point B), both in reference to ground; and E_{R_1} (the drop across R_1).

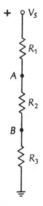

Figure 21-8 The voltage divider for Example 7.

$$E_B = \frac{V_s R_3}{R_t} = \frac{9 \times 40{,}000}{90{,}000} = 4.0 \text{ V};$$

$$E_A = \frac{V_s(R_2 + R_3)}{R_t} = \frac{9 \times 70{,}000}{90{,}000} = 7.0 \text{ V};$$

$$E_{R_1} = V_s - E_{R_2} - E_{R_3} = 9 - 4 - 3 = 2 \text{ V}.$$

Example 8. Refer to Figure 21-9. $+V_s = 9.0$ V, $-V_s = -9.0$ V, $R_1 = 20{,}000 \, \Omega$, $R_2 = 30{,}000 \, \Omega$, $R_3 = 40{,}000 \, \Omega$. Find the values of voltage at points A and B (E_A and E_B), both in reference to ground, and the drop across R_1. (Note that because neither end of the string is at ground, the divider formulas used in the previous example are not appropriate.)

$$I_t = \frac{(V_{s(\text{total})})}{R_t} = \frac{18}{90{,}000} = 0.2 \text{ mA} = 200 \, \mu\text{A}.$$

$$E_{R_1} = I_t R_1 = 200 \times 10^{-6} \times 20{,}000 = 4.0 \text{ V};$$
$$E_{R_2} = I_t R_2 = 200 \times 10^{-6} \times 30{,}000 = 6.0 \text{ V};$$
$$E_{R_3} = I_t R_3 = 200 \times 10^{-6} \times 40{,}000 = 8.0 \text{ V}.$$
$$E_B = -V_s + E_{R_3} = -9 + 8 = -1.0 \text{ V},$$

or

$$E_B = +V_s - E_{R_1} - E_{R_2} = 9 - 4 - 6 = -1.0 \text{ V}.$$
$$E_A = V_s - E_{R_1} = 9 - 4 = 5.0 \text{ V},$$

or

$$E_A = -V_s + E_{R_3} + E_{R_2} = -9 + 8 + 6 = 5.0 \text{ V}.$$

Example 9. Refer to Figure 21-10. $V_s = 12$ V, $R_1 = 200 \, \Omega$, $R_2 = 300 \, \Omega$, $R_3 = 400 \, \Omega$. Find the branch currents and I_t.

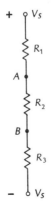

Figure 21-9 The voltage divider for Example 8.

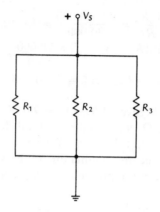

Figure 21-10 The current divider for Example 9.

$$I_{R_1} = \frac{V_s}{R_1} = \frac{12}{200} = 60 \text{ mA};$$
$$I_{R_2} = \frac{V_s}{R_2} = \frac{12}{300} = 40 \text{ mA};$$
$$I_{R_3} = \frac{V_s}{R_3} = \frac{12}{400} = 30 \text{ mA};$$
$$I_t = I_{R_1} + I_{R_2} + I_{R_3} = 130 \text{ mA}.$$

Example 10. Refer to Figure 21-11. $+V_s = 12$ V, $-V_s = -12$ V, $R_1 = 200 \, \Omega$, $R_2 = 300 \, \Omega$, $R_3 = 400 \, \Omega$. Find the branch currents and I_t.

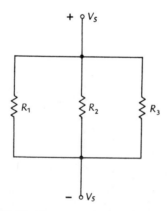

Figure 21-11 The current divider for Example 10.

$$I_{R_1} = \frac{V_{s(\text{total})}}{R_1} = \frac{24}{200} = 120 \text{ mA};$$
$$I_{R_2} = \frac{V_{s(\text{total})}}{R_2} = \frac{24}{300} = 80 \text{ mA};$$
$$I_{R_3} = \frac{V_{s(\text{total})}}{R_3} = \frac{24}{400} = 60 \text{ mA};$$
$$I_t = I_{R_1} + I_{R_2} + I_{R_3} = 120 \text{ mA} + 80 \text{ mA} + 60 \text{ mA} = 260 \text{ mA}.$$

Example 11. Refer to Figure 21-12. $+V_s = 6$ V, $-V_s = -12$ V, $R_1 = 100 \, \Omega$, $R_2 = 200 \, \Omega$, $R_3 = 470 \, \Omega$, $R_4 = 560 \, \Omega$, $R_5 = 820 \, \Omega$. Determine the value of E_A and E_B in reference to ground.

$$R_{\text{eq } 3,4,5} = \frac{1}{1/R_3 + 1/R_4 + 1/R_5} = \frac{1}{1/470 + 1/560 + 1/820}$$
$$= 194.8 \cong 195 \, \Omega;$$
$$R_t = R_1 + R_{\text{eq } 3,4,5} + R_2 = 100 + 195 + 200 = 495 \, \Omega.$$
$$I_t = \frac{V_{s(\text{total})}}{R_t} = \frac{18}{495} = 0.03636 \cong 0.0364 = 36.4 \text{ mA}.$$
$$E_{R_1} = I_t R_1 = 0.0364 \times 100 = 3.64 \text{ V};$$
$$E_A = +V_s - E_{R_1} = 6 - 3.64 = 2.36 \text{ V}.$$
$$E_{R_2} = I_t R_2 = 0.0364 \times 200 = 7.28 \text{ V}.$$
$$E_B = -V_s + E_{R_2} = -12 + 7.28 = -4.72 \text{ V}.$$

Example 12. Refer to Example 11. Using the values given, find the current I_{R_4} flowing in R_4.

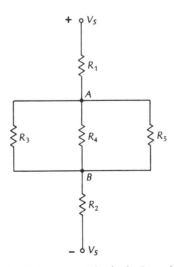

Figure 21-12 The current divider for Example 11.

$$E_{R_4} = V_s - E_{R_1} - (-V_s + E_{R_2}) = 6 - 3.64 - (-12 + 7.28)$$
$$= 7.08 \text{ V,}$$

or

$$E_{R_4} = V_{s(\text{total})} - E_{R_1} - E_{R_2} = 18 - 3.64 - 7.28 = 7.08 \text{ V.}$$
$$I_{R_4} = \frac{E_{R_4}}{R_4} = \frac{7.08}{560} = 0.0126 = 12.6 \text{ mA.}$$

21-2 KIRCHHOFF'S LAWS

One of the most powerful tools for solving problems in electrical circuitry are the relationships known as **Kirchhoff's laws**. In fact, you have been using them rather consistently; we simply have not identified them. In 1847 a German physicist, Gustav Robert Kirchhoff (1824–1887), discovered two very important relationships regarding the voltage and current in a closed, or complete, circuit. His observations led to major theoretical advances. We can state these laws as follows:

Kirchhoff's voltage law: The algebraic sum of all voltages around a closed loop is zero.

Kirchhoff's current law: The algebraic sum of all currents entering and leaving a junction, or node, is zero.

In order to apply the voltage law, you must first understand that all sources are given a *positive* sense while the voltage drops across loads are given a *negative* sense. (All sources must have the same polarity.) The current law requires a positive sense for all currents *entering* a node and a negative sense for currents *leaving* the node. Note that if the sources do not have the same polarity, the source having the largest voltage is considered *positive* and all others are treated as *drops*.

The series voltage divider shown in Figure 21-13 will be analyzed by use of Kirchhoff's voltage law. First, the individual voltage drops around the circuit must be found, using simple Ohm's law.

$$E_{R_1} = \frac{V_s \times R_1}{R_t} = \frac{6 \times 10}{10 + 20 + 30} = \frac{60}{60} = 1 \text{ V;}$$
$$E_{R_2} = \frac{V_s \times R_2}{R_t} = \frac{6 \times 20}{10 + 20 + 30} = \frac{120}{60} = 2 \text{ V;}$$
$$E_{R_3} = \frac{V_s \times R_3}{R_t} = \frac{6 \times 30}{10 + 20 + 30} = \frac{180}{60} = 3 \text{ V.}$$

Kirchhoff's voltage law allows us to equate the circuit to zero as follows:

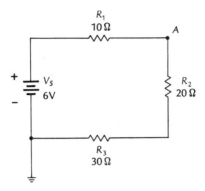

Figure 21-13 A series voltage divider to be analyzed by application of Kirchoff's laws.

$$V_s = E_{R_1} + E_{R_2} + E_{R_3},$$

so that

$$V_s - E_{R_1} - E_{R_2} - E_{R_3} = 0.$$

This, then, is the Kirchhoff statement for the circuit in question, and it is applied as shown below. Assume that the voltage at point A is required, with reference to ground.

$$E_A = V_s - E_{R_1} = +6 - 1 = 5 \text{ V},$$

or

$$E_A = 0 + E_{R_3} + E_{R_4} = 0 + 3 + 2 = 5 \text{ V}.$$

Such an application has been used before in this book; for instance, you will discover it more than once when you restudy Examples 7 through 12 above.

Kirchhoff's voltage law is most useful in series circuits in obtaining the value of voltage at an intermediate point. Parallel circuits require the use of the current law, and a circuit to be used as an example is given in Figure 21-14. The current law states that the current flowing toward point A must equal the current flowing away from point B; or, stated another way, the algebraic sum of I_1, I_2, I_3, and I_t must equal zero. We assign a + sign to currents flowing toward a node and a − sign to those flowing away from the node. Therefore, in the illustrated example,

$$I_t - I_1 - I_2 - I_3 = 0,$$

or

$$60 \text{ mA} - 10 \text{ mA} - 20 \text{ mA} - 30 \text{ mA} = 0.$$

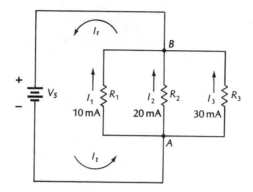

Figure 21-14 A current divider to be analyzed by application of Kirchoff's laws.

This relationship can be solved for any one unknown if the other values are known.

$$I_1 = I_t - I_2 - I_3,$$
$$I_2 = I_t - I_1 - I_3,$$
$$I_3 = I_t - I_2 - I_1.$$

Again, these relationships have been used before, but without identifying them as Kirchhoff's laws.

As a further example of how Kirchhoff's laws are applied, consider Figure 21-15. Here, two sources operate into a somewhat complex network, and it is not possible to analyze the circuit using only Ohm's law. While other somewhat simpler methods do exist, we will show the complete Kirchhoff procedure to help you become familiar with this key method.

Note first that there are three values of current to concern us: I_1, I_2, and I_3. Using the current law with respect to points A or B,

$$I_3 - I_1 - I_2 = 0 \quad \text{or} \quad I_3 = I_1 + I_2.$$

To determine the value of current flowing in R_3, the voltage equations for each loop are written to obtain interim values. Loop 1 consists of the path C, B, A, and D. Loop 2 consists of path E, B, A, and F.

Loop 1 Voltage Equation:

$$V_{s_1} - E_{R_1} - E_{R_3} = 0,$$
$$V_{s_1} - I_1 \times R_1 - I_3 \times R_3 = 0.$$

When this expression is evaluated it will *not* give a value in amperes, but instead is only an interim result to be used later in solving for the true value of current in R_3. Substituting known values we have

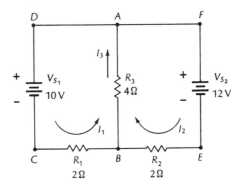

Figure 21-15 A complex multisource network to be analyzed by Kirchoff's laws.

$$V_{s_1} - I_1 \times R_1 - I_3 \times R_3 = 0,$$
$$10 - (I_1 \times 2) - (I_3 \times 4) = 0,$$

or

$$10 - (2I_1) - (4I_3) = 0.$$

Solve for I_1, which is the current (interim value) flowing in Loop 1:

$$10 - 4I_3 = 2I_1.$$

Now divide both terms by 2 to obtain I:

$$5 - 2I_3 = I_1.$$

This is the value of the Loop 1 interim solution.

We now follow a similar procedure for Loop 2.

Loop 2 Voltage Equation:

$$V_{s_2} - E_{R_2} - E_{R_3} = 0.$$

First, substitute known values:

$$12 - (I_2 \times R_2) - (I_3 \times R_3) = 0,$$

or

$$12 - (2I_2) - (4I_3) = 0.$$

Solve for I_2, which is the current (interim value) in Loop 2:

$$12 - (4I_3) = 2I_2.$$

Solve for I_2 by dividing by 2:

$$6 - (2I_3) = I_2.$$

This is the interim value for Loop 2.

Now combine the two loop equations. Since $I_3 - I_1 - I_2 = 0$, then $I_1 + I_2 = I_3$. Recall that the current law states that the sum of the two currents arriving at a node (A or B) and the current leaving the node must equal zero. Now, substitute the interim values for the two loops in the equation:

$$I_1 + I_2 = I_3,$$
$$(5 - 2I_3) + (6 - 2I_3) = I_3.$$

Finally, combine terms:

$$11 = 2I_3 + 2I_3 + I_3;$$
$$11 = 5I_3;$$
$$2.2 \text{ A} = I_3.$$

This is the true value of current through R_3. It is now possible to find the remaining values by conventional means.

$$E_{R_3} = I_3 \times R_3 = 2.2 \times 4 = 8.8 \text{ V};$$
$$E_{R_1} = V_{s_1} - E_{R_3} = 10 - 8.8 = 1.2 \text{ V};$$
$$E_{R_2} = V_{s_2} - E_{R_3} = 12 - 8.8 = 3.2 \text{ V};$$
$$I_{R_1} = I_1 = \frac{E_{R_1}}{R_2} = \frac{1.2}{2} = 0.6 \text{ A};$$
$$I_{R_2} = I_2 = \frac{E_{R_2}}{R_2} = \frac{3.2}{2} = 1.6 \text{ A}.$$

Proof:

$0.6 + 1.6 = 2.2$ A, which agrees with the computed value.

Highly complex multisource circuits can be solved using Kirchhoff's laws, as we have just shown. Such circuits often require the use of simultaneous equations, so that most of these circuits can be solved more easily by the use of one or more of the theorems to be discussed subsequently.

21-3 THEVENIN'S THEOREM

Thevenin's theorem allows us to solve for certain circuit values that we could not find with the methods discussed thus far. As has been mentioned earlier, *all* sources of electrical energy have an internal impedance to a greater or lesser degree. We have thus far considered the source to be a perfect voltage generator. Thus, no matter how much or how little current is drawn from the source, we have considered the source to provide a given, fixed value of terminal voltage. However, all practical sources provide a voltage that *decreases* as more current is drawn, due to the internal impedance. Thevenin's theorem allows us to construct an equivalent circuit that accounts for the internal impedance and thus gives much more accurate results.

A Thevenin-equivalent circuit is shown in Figure 21-16, with three different loads that may be connected to the source. The source has been *Thevenized*, which simply means that the internal resistance R_i or R_{th} has been accounted for, and the open-circuit (no-load) voltage V_{oc}, or V_{th}, is given. V_{th} stands for Thevenin equivalent source voltage, and R_{th} is the equivalent source internal resistance.

To solve for the current and voltage values, simply treat the circuit as a normal series circuit, *but* include the source's internal resistance where applicable. For example, find the total current I_t and the true drop across R_1 when it is connected into the circuit. If no load at all is connected, the terminal open-circuit voltage V_{th} of the battery is 12 V, since if no current flows in R_{th}, there can be no drop across it. However, when a load is connected, say R_1, then R_{th} must be considered:

$$I_t = \frac{V_{th}}{R_{th} + R_1} = \frac{12}{5 + 100} = 0.1143 \text{ A} = 114.3 \text{ mA}.$$

Now $E_{R_{th}}$ and E_{R_1} can be found:

$$E_{R_{th}} = I_t \times R_{th} = (114.3 \times 10^{-3})(5) = 0.57 \text{ V}.$$
$$E_{R_1} = I_t \times R_1 = (114.3 \times 10^{-3})(100) = 11.43 \text{ V}.$$

The voltage across R_1 is therefore *not* 12.0 V, but instead about 0.5 V less.

If R_2 is connected instead of R_1, the resistance of the load is less, and therefore the load current is greater.

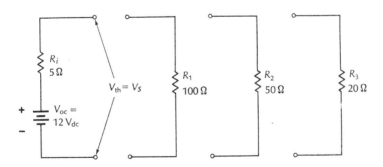

Figure 21-16   A Thevenin equivalent circuit.

$$I_t = \frac{V_{th}}{R_{th} + R_2} = \frac{12}{5 + 50} = 0.2182 \text{ A} = 218.2 \text{ mA}.$$
$$E_{R_{th}} = I_t \times R_{th} = 0.2182 \times 5 = 1.1 \text{ V}.$$
$$E_{R_2} = I_t \times R_2 = 0.2182 \times 50 = 10.9 \text{ V}.$$

Note how the voltage across the load decreases as the load current is increased. Finally, if R_3 is connected, the load current is even greater, and the load voltage decreases even more.

$$I_t = \frac{E_{oc}}{R_{th} + R_3} = \frac{12}{5 + 20} = 0.48 \text{ A} = 480 \text{ mA}.$$
$$E_{R_{th}} = I_t \times R_{th} = 0.48 \times 5 = 2.4 \text{ V}.$$
$$E_{R_3} = I_t \times R_3 = 0.48 \times 20 = 9.6 \text{ V}.$$

By constructing the Thevenin-equivalent circuit, which includes the internal impedance or resistance, the true circuit action is described and the true values easily determined.

The preceding example was analyzed by simple Ohm's law to illustrate the basic principles of Thevenin's theorem. A slightly different circuit is shown in Figure 21-17. Here, a 100-V source has a voltage divider connected across it to drop the total supply to something less than 100 V. The load R_3 is then connected to the tap between R_1 and R_2. This circuit is a perfect example for the application of Thevenin's theorem. In Figure 21-17a the basic circuit is shown, followed by an equivalent circuit with values given. Such a circuit configuration is not impossible to solve using Ohm's law, but when Thevenin's theorem is applied, more complex configurations are much easier to deal with than would be the case using only Ohm's law.

The first step in Thevenizing the circuit is to consider V_s, R_1, and R_2 as the complete source, as shown in part b of the figure. Hence V_s provides a total of 100 V, while R_1 and R_2 comprise the source's internal resistance. Drawing c shows how to find the overall source internal resistance R_{th}. V_s is mentally disconnected and replaced with its internal resistance, which we will consider to be zero. Now, the total resistance is that which is seen by the disconnected load *looking back* into the two terminals. From this viewpoint, the two resistors R_1 and R_2 are simply in parallel, as far as the load is concerned. R_{th} is therefore determined as follows:

$$R_{th} = \frac{R_1 \times R_2}{R_1 + R_2} = \frac{(30 \times 10^3) \times (20 \times 10^3)}{(30 \times 10^3) + (20 \times 10^3)} = 12,000 \text{ }\Omega.$$

The load, then, will perform as though the source had a total internal impedance of 12,000 Ω.

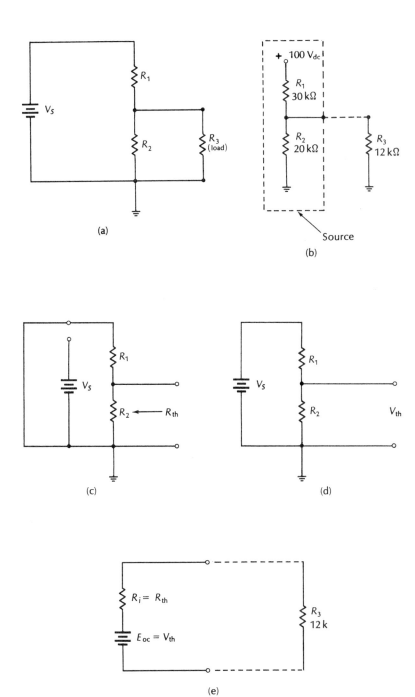

Figure 21-17 Developing the Thevenin equivalent circuit.

Next, find the equivalent source voltage V_{th}, as indicated in d. Here, the load is mentally disconnected and the result is a simple voltage divider. This is the source's open-circuit (load-disconnected) voltage V_{th}:

$$V_{th} = \frac{R_2}{R_1 + R_2} \times V_s = \frac{20 \times 10^3}{(20 \times 10^3) + (30 \times 10^3)} \times 100$$
$$= 40 \text{ V}.$$

Finally, the equivalent circuit in part e of the figure allows the calculations to be completed. This is the Thevenin-equivalent circuit for the original configuration. The equivalent circuit has been reduced to a simple series circuit. Inserting the calculated values from above, the voltage across the load and the current through it can now be found rather easily:

$$I_t = \frac{V_{th}}{R_{th} + R_3} = \frac{40}{(12 \times 10^3) + (12 \times 10^3)} = 1.67 \text{ mA}.$$
$$E_{R_3} = I_t \times R_3 = (1.67 \times 10^{-3})(12{,}000) = 20 \text{ V}.$$

Also, the drop across R_{th} is

$$E_{R_{th}} = I_t \times R_3 = (1.67 \times 10^{-3})(12{,}000) = 20 \text{ V}.$$

While simple Ohm's law could have been used to solve for the unknown values in the preceding example, the next example requires something more.

The circuit in Figure 21-18a has three voltage sources, so in order to determine the true current through and the voltage across the load resistor, as well as the voltage at point A in reference to ground, Thevenin's theorem will be used. First, the Thevenin-equivalent source resistance as "seen" by the load will be determined. R_L is therefore first considered to be disconnected from the rest of the circuit and the entire source replaced with a short circuit; R_{th} is, as before, simply R_1 and R_2 in parallel.

$$R_{th} = \frac{R_1 \times R_2}{R_1 + R_2} = \frac{6000 \times 3000}{9000} = 2000 \text{ }\Omega.$$

Next, V_{th} is determined with R_L out of the circuit.

$$E_{R_1} = \left(\frac{R_1}{R_1 + R_2}\right) V_{total} = \left(\frac{6000}{9000}\right) 24 = 16 \text{ V};$$
$$V_{th} = 12 - 16 = -4 \text{ V}.$$

Enough information is now at hand to draw the Thevenin-equivalent circuit, shown in Figure 21-18b. The open-circuit voltage V_{th} at the junction of R_1 and R_2 is represented by the 4-V source in series with R_{th}. Connected to this is the load resistor R_L, which is returned to a

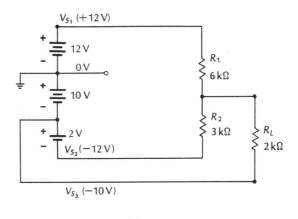

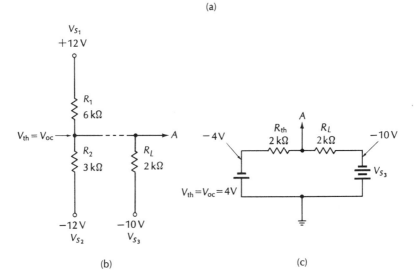

Figure 21-18 A circuit with three sources to be solved by Thevenin's theorem.

−10-V source. The circuit is now simplified to the point that Ohm's law will specify the unknown values, as can be seen in part c of the figure.

$$I_t = \frac{-V_{s_3} + V_{th}}{R_{th} + R_L} = \frac{(-10 + 4)}{4000}$$

$$= \frac{-6}{4000} = -1.5 \times 10^{-3} = -1.5 \text{ mA}.$$

The negative quotient simply indicates that the predominate voltage is negative with respect to ground.

The voltage drop across R_L is simply $I_t \times R_L$.

$$E_{R_L} = -I_t \times R_L = (-1.5 \times 10^{-3})(2 \times 10^3) = -3 \text{ V}.$$

Finally, the voltage at point A in reference to ground is

$$E_A = -V_{s_a} + E_{R_L} = -10 + 3 = -7 \text{ V}.$$

As you can see, proper use of Thevenin's theorem greatly simplifies what would otherwise be a complex problem.

21-4 NORTON'S THEOREM

While Thevenin's theorem is used with a constant-voltage source (R_L much larger than the internal resistance of the source), Norton's theorem is used when the source is a constant-current generator (R_L much smaller than the internal resistance). Before discussing Norton's theorem, however, a brief review of constant-voltage and constant-current sources will help you apply both Thevenin's and Norton's theorems.

Figure 21-19 illustrates a constant-voltage generator and a constant-current generator. The constant-voltage generator is such that R_L is at least five times greater than R_i, and when this is true, the voltage across the load remains essentially constant, even though the load is varied. As an example, assume in Figure 21-19a that $V_s = 10$ V, $R_i = 5$ Ω, and $R_L = 500$ Ω (note that R_L can be varied to some extent). The voltage across the load is then

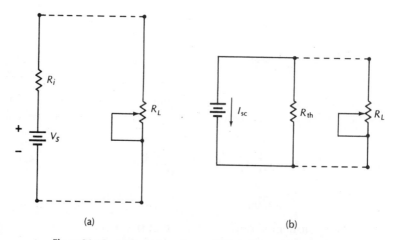

Figure 21-19 (a) Constant-voltage and (b) constant-current sources.

$$E_{R_L} = I_t \times R_L = \frac{V_s}{R_i + R_L} \times R_L = \frac{10}{505} \times 500 = 9.9 \text{ V}.$$

If the resistance of the load is now changed to 600 Ω, how is the load voltage altered?

$$E_{R_L} = I_t \times R_L = \frac{V_s}{R_i + R_L} \times R_L = \frac{10}{605} \times 600 = 9.92 \text{ V}.$$

If these two circuits were measured with a standard voltmeter, the 20-mV difference would not be detected; therefore we consider it negligible. Such a source is therefore a constant-voltage source.

Now consider Figure 21-19*b*, which illustrates a constant-current generator which has a *very high* internal resistance R_{th} compared to R_L. This is the Norton-equivalent circuit, where V_s has an *infinite* internal resistance shunted by the Thevenin-equivalent resistance of the actual source. (If a source having an infinite resistance were possible, which it is not, the current through R_L would remain *absolutely constant* no matter how high the value of R_L were set. Thus R_{th} is a measure of how much the source deviates from the ideal, or perfect, current source.) If $V_s = 10$ V, $R_{th} = 500{,}000$ Ω, and $R_L = 5000$ Ω, find the load current. First, find the current I_{sc} that would flow if the source terminals were short circuited:

$$I_{sc} = \frac{V_s}{R_{th}} = \frac{10}{500{,}000} = 20 \,\mu\text{A}.$$

Now I_{R_L} can be found by using the current-divider relationship:

$$I_{R_L} = I_{sc}\left(\frac{R_{th}}{R_{th} + R_L}\right) = 20\left(\frac{500{,}000}{505{,}000}\right) = 19.8 \,\mu\text{A}.$$

Note that the current in the load differs from I_{sc} by only 0.2 μA. The load is changed to 2500 Ω. Find the new value of load current.

$$I_{R_L} = I_{sc}\left(\frac{R_{th}}{R_{th} + R_L}\right) = 20\left(\frac{500{,}000}{502{,}500}\right) = 19.9 \,\mu\text{A}.$$

Both conditions yield a result that for all practical purposes is 20 μA.

The foregoing example demonstrates the value of Norton's theorem, which applies whenever the load resistance (or impedance) is one hundred or more times *less* than the source impedance.

21-5 SUPERPOSITION THEOREM

The superposition theorem is a most useful device for simplifying complex circuits. Basically, the theorem states that the current flowing at any point in a complex circuit can be found by considering one source

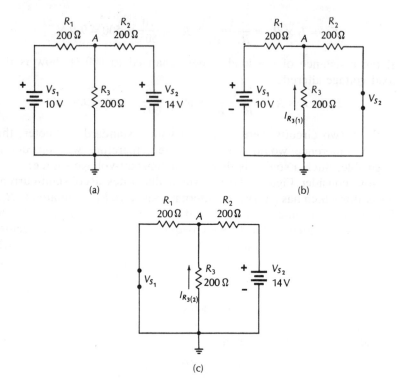

Figure 21-20 A complex network to be solved by the superposition theorem.

at a time and replacing all others with their internal resistances. The individual currents are then algebraically summed to find the actual value.

Consider Figure 21-20a, which shows a complex network of three resistors and two sources. According to the theorem, the current flowing in R_3, for example, can be determined by considering only one source at a time, with all others replaced by their internal resistances. If we begin with V_{s_1}, then V_{s_2} is replaced by a short circuit (if R_i is essentially zero). This is illustrated in Figure 21-20b, where V_{s_2} is replaced by a short circuit. Now the circuit is a simple series-parallel arrangement. Next, V_{s_1} is replaced by a short circuit, and the result is again an easily analyzed circuit. Finally, the two interim values of current are summed to find the true value.

Using the values given in Figure 21-20a, we will find the unknown values by the use of the superposition theorem.

Using the equivalent circuit of Figure 21-20b, the current flowing in R_3 is found by simple Ohm's law:

$$R_{eq\ 2,3} = \frac{R_2 \times R_3}{R_2 + R_3} = \frac{200 \times 200}{200 + 200} = 100\ \Omega;$$
$$R_t = R_1 + R_{eq\ 2,3} = 200 + 100 = 300\ \Omega;$$
$$I_t = \frac{V_{s1}}{R_t} = \frac{10}{300} = 0.0333 = 33.3\ \text{mA};$$
$$I_{R3} = I_t \left(\frac{R_2}{R_2 + R_3}\right) = 33.3 \times 0.5 = 16.67\ \text{mA}.$$

This value is labeled $I_{R_{3(1)}}$ on the drawing, and its direction is noted by the arrowhead.

Next, use V_{s2} to find the current for the second condition. Note in Figure 21-20c that V_{s1} is now short circuited.

$$R_{eq\ 1,3} = \frac{R_1 \times R_3}{R_1 + R_3} = \frac{200 \times 200}{200 + 200} = 100\ \Omega;$$
$$R_t = R_2 + R_{eq\ 1,3} = 200 + 100 = 300\ \Omega;$$
$$I_t = \frac{V_{s2}}{R_t} = \frac{14}{300} = 0.0467 = 46.7\ \text{mA};$$
$$I_{R3} = I_t \left(\frac{R_1}{R_3 + R_1}\right) = 46.7\ \text{mA} \left(\frac{200}{200 + 200}\right) = 23.33\ \text{mA}.$$

This value is indicated on the drawing as $I_{R_{3(2)}}$, and note that it flows upward, as does $I_{R_{3(1)}}$. The two currents are therefore additive:

$$I_{R_{3(true)}} = I_{R_{3(1)}} + I_{R_{3(2)}} = 16.67\ \text{mA} + 23.33\ \text{mA}$$
$$= 40\ \text{mA}$$

The voltage at junction A can now be found if required:

$$E_A = I_{R_{3(true)}} \times R_3 = 0.04 \times 200 = 8\ \text{V}.$$

Since the current through R_3 is in the direction indicated, A is more positive than ground by 8 V.

The values of currents through R_1 and R_2 can be determined:

$$I_{R1} = \frac{V_{s1} - E_A}{R_1} = \frac{10 - 8}{200}$$
$$= \frac{2}{200} = 0.01 = 10\ \text{mA};$$
$$I_{R2} = \frac{V_{s2} - E_A}{R_2} = \frac{14 - 8}{200}$$
$$= \frac{6}{200} = 0.03 = 30\ \text{mA}.$$

A slightly different situation is shown in Figure 21-21. Here, V_{s2} is reversed, and now the current in R_3 produced by V_{s2} is *opposite* to

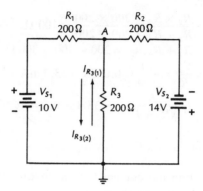

Figure 21-21 A complex network whose sources have opposite polarity.

that produced by V_{s_1}. This, of course, greatly influences the current through R_3, since now the two interim currents are *subtractive*. The total current through R_3 and the voltage at A are determined as follows, substituting a short circuit for V_{s_2}:

$$R_{eq\ 2,3} = \frac{R_2 \times R_3}{R_2 + R_3} = 100\ \Omega;$$
$$R_t = R_1 + R_{eq\ 2,3} = 300\ \Omega;$$
$$I_t = \frac{V_{s_1}}{R_t} = \frac{10}{300} = 0.0333 = 33.3\ \text{mA};$$
$$I_{R_3} = I_t \left(\frac{R_2}{R_2 + R_3}\right) = 0.01667 = 16.67\ \text{mA}.$$

Note that this is the same result as in the previous problem, since the fact that V_{s_2} is reversed has no bearing on the result because it is temporarily shorted.

Now, using V_{s_2} as the source *will* result in different conditions:

$$R_{eq\ 1,3} = \frac{R_1 \times R_3}{R_1 + R_3} = \frac{200 \times 200}{200 + 200} = 100\ \Omega;$$
$$R_t = R_2 + R_{eq\ 1,3} = 200 + 100 = 300\ \Omega;$$
$$I_t = \frac{V_{s_2}}{R_t} = \frac{14}{300} = 0.0467 = 46.7\ \text{mA};$$
$$I_{R_3} = I_t \left(\frac{R_1}{R_3 + R_1}\right) = 0.0467 \left(\frac{200}{200 + 200}\right) = 0.0233 = 23.3\ \text{mA}.$$

At this point, the circuit results are different than in the preceding example although the values are the same. Because V_{s_2} is reversed relative to the first example, $I_{R_{3(2)}}$ flows in the opposite direction through R_3 than does $I_{R_{3(1)}}$. These currents, then, are subtractive, and the true current in I_{R_3} is the difference between the two values. Subtracting the smaller from the larger gives:

$$I_{R_{3(true)}} = I_{R_{3(2)}} - I_{R_{3(1)}} = 23.3 \text{ mA} - 16.67 \text{ mA} = 6.67 \text{ mA}.$$

$I_{R_{3(2)}}$ is the larger of the two, so it determines the polarity across R_3, as shown. Now, knowing the current through R_3 allows the voltage drop to be found:

$$E_{R_3} = I_{R_3} \times R_3 = 6.67 \text{ mA} \times 200 = 1.33 \text{ V}.$$

Point A is therefore at a potential that is 1.33 V more negative than ground.

Next, the current in R_1 and R_2 can be determined.

$$I_{R_1} = \frac{V_{s_1} - E_{R_3}}{R_1} = \frac{10 - 1.33}{200} = 0.0433 = 43.3 \text{ mA};$$

$$I_{R_2} = \frac{-V_{s_2} + E_{R_3}}{R_2} = \frac{-14 + 1.33}{200} = -0.0634 = -6.34 \text{ mA}.$$

Hence, the voltage at point A is indicative of the difference between the two circuits.

21-6 MILLMAN'S THEOREM

Millman's theorem combines Norton's and Thevenin's theorems to provide a simple expression to resolve unknown values in complex circuits having many sources. A sample circuit is given in Figure 21-22, drawn in two different ways. In Figure 21-22b, note the line labeled I_{sc}. This represents an imaginary short circuit between X and Y, which is used to find the value of I_{sc} (Norton's theorem) that would flow here if the short circuit actually existed. Also, points X and Y are used to find the Thevenin-equivalent resistance between these two points.

Now, Millman's theorem states that in a circuit such as this, the voltage between points X and Y is equal to the product of I_{sc} and R_{th}. That is, the Norton short-circuit current multiplied by the Thevenin-equivalent resistance seen by the two terminals X and Y looking back into the circuit (no short circuit) equals the potential difference across the terminals. (The sources are replaced by their internal resistances, as in Thevenin's theorem.)

To illustrate Millman's theorem, the values given in the illustration will be used to find the voltage at A relative to ground. First, the individual short-circuit currents are determined:

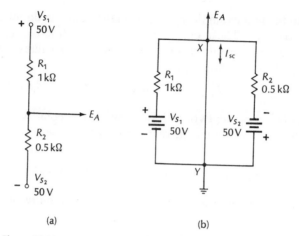

Figure 21-22 (a) A two-source circuit; (b) the circuit redrawn with the dashed line representing an imaginary short circuit.

$$I_{R_1} = \frac{V_{s_1}}{R_1} = \frac{50}{1000} = 0.05 = 50 \text{ mA};$$

$$I_{R_2} = \frac{-V_{s_2}}{R_2} = \frac{-50}{500} = -0.1 = -100 \text{ mA}.$$

Depending on the directions of these currents, they are either additive (both flowing in the same direction) or subtractive (flowing in opposite directions). In this instance they are flowing in opposite directions.

$$I_{sc} = I_{R_2} - I_{R_1} = 100 \text{ mA} - 50 \text{ mA} = 50 \text{ mA}.$$

Because I_{R_2} is larger, it predominates, determining the polarity of E_A.

Now the Thevenin-equivalent resistance is determined. If the sources are assumed to have zero internal resistance, they are replaced with short circuits. Between X and Y, then R_{th} is simply R_1 and R_2 in parallel:

$$R_{th} = \frac{1}{1/R_1 + 1/R_2} = \frac{1}{1/1000 + 1/500} = 333.3 \text{ }\Omega.$$

The voltage at A is therefore the product of I_{sc} and R_{th}:

$$E_A = I_{sc} \times R_{th} = 0.05 \times 333.3 = 16.67 \text{ V}.$$

Because I_{R_2} is the larger current, the sign of E_A is negative, and

$$E_A = -16.67 \text{ V}.$$

With the aid of a little algebraic manipulation, *all* of the foregoing steps can be combined into a single equation, the Millman equation:

Circuit Analysis Techniques 535

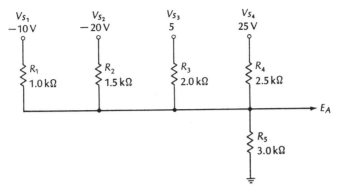

Figure 21-23 Millman's theorem can be used to solve this four-source circuit.

$$E_A = \left(\frac{V_{s1}}{R_1} + \frac{V_{s2}}{R_2}\right)\left(\frac{1}{1/R_1 + 1/R_2}\right) = \frac{(50/1000) + (-50/500)}{1/R_1 + 1/R_2}$$
$$= \frac{0.05 + (-0.1)}{0.001 + 0.002} = \frac{-0.05}{0.003} = -16.67 \text{ V.}$$

Note that the polarity sign of the sources must be used in the numerator.

These results can be arrived at by other means, of course; even simple Ohm's law would suffice. In Figure 21-23, however, a circuit ideally suited for Millman's theorem is shown. Here, four sources provide a rather complicated network feeding current into R_5, across which the voltage E_A is developed. The Millman equation is first written; then given values are substituted to find the voltage at the junction of all resistors:

$$E_A = \frac{(-V_{s1}/R_1) + (-V_{s2}/R_2) + (V_{s3}/R_3) + (V_{s4}/R_4) + (0/R_5)}{1/R_1 + 1/R_2 + 1/R_3 + 1/R_4 + 1/R_5}$$
$$= \frac{(-10/1000) + (-20/1500) + (5/2000) + (25/2500)}{1/1000 + 1/1500 + 1/2000 + 1/2500 + 1/3000}$$
$$= \frac{(-0.01) + (-0.0133) + (0.0025) + (0.01)}{0.001 + 0.000667 + 0.0005 + 0.0004 + 0.000333}$$
$$= \frac{-0.0104333}{0.0029} = -3.7356 \cong -3.74 \text{ V.}$$

Note that the term $0/R_5$ is dropped from the numerator, since R_5 is not returned to a voltage source; however, R_5 *is* included in the denominator, since it certainly contributes to the total Thevenin resistance.

21-7 DELTA-WYE TRANSFORMATIONS

In Figure 21-24a, a circuit is shown that contains a delta configuration. That is, R_1, R_2, and R_3 can be redrawn, as in part b of the figure, to resemble the Greek letter Δ. Because this type of connection has multiple parallel paths, it is difficult to analyze for current and voltage values. One method of simplifying the analysis is to convert the delta connection to an equivalent *wye* or *tee* configuration that is electrically identical but consists of series rather than parallel elements. In Figure 21-24c the wye connection is shown, and its resemblance to a Y is clear. If the arms of the wye were straightened, then the shape of the T would be apparent, and the circuit would be a tee connection.

The relationship between the delta and wye configurations is illustrated in Figure 21-25. By applying a specific set of equations, either

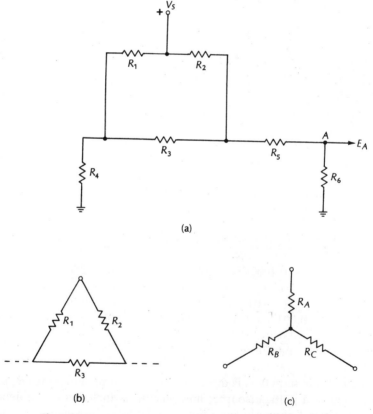

Figure 21-24 (a) A circuit with a "delta" configuration; (b) the delta portion of the circuit; (c) the "wye" equivalent.

configuration can be converted to the other, often simplifying the determination of unknown values.

It is important to realize that either configuration must be electrically equivalent to the other. That is, the resistance measured between any two terminals (A, B, C) must be the same for either circuit. As shown in Figure 21-25, both the delta and wye are shown connected to the three terminals A, B, and C. In practice, of course, only one connection actually exists; the other is simply a mathematical model used to simplify calculations.

Delta-to-Wye Transformation

If the original circuit is a delta configuration, and it is desired to change it to a wye, the following equations are used:

$$R_A = \frac{R_1 R_2}{R_1 + R_2 + R_3};$$
$$R_B = \frac{R_1 R_3}{R_1 + R_2 + R_3};$$
$$R_C = \frac{R_2 R_3}{R_1 + R_2 + R_3}.$$

Note that to find R_A, the two delta resistors lying on either side are used in the numerator, and so on around the circuit. Also, once the denominator value is found for the first equation, the same number is used in both other equations. Therefore if you are using an addressable-memory calculator, you can store and recall this value several times.

As a numerical example, assume that in Figure 21-25, $R_1 = 800\ \Omega$, $R_2 = 1000\ \Omega$, and $R_3 = 1200\ \Omega$. Determine the equivalent values of the wye circuit.

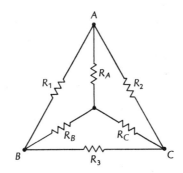

Figure 21-25 Delta-wye interrelationships.

$$R_A = \frac{R_1 R_2}{R_1 + R_2 + R_3} = \frac{800 \times 1000}{800 + 1000 + 1200} = \frac{800{,}000}{3000} = 267\ \Omega;$$

$$R_B = \frac{R_1 R_3}{R_t} = \frac{800 \times 1200}{3000} = 320\ \Omega;$$

$$R_C = \frac{R_2 R_3}{R_t} = \frac{1000 \times 1200}{3000} = 400\ \Omega.$$

To prove that each configuration, both of which are shown in Figure 21-26, is the electrical equivalent of the other, check the resistance between any two terminals, say B and C. Total resistance between B and C in the wye circuit is simply $400 + 320 = 720\ \Omega$. The resistance between terminals B and C in the delta circuit is R_3, *in parallel* with $R_1 + R_2$:

$$R_{B-C} = \frac{1200 \times 1800}{1200 + 1800} = 720\ \Omega.$$

If you check any other two terminals, you will find that the resistance is the same for either circuit configuration, all the way around the circuit.

Wye-to-Delta Transformation

To convert from a wye to a delta configuration, a different set of equations is used:

$$R_1 = \frac{R_A R_B + R_A R_C + R_B R_C}{R_C};$$

$$R_2 = \frac{R_A R_B + R_A R_C + R_B R_C}{R_B};$$

$$R_3 = \frac{R_A R_B + R_A R_C + R_B R_C}{R_A}.$$

Note that in this case the numerators are identical, and that the denominator is the resistor *opposite* to the one being solved for. For example, assume that the wye circuit values in Figure 21-26b are given, and that the delta equivalent is to be found. Then

$$R_1 = \frac{R_A R_B + R_A R_C + R_B R_C}{R_C} = \frac{(267 \times 320) + (267 \times 400) + (320 \times 400)}{400}$$

$$= \frac{320{,}240}{400} = 800\ \Omega.$$

Now, we can simply use the computed numerator, thus streamlining the remaining calculations.

$$R_2 = \frac{R_A R_B + R_A R_C + R_B R_C}{R_B} = \frac{320{,}240}{320} = 1000\ \Omega;$$

$$R_3 = \frac{R_A R_B + R_A R_C + R_B R_C}{R_A} = \frac{320{,}240}{267} = 1200\ \Omega.$$

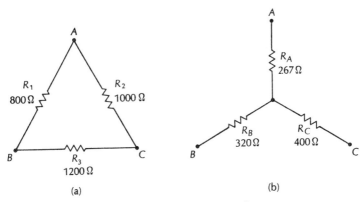

Figure 21-26 Delta-wye equivalents.

Thus, the wye-delta conversions are seen to be relatively simple.

To illustrate the application of delta-wye conversion, consider Figure 21-24a. This is a circuit within which is a delta circuit: $R_1, R_2,$ and R_3. Finding the voltage would actually be rather difficult by other means, but the circuit can be simplified by changing this delta configuration into the equivalent wye circuit. The circuit is repeated in Figure 21-27, along with the simplified results. As a numerical example, assume all resistors in Figure 21-27a are 100 Ω. Find the voltage E_X at point X.

The first step is to redraw the circuit with the wye substituted for the delta. This is done in part b of the figure. Then, transform from delta to wye:

$$R_A = \frac{R_1 R_2}{R_t} = \frac{100 \times 100}{300} = 33.3 \ \Omega;$$

$$R_B = \frac{R_1 R_3}{R_t} = \frac{100 \times 100}{300} = 33.3 \ \Omega;$$

$$R_C = \frac{R_2 R_3}{R_t} = \frac{100 \times 100}{300} = 33.3 \ \Omega.$$

These values are illustrated.

Next, observe in part c of the illustration that $R_B + R_4$ is in parallel with $R_C + R_5 + R_6$. This network can be resolved into a single equivalent resistance R_{eq}.

$$R_{eq} = \frac{(R_B + R_4)(R_C + R_5 + R_6)}{R_B + R_4 + R_C + R_5 + R_6} = 93.3 \ \Omega.$$

Total current is simply

$$I_t = \frac{V_s}{R_A + R_{eq}} = \frac{48}{33.3 + 93.3} = \frac{48}{126.6} = 0.38 \ \text{A}.$$

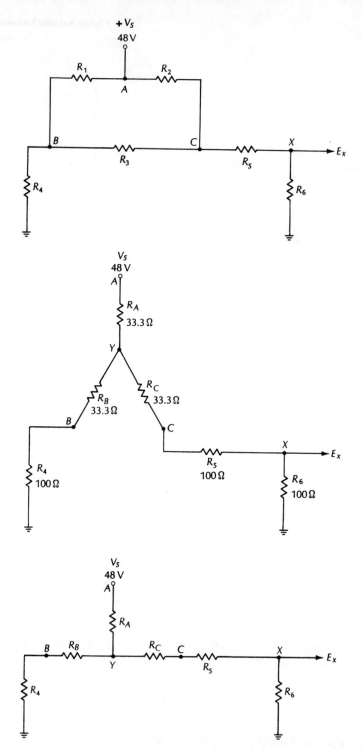

Figure 21-27 A practical example of a delta-wye transformation.

Now the voltage across R_A, which is $I_t R_A$, can be determined:

$$E_Y = V_s - (I_t R_A) = 48 - (0.38 \times 33.3) = 48 - 12.62$$
$$= 35.37 \text{ V}.$$

Now, with 35.37 V across R_C, R_5, and R_6, a simple voltage divider exists, and

$$E_X = \frac{R_6}{R_C + R_5 + R_6}(35.37) = \frac{100}{233.3}(35.37) = 15.16 \text{ V}.$$

Thus, by changing from a delta to a wye configuration, the circuit becomes a simple Ohm's law problem.

SUMMARY

- In a series string, the voltage across any one resistor is determined by the ratio of the resistor's value to the total resistance.
- In parallel branches, total current divides in amounts that are inversely proportional to the branch resistances.
- Kirchhoff's voltage law states that the algebraic sum of all voltages around a closed loop is zero, providing polarities are observed.
- Kirchhoff's current law states that the sum of all currents entering a node and leaving a node is zero.
- Thevenin's theorem is a method of replacing a complex source with an equivalent open-circuit voltage V_{th} having an internal resistance R_{th}. The load will then perform as though V_{th} were applied to it through R_{th}.
- Norton's theorem is similar to Thevenin's theorem, except that a *constant-current* generator is the source instead of a *constant-voltage* generator.
- The superposition theorem states that a complex circuit can be solved by replacing all sources but one with their internal resistances (usually zero) and then by using only one source at a time. The individual currents are then algebraically summed to arrive at the true current.
- Millman's theorem states that the product of I_{sc} and R_{th} at a given point in a circuit will specify the voltage at that point.
- Delta-wye transformations allow a complex circuit of a certain configuration to be changed to an equivalent series circuit which can be analyzed by conventional means.

PROBLEMS

For the following problems, refer to Figure 21-28.

1. $V_s = 9 \text{ V}, R_1 = 60 \text{ k}\Omega, R_2 = 80 \text{ k}\Omega,$ and $R_3 = 100 \text{ k}\Omega$. Determine the voltage at point A; at point B.

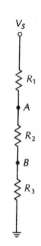

Figure 21-28

2. $V_s = 12$ V, $R_1 = 720$ Ω, $R_2 = 960$ Ω, and $R_3 = 1200$ Ω. Determine the voltage at point A; at point B.
3. $V_s = 12$ V, $R_1 = 1$ kΩ, $E_A = 11$ V, $E_B = 6$ V. Find R_1 and R_2.
4. $V_s = 12$ V, $R_1 = 2500$ Ω, $E_A = 11$ V, and $E_B = 6$ V. Find R_1 and R_2.

For the following problems, refer to Figure 21-29.

5. $R_1 = 2000$ Ω, $R_2 = 3000$ Ω, and $V_s = 6$ V. Find the branch currents.
6. $R_1 = 2000$ Ω, $R_2 = 6000$ Ω, and $V_s = 12$ V. Find the branch currents.
7. $R_1 = 2000$ Ω, $R_2 = 3000$ Ω, and $I_t = 20$ mA. Find the branch currents and V_s.
8. $R_1 = 4000$ Ω, $R_2 = 6000$ Ω, and $I_t = 60$ mA. Find the branch currents and V_s.

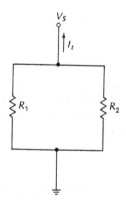

Figure 21-29

9. $V_s = 6$ V, $I_t = 10$ mA, $I_{R_2} = 4$ mA. Find R_t, R_1, R_2, and I_{R_1}.
10. $V_s = 12$ V, $I_t = 0.5$ mA, $I_{R_2} = 0.1$ mA. Find R_t, R_1, R_2, and I_{R_1}.

For the following problems, refer to Figure 21-30.

11. $V_s = 9$ V, $R_1 = 100$ Ω, $R_2 = 200$ Ω, and $R_3 = 300$ Ω. Find the branch currents and I_t.
12. $V_s = 24$ V, $R_1 = 10$ kΩ, $R_2 = 20$ kΩ, and $R_3 = 30$ kΩ. Find the branch currents and I_t.
13. Given that $I_t = 330$ mA, $I_{R_2} = 90$ mA, $I_{R_3} = 60$ mA, and $R_1 = 1000$ Ω, find V_s, R_2, and R_3.
14. Given that $I_t = 33$ mA, $I_{R_2} = 9$ mA, $I_{R_3} = 6$ mA, and $R_2 = 200$ Ω, find V_s, R_1, and R_3.
15. Given that $I_t = 150$ mA, $R_2 = 1000$ Ω, $I_{R_1} = 40$ mA, and $I_{R_3} = 50$ mA, find V_s, R_1, and R_3.
16. Given that $I_t = 225$ mA, $R_2 = 1500$ Ω, $I_{R_1} = 60$ mA, and $I_{R_3} = 75$ mA, find V_s, R_1, and R_3.
17. A battery has an open-circuit terminal voltage of 12.0 V. When a 100-Ω load is connected, E_{R_L} is 10 V. Find R_{th}.
18. A battery has an open-circuit terminal voltage of 24 V. When a 10-Ω load is connected, E_{R_L} is 22 V. Find the internal Thevenin resistance.

For the following problems, refer to Figure 21-31.

19. $+V_s = 6$ V, $-V_s = -12$ V, $R_1 = 1000$ Ω, $R_2 = 2500$ Ω, and $R_3 = 4700$ Ω. Find the equivalent Thevenin resistance R_{th} and the voltage E_A at point A.
20. $+V_s = 6$ V, $-V_s = -12$ V, $R_1 = 3$ kΩ, $R_2 = 4$ kΩ, and $R_3 = 6.8$ kΩ. Find the values of the equivalent Thevenin resistance R_{th} and the voltage E_A at point A.

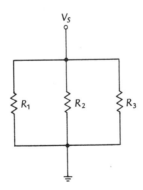

Figure 21-30

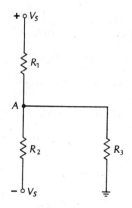

Figure 21-31

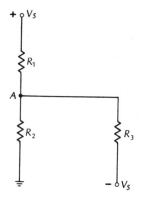

Figure 21-32

For the following problems, refer to Figure 21-32.

21. $+V_s = 12$ V, $-V_s = -6$ V, $R_1 = 1000$ Ω, $R_2 = 2500$ Ω, and $R_3 = 4700$ Ω. Find the equivalent Thevenin resistance R_{th} and the voltage E_A at point A.

22. $+V_s = 12$ V, $-V_s = -6$ V, $R_1 = 2000$ Ω, $R_2 = 3000$ Ω, and $R_3 = 4000$ Ω. Find the equivalent Thevenin resistance R_{th} and the voltage E_A at point A.

For the following problems, refer to Figure 21-33.

23. $V_{s_1} = +9$ V, $V_{s_2} = +6$ V, $V_{s_3} = +3$ V, $V_{s_4} = -12$ V, $R_1 = 4000$ Ω, $R_2 = 3000$ Ω, $R_3 = 2000$ Ω, $R_4 = 1000$ Ω, and $R_5 = 500$ Ω. Determine the voltage E_A at point A.

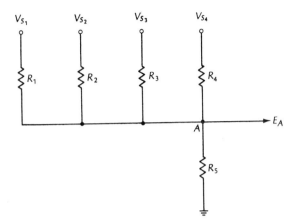

Figure 21-33

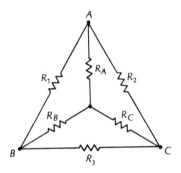

Figure 21-34

24. $V_{s_1} = +9$ V, $V_{s_2} = +6$ V, $V_{s_3} = +3$ V, $V_{s_4} = -12$ V, $R_1 = 1000$ Ω, $R_2 = 2000$ Ω, $R_3 = 3000$ Ω, $R_4 = 4000$ Ω, and $R_5 = 500$ Ω. Determine the voltage E_A at point A.

For the following problems, refer to Figure 21-34.

25. $R_1 = 200$ Ω, $R_2 = 300$ Ω, and $R_3 = 400$ Ω. Determine the equivalent-wye values of R_A, R_B, and R_C.
26. $R_1 = 1000$ Ω, $R_2 = 2000$ Ω, and $R_3 = 3000$ Ω. Determine the equivalent-wye values of R_A, R_B, and R_C.
27. $R_A = 800$ Ω, $R_B = 600$ Ω, and $R_C = 400$ Ω. Determine the equivalent-delta values of R_1, R_2, and R_3.
28. $R_A = 1200$ Ω, $R_B = 900$ Ω, and $R_C = 600$ Ω. Determine the equivalent-delta values of R_1, R_2, and R_3.

ANSWERS TO ODD-NUMBERED QUESTIONS AND PROBLEMS

Chapter 2 Answers to Odd-Numbered Questions

1. True
3. Electron
5. Equal; opposite
7. True
9. Repel
11. True
13. False
15. False
17. 16

Chapter 3 Answers to Odd-Numbered Questions

1. The simple circuit must have at least a source of EMF, such as a battery, conductors (wires), and a load (device which uses the electrical energy).
3. Protects
5. Slow
7. True
9. Resistance
11. Resistance
13. The current varies directly with the voltage and inversely with the resistance.
15. •————•
17. ————|⊢|⊢————
19. 10,000
21. 100,000
23. Heat
25. Voltmeter
27. Ohmmeter

Chapter 3 Answers to Odd-Numbered Problems

1. $E = IR$
   $I = E/R$
   $R = E/I$
3. 0.0033 A
5. 12,000 Ω
7. 0.000025 W

9. 25,000 V
11. 0.000350 s
13. 0.255 A
15. 400 Ω
17. 11.88 V
19. 0.00006 A

Chapter 4   Answers to Odd-Numbered Questions

1. In a series circuit the current has the same value at any instant, the value being determined by the supply voltage and the total resistance.
3. Sum
5. Is the same
7. A voltage drop is that voltage appearing across any *load*, as opposed to a voltage appearing across a source.
9. Is
11. Is
13. Zero

Chapter 4   Answers to Odd-Numbered Problems

1. 0.3 A
3. 0.3 A
5. 0.3 A
7. 0.1 A
9. 0.067 A
11. 0.333 A
13. 20 μA
15. 20 μA
17. $R_t = 40$ Ω
    $I_t = 0.1$ A
    $E_{R_1} = 1$ V
    $E_{R_2} = 3$ V
    $P_{R_1} = 0.1$ W
    $P_{R_2} = 0.3$ W
    $P_t = 0.4$ W
19. $R_t = 32,000 = 32$ kΩ
    $I_t = 0.375$ mA
    $E_{R_1} = 3.75$ V
    $E_{R_2} = 8.25$ V
    $P_{R_1} = 0.0014 = 1.4$ mW
    $P_{R_2} = 0.00309 = 3.1$ mW
    $P_t = 0.0045 = 4.5$ mW
21. $R_t = 110,000 = 110$ kΩ
    $I_t = 0.001 = 1$ mA
    $E_{R_1} = 88$ V
    $E_{R_2} = 22$ V
    $P_{R_1} = 0.088 = 88$ mW
    $P_{R_2} = 0.022 = 22$ mW
    $P_t = 110$ mW

23. $R_t = 60\ \Omega$
    $I_t = 0.1\ A$
    $E_{R_1} = 1.0\ V$
    $E_{R_2} = 2.0\ V$
    $E_{R_3} = 3.0\ V$
    $P_{R_1} = 0.1\ W$
    $P_{R_2} = 0.2\ W$
    $P_{R_3} = 0.3\ W$
    $P_t = 0.6\ W$

25. $R_t = 9000\ \Omega$
    $I_t = 0.002 = 2\ mA$
    $E_{R_1} = 3.0\ V$
    $E_{R_2} = 6.0\ V$
    $E_{R_3} = 9.0\ V$
    $P_{R_1} = 0.006 = 6\ mW$
    $P_{R_2} = 0.012 = 12\ mW$
    $P_{R_3} = 0.018 = 18\ mW$
    $P_t = 0.036 = 36\ mW$

27. $R_t = 900\ \Omega$
    $I_t = 0.00667 = 6.67\ mA$
    $E_{R_1} = 1.333\ V$
    $E_{R_2} = 1.999 \cong 2\ V$
    $E_{R_3} = 2.666\ V$
    $P_{R_1} = 0.00888 = 8.88\ mW$
    $P_{R_2} = 0.0133 = 13.3\ mW$
    $P_{R_3} = 0.0177 = 17.7\ mW$
    $P_t = 0.0398 = 40\ mW$

29. $R_t = 148{,}000 = 148\ k\Omega$
    $I_t = 0.1216\ mA$
    $E_{R_1} = 4.01\ V$
    $E_{R_2} = 5.72\ V$
    $E_{R_3} = 8.27\ V$
    $P_{R_1} = 0.488\ mW$
    $P_{R_2} = 0.695\ mW$
    $P_{R_3} = 0.001 = 1\ mW$
    $P_t = 0.00218 = 2.18\ mW$

31. $R_t = 60{,}000 = 60\ k\Omega$
    $I_t = 0.0001 = 100\ \mu A$
    $E_{R_1} = 1\ V$
    $E_{R_2} = 2\ V$
    $E_{R_3} = 3\ V$
    $E_A = +6\ V$
    $E_B = +6 - 1 = +5\ V$
    $E_C = +6 - 1 - 2 = +3\ V$

33. $R_t = 20{,}600 = 20.6\ k\Omega$
    $I_t = 0.000583 = 583\ \mu A$
    $E_{R_1} = 2.74\ V$
    $E_{R_2} = 3.96\ V$
    $E_{R_3} = 5.30\ V$

$E_A = +12$ V
$E_B = +9.26$ V
$E_C = +5.3$ V
35. (a) $E_C = +1$ V
   $E_D = -5$ V
   (b) $E_C = -5$ V
   $E_D = -11$ V
   (c) $E_B = +5$ V
   $E_D = -6$ V
37. $R_1 = 700$ Ω
   $R_3 = 600$ Ω
   $E_{R_2} = 5$ V
   $E_{R_3} = 6$ V
   $R_t = 1800$ Ω
39. $E_{R_1} = 1$ V
   $R_2 = 20,000$ Ω
   $E_{R_4} = 4.0$ V
   $R_5 = 50,000$ Ω
   $R_t = 150,000$ Ω
   $R_3 = 30,000$ Ω
   $E_{R_3} = 3$ V

Chapter 5 Answers to Odd-Numbered Questions

1. Will not
3. False
5. False
7. Is not
9. Is
11. Because R_3 is now open, there is no current flow in this branch. Total current is reduced, and total resistance is increased.
13. The individual branch currents in a parallel circuit are a function of the applied voltage and the total resistance in the branch. The sum of all branch currents equals total source current.

Chapter 5 Answers to Odd-Numbered Problems

1. $R_t = 180$ Ω
   $I_{R_1} = 0.025$ A $= 25$ mA
   $I_{R_2} = 0.025$ A $= 25$ mA
   $I_t = 0.05$ A $= 50$ mA
   $P_{R_1} = 0.225$ W
   $P_{R_2} = 0.225$ W
   $P_t = 0.45$ W
3. $R_t = 1440$ Ω
   $I_{R_1} = 0.005 = 5$ mA
   $I_{R_2} = 0.00333 = 3.33$ mA
   $I_t = 0.00833 = 8.33$ mA
   $P_{R_1} = 0.06 = 60$ mW
   $P_{R_2} = 0.0399 \cong 40$ mW
   $P_t = 0.0999 \cong 100$ mW

5. $R_t = 54.545 \cong 54.55 \, \Omega$
$I_{R_1} = 0.09 \, A = 90 \, mA$
$I_{R_2} = 0.045 \, A = 45 \, mA$
$I_{R_3} = 0.03 \, A = 30 \, mA$
$I_t = 0.165 \, A = 165 \, mA$

7. $R_t = 721 \, \Omega$
$I_{R_1} = 0.00333 \, A = 3.33 \, mA$
$I_{R_2} = 0.00208 \, A = 2.08 \, mA$
$I_{R_3} = 0.001515 \, A \cong 1.52 \, mA$
$I_t = 0.00693 \, A = 6.93 \, mA$
$G_t = 0.001386 \cong 1.39 \times 10^{-3} \, mho$

9. $R_t = 150,863 \cong 151 \, k\Omega$
$I_{R_1} = 0.455 \, mA = 455 \, \mu A$
$I_{R_2} = 0.319 \, mA = 319 \, \mu A$
$I_{R_3} = 0.221 \, mA = 221 \, \mu A$
$I_t = 0.995 \, mA = 995 \, \mu A$

Chapter 6   Answers to Odd-Numbered Questions

1. Is not
3. Are not
5. Are, is
7. A voltage divider is a series connection that divides source voltage in proportion to resistor values; a current divider is a parallel connection that divides current in inverse proportion to resistor values.
9. Do
11. Do not

Chapter 6   Answers to Odd-Numbered Problems

1. $R_t = 150 \, \Omega$
$I_t = 0.08 = 80 \, mA$
$I_{R_3} = 0.048 = 48 \, mA$
$I_{R_4} = 0.032 = 32 \, mA$
$E_{R_1} = 2.4 \, V$
$E_{R_2} = 4.8 \, V$
$E_{R_3} = 4.8 \, V$
$E_{R_4} = 4.8 \, V$

3. $R_t = 1500 \, \Omega$
$I_t = 0.08 = 80 \, mA$
$I_{R_3} = 0.048 = 48 \, mA$
$I_{R_4} = 0.032 = 32 \, mA$
$E_{R_1} = 48 \, V$
$E_{R_2} = 24 \, V$
$E_{R_3} = 48 \, V$
$E_{R_4} = 48 \, V$

5. $R_t = 1200 \, \Omega$
$I_t = 0.01 = 10 \, mA$
$I_{R_{1,2}} = 3.33 \, mA$

$I_{R_{3,4}} = 3.33$ mA
$I_{R_{5,6}} = 3.33$ mA
$E_{R_3} = 6.0$ V
7. $R_t = 125\ \Omega$
$I_t = 0.04 = 40$ mA
$E_{R_5} = 2$ V
9. The value of each branch current is determined by V_s and the total resistance in that branch. Total current is the sum of all the branch currents.
11. $R_t = 4950 = 4.95$ kΩ
$I_t = 0.00182 = 1.82$ mA
$E_{R_3} = 3$ V
13. $R_t = 18{,}000 = 18$ kΩ
$I_t = 0.666$ mA
$E_{R_3} = 4$ V
15. $R_t = 4500\ \Omega$
$I_t = 0.001 = 1$ mA
$I_{R_3} = 0.5$ mA
$I_{R_6} = 0.5$ mA
$E_{R_6} = 1.5$ V
17. $R_t = 5650$
$I_t = 1.77$ mA
$I_{R_3} = 0.74$ mA
$I_{R_6} = 1.03$ mA
$E_{R_3} = 0.148$ V
19. $R_t = 4750\ \Omega$
$I_t = 0.02 = 20$ mA
$I_{R_3} = 0.005 = 5.0$ mA
$I_{R_6} = 0.01 \cong 10$ mA
$E_{R_7} = 15$ V
21. (a) Unbalanced; (b) $E_{A-B} = 0.153$ V
23. (a) 393.2 Ω; (b) $E_A = E_B = 5.437$ V

Chapter 7 Answers to Odd-Numbered Questions

1. Coil current
3. Magnetic field
5. Friction
7. R_m is the internal resistance of the movement, and is the resistance of the copper wire used in the coil.
9. 82.5 mA
11. (a) Make a break (open) in the circuit; (b) observe polarity; (c) set the full-scale reading to a value larger than the expected measurement.
13. (a) Ensure that power is off; (b) "zero" the meter at full scale; (c) estimate the range so that the reading will be in the upper two-thirds of the scale; (d) multiply the meter reading by the switch factor; (e) beware of "sneak" circuits, and always disconnect one end of the component being measured; (f) if semiconductors are present, check that polarity does not affect the reading.

15. High
17. 10,000 Ω/V
19. 10 μA (0.01 mA)
21. So that circuit current will flow through the meter and thus be able to be measured

Chapter 7 Answers to Odd-Numbered Problems

1. 10,000 Ω/V
3. 30.3 Ω
5. 0.05 V
7. 2.493 MΩ
9. 1473 Ω (1500 − 27)

Chapter 8 Answers to Odd-Numbered Questions

1. Are
3. Is not
5. True
7. 404 circular mil
9. Larger
11. Does not
13. Is not
15. May
17. Is
19. True
21. True
23. False
25. True
27. True
29. False

Chapter 8 Answers to Odd-Numbered Problems

1. 20.075 ≅ 20.1 V
3. 8.03 V
5. 29.1 ft (total wire length = 58.2 ft)
7. 1.02 Ω
9. 50.75 Ω
11. 19.72 Ω
13. 13.4 Ω
15. $I_t = 23$ mA; $E_R = 11.5$ V
17. $R = 5400$ Ω

Chapter 9 Answers to Odd-Numbered Questions

1. Are not
3. True
5. Less
7. Greater
9. Larger

Answers to Odd-Numbered Questions and Problems   553

Chapter 9   Answers to Odd-Numbered Problems

1. Orange, orange, black
3. 2400 Ω
5. Brown, brown, yellow, silver
7. 2430 Ω; 2970 Ω
9. 460 kΩ
11. ⅛ W
13. (a) 100 Ω, (b) 1 W, (c) carbon composition

Chapter 10   Answers to Odd-Numbered Questions

1. Is
3. Is not
5. False
7. Are not
9. False
11. Are not
13. Are
15. Has not
17. Is not
19. Has

Chapter 10   Answers to Odd-Numbered Problems

1. 0.8475 Ω
3. 11 V
5. 5 V
7. 750 A
9. $R_i = 10\ \Omega$, $R_l = 230\ \Omega$, $R_{match} = 10\ \Omega$

Chapter 11   Answers to Odd-Numbered Questions

1. Lodestone; magnetite
3. Any material exhibiting attraction or repulsion of other ferromagnetic materials
5. Any material having magnetic properties slightly less than air
7. Attract
9. Flux
11. True
13. False
15. False
17. Degauss
19. Current

Chapter 11   Answers to Odd-Numbered Problems

1. 225 AT
3. 125.7 Gb
5. 198.9 AT
7. 8.48 H
9. 1416 H
11. 50

Chapter 12 Answers to Odd-Numbered Questions

1. Wire would tend to move downward.
3. Wire would tend to move upward.
5. Current would be zero.
7. The signs encircled represent the normal *voltage drop* across the dc resistance of the coil; those in squares are a generated voltage and represent a *source* voltage.
9. Because the coil and slip rings rotate, the static brush provides a low-friction means of contacting the slip ring.

Chapter 12 Answers to Odd-Numbered Problems

1. 500 V
3. 22.5 V
5. 1750 turns
7. 0.04 webers per s.
   4×10^6 gilberts per s.

Chapter 13 Answers to Odd-Numbered Questions

1. 0.333 A
3. 20 W
5. A vector is a graphical construction, or line with arrowhead, showing *both* amplitude and direction.
7. $e = E \sin \theta$. The instantaneous value is proportional to the sine of the angle measured from the point of origin to the point of interest.
9. Radio frequency
11. π rad = 3.1416 rad = 180°
13. 4 msec; 250 Hz

Chapter 13 Answers to Odd-Numbered Problems

1. $V_{gen} = Vs = 220$ Vac @ 10 A
3. 198 V_{ave} at 8.99 A_{ave}
5. 212 V_{pk} at 3.25 A_{pk}
7. .002 = 2 msec
9. 66.7 μsec
11. 240 V
13. 153.8 V
15. 876.9 ft
17. 2 rads
19. 114.59 degrees

Chapter 14 Answers to Odd-Numbered Questions

1. The varying current induces a counter EMF that opposes current and that acts at a 90° angle to resistive opposition.
3. Induced voltage is generated as a function of the rate-of-change of current. Current rate-of-change is displaced 90° from the current waveform itself.

Answers to Odd-Numbered Questions and Problems 555

5. Core losses occur in the transformer due to I^2R, eddy-current, and hysteresis losses and are in the amount of:
$$\text{losses} = P_{input} - P_{output}.$$
7. Hysteresis loss occurs in the iron core because of the necessity of remagnetizing the iron to overcome the residual magnetism.
9. Laminated, powdered iron, ferrite
11. $k = 1.0$.

Chapter 14 Answers to Odd-Numbered Problems

1. 1.0 mH
3. 20 V
5. 565.5 Ω
7. 9425 Ω
9. $X_L = 94.3$ Ω
   $Z = 210$ Ω
   $\theta = 26.6°$
11. 45°
13. 2.5 mV
15. $E_{XL} = 23.4$ V
17. $\theta = 51.34°$
19. $L_T = 30$ mH
21. $L_T = 150$ μH
23. $X_L = 188.5$ Ω
25. $L_M = 0.25$ mH
    $k = 0.333$
27. $E_{sec} = 12$ V
    $I_{sec} = 0.25$ A
29. 93.6%

Chapter 15 Answers to Odd-Numbered Questions

1. A complex number is one having both magnitude and either direction (vector) or angular rotation (phasor).
3. 75/53.1 means that the circuit has an impedance of 75 Ω and the phase angle is +53.1°.
5. These values represent real, positive numbers. Those points on the abscissa represent magnitude only.
7. These values represent the magnitude of the phasor quantity that is rotated 90° CCW from the real, positive values.
9. J^2 indicates CCW rotation of 180°, and is equal to -1.
11. Apparent power in ac circuits is the simple product of voltage and current, without regard for any phase differences.

Chapter 15 Answers to Odd-Numbered Problems

1. $Z = 20.15$ Ω
   $\theta = 31°$
   Polar form: 29.15/31°
3. $52.5 + j\,70$

5. Rectangular: $Z = 2500 + j\,3770$
   Polar: $Z = 4524\underline{/56.45°}$
7. Rectangular: $Z = 333 + j\,501.4$
   Polar: $Z = 602\underline{/56.4°}$
   $i_T = 0.03986 \cong 0.04 = 40$ mA
9. $E_R = 13.26$ V
   $E_L = 20$ V
11. $135.5 + j\,63.2$
13. $2176\underline{/81.4°}$
15. $P_{app} = 0.955$ W
    $P_{true} = 0.745$ W

Chapter 16 Answers to Odd-Numbered Questions

1. Two conductors separated by a dielectric (insulator)
3. B
5. B
7. 0.10 F
9. Voltage lags the current, or current leads the voltage, by less than 90°. The more resistance, the smaller the phase angle; the greater the capacitance, the greater the phase angle.

Chapter 16 Answers to Odd-Numbered Problems

1. 0.0025×10^{-6}
   or 0.0025 μF
3. 47,000 pF
5. 8.44 μJ
7. 132.6 kΩ
9. 90.5 μA
11. 0.333 μF
13. 3 μF
15. 100 Ω
17. 25 Ω

Chapter 17 Answers to Odd-Numbered Questions

1. I_T
3. 0.0235×10^{-3} s = 23.5 μsec
5. 63.2%

Chapter 17 Answers to Odd-Numbered Problems

1. $X_c = 159$ Ω; $\theta = 57.8°$
3. $i_T = 0.638$ A
5. $Z = 100 + j\,159$ (rect); $Z = 188\underline{/57.8°}$ (polar)
7. $X_c = 1000$ Ω
9. $Z = 1414$ Ω
11. $e_{R_1} = 8.5$ V$_{ac}$
13. $12 = \sqrt{8.5^2 + 8.5^2} = 12$
15. $Z_1 = 10{,}034$ Ω; $Z_2 = 5927$ Ω

17. $Z_T = 3870\ \Omega$
19. $i_T = 1.55\ \text{mA};\ \theta_T = 37.5°$
21. $e_{R_2} = 5.06\ \text{V};\ e_{c_2} = 3.22\ \text{V}$
23. $6 = \sqrt{5.06^2 + 3.22^2}$
25. $e_{R_1} = E_{out} = 5.4\ \text{V}$
27. $e_{R_1} = E_{out} = 5.68\ \text{V}$
29. $t_c = 0.221\ \mu\text{sec}$
31. $E_c = 0.5067 \cong 0.51\ \text{V}$

Chapter 18 Answers to Odd-Numbered Questions

1. Because the phase angle is zero.
3. In an inductor, current *lags* voltage, while in a capacitor, current *leads* voltage. When both exist in the same circuit, this leads to opposite effects.
5. $R = \dfrac{1}{G}$, where G = conductance

   $X = \dfrac{1}{B}$, where B = susceptance

   $Z = \dfrac{1}{Y}$, where Y = admittance
7. The bel and decibel are logarithmic functions of the power ratios, and the human ear responds logarithmically to sound intensity.

Chapter 18 Answers to Odd-Numbered Problems

1. $X_C = 39.997 \cong 40\ \Omega$
   $X_L = 30\ \Omega$
   $X_T = -10\ \Omega$
   $Z = X_T = 10\ \Omega$
   $\theta = -90°$
3. $E_{XL} = 15\ \text{V}\ \underline{/90°}$
   $E_{XC} = 20\ \text{V}\ \underline{/-90°}$
5. $I_L = 19.11\ \text{mA}$
   $I_C = 7.54\ \text{mA}$
   $I_T = 11.57\ \text{mA}$
   $Z = 1037\ \Omega$
7. $X_T = 45\ \Omega$
   $Z = 75\ \Omega$
   $\theta = 36.9°$
   $I_T = 120\ \text{mA}$
   $E_R = 7.2\ \text{Vac}$
   $E_L = 10.8\ \text{Vac}$
   $E_C = 5.4\ \text{Vac}$
   $V_s = \sqrt{7.2^2 + 5.4^2} = 9\ \text{Vac}$
9. $I_{R_1} = 593\ \text{mA}$
   $I_{XC} = 200\ \text{mA}$
   $I_{XL} = 400\ \text{mA}$
   $I_{XT} = 200\ \text{mA}$
   $I_T = 626\ \text{mA}$
   $Z = 25.6\ \Omega$
   $\theta = 18.64°$

11. $C = 2.0\ \mu F$
    $L = 0.0064\ H = 6.4\ mH$
13. $Z = 16.8\ K\ \underline{/1.43°}$
15. Power gain = 30 dB
17. 24.3 A

Chapter 19   Answers to Odd-Numbered Questions

1. $X_L = X_C$
3. There is always some resistance to limit current.
5. X_L decreases and X_C increases.
7. Current is limited only by R_L since $X_L - X_C = 0$, and $I_T\ (X_L)$ may be several times V_s.
9. Series: Z is minimum and I_T is maximum.
   Parallel: Z is maximum and I_T is minimum.
11. To achieve resonance, I_{XL} and I_{XC} must be equal (but of opposite sign). When R is large, the value of I_{XL} does not equal I_{XC} at $1/(2\pi\sqrt{LC})$.
13. Passband is the range of frequencies extending from f_1 (lower 3dB, or 70.7%) to f_2 (upper 3dB, or 70.7%) that are permitted to get through a filter with minimum attenuation.

Chapter 19   Answers to Odd-Numbered Problems

1. 503 kHz
3. 15.9 kHz
5. $f_r = 734$ kHz
   $i_T = 0.606$ A
7. $i_{fr} = 0.24$ A
   $Z = R = 50\ \Omega$
   $Q = 6.32$
   $E_L = 75.84$ Vac
9. $f_r = 919$ kHz
   $Q = 115.5$
11. $f_r = 822$ kHz
13. $f = 100$ kHz
    $f_1 = 950$ kHz (0.95 MHz)
    $f_2 = 1.05$ MHz

Chapter 20   Answers to Odd-Numbered Questions

1. The rise time occurs on the leading edge of a pulse and is measured between the 10 and 90% points.
3. Pulse period is the time period encompassed by one complete cycle.
5. Repetition rate is the number of pulses per second, or frequency.
7. The harmonic waveforms must be odd-order (3rd, 5th, 7th, etc.); they must be in phase; and they must be progressively smaller in amplitude.
9. Very long time constant

Answers to Odd-Numbered Questions and Problems

Chapter 20 Answers to Odd-Numbered Problems

1. Period is 120 μsec; frequency is 8333 Hz
3. $t_p = 0.001 = 1.0$ msec
5. $F = 167$ Hz
   $t_p = 6$ msec
   $t_d = 2$ msec
   $E_{ave} = 1.0$ V
7. $F = 167$ Hz
   $t_d = 1$ msec
   $t_p = 6$ msec
   $E_{ave} = 0.833$ V
9. $t_c = 0.005 = 5$ msec
11. $R = 33$ K
13. $C_1 = 30$ pF (30×10^{-12})

Chapter 21 Answers to Odd-Numbered Questions and Problems

1. $E_A = 6.75$ V; $E_B = 3.75$ V
3. $R_1 = 5000 \, \Omega$; $R_2 = 6000 \, \Omega$
5. $I_{R_1} = 3$ mA; $I_{R_2} = 2$ mA
7. $I_{R_1} = 12$ mA; $I_{R_2} = 8$ mA; $V_s = 24$ V
9. $R_T = 600 \, \Omega$; $R_1 = 1500 \, \Omega$; $R_2 = 1000 \, \Omega$; $I_{R_1} = 6$ mA
11. $I_{R_1} = 0.09 = 90$ mA; $I_{R_2} = 0.045 = 45$ mA; $I_{R_3} = 0.030 = 30$ mA;
    $I_T = 165$ mA
13. $V_s = 180$ V; $R_2 = 2000 \, \Omega$; $R_3 = 3000 \, \Omega$
15. $V_s = 60$ V; $R_1 = 1500 \, \Omega$; $R_2 = 1200 \, \Omega$
17. $R_i = 20 \, \Omega$
19. $R_i = 714 \, \Omega$; $E_A = 0.74$ V
21. $R_i = 714 \, \Omega$; $E_A = +6.65$ V
23. $E_A = -1.53$ V
25. $R_A = 75 \, \Omega$; $R_B = 100 \, \Omega$; $R_C = 150 \, \Omega$
27. $R_1 = 2600 \, \Omega$; $R_2 = 1733 \, \Omega$; $R_3 = 1300 \, \Omega$

APPENDIX 1

STANDARD GRAPHIC SYMBOLS FOR ELECTRICAL AND ELECTRONICS DIAGRAMS

A huge variety of graphic symbols for use in schematic diagrams exists. As you continue your studies and enter the working world, you will no doubt encounter many different symbols used to represent the same circuit function or component. In order to improve graphic communications among those who work in electricity and electronics, the Institute of Electrical and Electronics Engineers (IEEE) and the American National Standards Institute (ANSI) maintain a list of *standard* graphic symbols. This list is lengthy and detailed. For instance, under switches, contacts, relays, and related items, there are more than one hundred separate symbols. It is far beyond the scope of this text to reproduce the entire list of standard symbols. Following are the symbols used in this text, plus other selected common symbols.

Note that in some instances, alternate symbols are given (for example, the junction of two conductors). Alternate symbols indicate that no agreement on a single symbol exists *at this time*.

1. Transmission Path Elements
   Guided path, general

Note: This symbol can be used to represent any wire or cable. Details (such as size, type, rating, or length) may be placed next to the symbol if needed.

Two conductors or paths

Crossing of paths, no connection

Note: Crossings can occur at any angle.

Junction

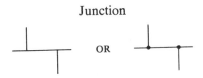

If space requires it, use

Three alternative paths

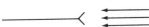

Shielded single conductor

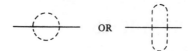

Shielded two-conductor cable with shield grounded

Note: Shielding around a component or assembly is indicated by long dashes.

— — — — —

Open circuit

Short circuit

Earth ground

Chassis or frame connection

Common returns at same level of potential

Note: * is not part of the symbol and is replaced by identifying values or letters.

2. Fundamental Items

Resistor, general

Tapped resistor

Resistor with adjustable contact

Adjustable or continuously adjustable resistor, rheostat

Shunt resistor

Capacitor, general

Adjustable or variable capacitor

Adjustable or variable capacitors, ganged

Standard Graphic Symbols for Electrical and Electronics Diagrams 563

Shunt capacitor

Antenna, general

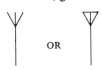

OR

Dipole

Loop

Direct-current source, general

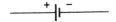

Note: The long line is always positive, so the polarity signs are optional.

One-cell direct-current source

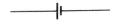

Multicell direct-current source

Generalized alternating-current source

3. Switching Functions

Fixed contact for switch

○   OR   ⟶

Adjustable or sliding contact for resistor, inductor, and so on

⟶   OR   ⌐

Closed contact (break)

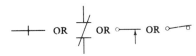

Open contact (make)

─✕─ OR ⊥ OR ○─── OR ○╲○

Single-throw switch, general

Double-throw switch, general

Double-pole, double-throw switch, with contacts shown

Nonlocking switch, circuit closing (make)

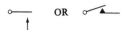

Nonlocking switch, circuit opening (break)

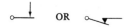

Locking switch, circuit closing (make)

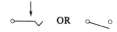

Locking switch, circuit opening (break)

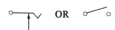

Selector or multiposition switch, general

Note: Any number of transmission paths may be shown.

4. Terminals and Connectors

Circuit terminal

o

Female contact, jack

———⟨

Male contact, plug

———→

Switchboard-type two-conductor jack

Switchboard-type two-conductor plug

5. Transformers, Inductors, and Windings

Inductor or winding, general

⌒⌒⌒ OR ⌒⌒⌒⌒⌒

Note: Either of these symbols may be used in the following items; we have chosen the right-hand one.

Magnetic-core inductor (coil)

Tapped inductor

Adjustable inductor

Adjustable or continuously adjustable inductor

Magnetic core

=
=

Standard Graphic Symbols for Electrical and Electronics Diagrams   565

Air core

NO SYMBOL

Note: If it is necessary to identify an air core, place a note next to the transformer or inductor symbol.

Transformer, general

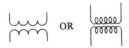

Note: Either symbol may be used in specific transformer symbols.

6. Semiconductor devices

Diode

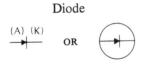

Capacative diode (varactor)

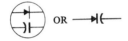

7. Circuit Protectors

Fuse

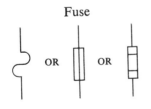

Lightning arrester (air-gap type)

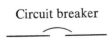

Circuit breaker

8. Acoustic Devides

Bell

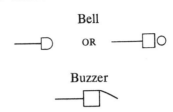

Buzzer

Audible-signaling device, general

Audible-signaling device, function specified

Note: † and * are not part of this symbol. When possible, the general symbol alone is used. If specific identification is necessary place the correct letter combination into the appropriate space:

*HN Horn
*LS Loudspeaker
*SN Siren
†EM Electromagnetic with moving coil
†MG Magnetic armature
†PM Permanent magnet with moving coil

Microphone

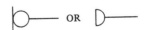

Light-emitting diode (LED)

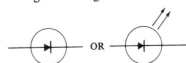

Headset

9. Lamps

Lamp, general

Note: If necessary, the following specific identifying letter combinations may be placed into the general symbol:

ARC   Arc
FL   Fluorescent
HG   Mercury vapor
IN   Incandescent
IR   Infrared
NE   Neon

Glow lamp, neon lamp, alternating-current type

Glow lamp, neon lamp, direct-current type

Incandescent lamp

Indicating lamp

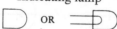

10. Metering Devices

Meter, general

Note: Always replace the asterisk with one of the following combinations, depending on meter function.

A	Ammeter
CRO	Oscilloscope
DB	Decibel meter
μA or UA	Microammeter
MA	Milliammeter
OHM	Ohmmeter
V	Voltmeter
VA	Volt-ammeter
W	Wattmeter

11. Rotating Machinery

Basic

◯

Generator, general

Ⓖ OR (GEN)

Generator, direct-current

(G̲)

Generator, alternating-current

(G̰)

Motor, general

 OR

Motor, direct-current

Motor, alternating-current

12. Composite Assemblies

General

Note: This symbol is used in block diagrams or to substitute for complex elements in schematic diagrams when the internal operation of the element is unimportant to the purpose of the diagram. The asterisk must be replaced by one of the following letter combinations (not all existing letter combinations are given here).

    CLK   Clock
      FL   Filter
  FL-BP   Filter, bandpass
  FL-HP   Filter, high-pass
  FL-LP   Filter, low-pass
    IND   Indicator
     PS   Power Source

Amplifier, general

Note: The triangle points in the direction of transmission.

Amplifier with two inputs

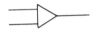

Amplifier with two outputs

Bridge-type rectifier

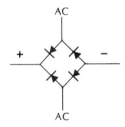

Appendix 2

COLOR CODES

Wiring Color Code. Table A-1 illustrates the color of wires used in wiring individual chassis and in interconnecting units. Colors may either be solid or helical stripes on a white background.

Resistors. Carbon resistors (2 W or less) are generally coded for resistance in ohms and percent of tolerance. Two methods may be encountered: the obsolete body-end-dot (BED) system, and the current circular-stripe method. These are illustrated in Table A-2, along with the universal color code.

Note that wire-wound resistors with axial leads have a double-wide A band. Carbon resistors with axial leads have a neutral tan body, which is not considered a color. The values of film-type resistors are more accurate, and thus such resistors have five stripes: bands A, B, and C are the first, second, and third significant figures respectively, band D is the multiplier, and band E is the tolerance. Only gold and silver are used for carbon-resistor tolerance, but all colors are used for tolerance of capacitors and film resistors. If a resistor has a *third* stripe of gold or silver, the multipliers are 0.1 and 0.01 respectively.

The preferred values for carbon resistors are given in Table A-3. These are values between 10 and 100, which must be multiplied by the correct multiple of 10 to obtain the actual value of the resistor. For example, in the ±5% column, the 12 ohms might represent 1.2, 12, 120, 1200, 12,000, and so on. Note that, with the exception of the ±1 percent column, Table A-3 is also used for capacitors.

TABLE A-1 WIRING COLOR CODE

Color	Connected to
Red	Ungrounded side of voltage source
Blue	Vacuum-tube plate, transistor collector, FET drain, ungrounded antenna connection
Green	Tube control grid, transistor base, FET gate, input of diode detector
Yellow	Tube cathode, transistor emitter, FET source
Orange	Screen grid of tube, second base of transistor
Brown	Vacuum-tube heaters or filaments
Black	Chassis ground return
White	Return for control grid (AVC bias) or base of transistor
Gray	Ac power line

TABLE A-2 RESISTOR AND CAPACITOR COLOR CODES

Color	Significant Figure	Decimal Multiplier	Tolerance* (%)	Voltage Rating*	Temperature Coefficient (ppm/°C)
Black	0	1	20		0
Brown	1	10	1	100	−30
Red	2	10^2	2	200	−80
Orange	3	10^3	3	300	−150
Yellow	4	10^4	4	400	−220
Green	5	10^5	5	500	−330
Blue	6	10^6	6	600	−470
Violet	7	10^7	7	700	−750
Gray	8	10^8	8	800	30
White	9	10^9	9	900	500
Gold		0.1	5	1000	
Silver		0.01	10	2000	
No color			20	500	

* Tolerance colors other than gold and silver and colors for voltage rating apply only to capacitors.

Axial Leads	Color	Radial Leads
Band A	First significant figure	Body A
Band B	Second significant figure	End B
Band C	Decimal multiplier	Dot C
Band D	Tolerance	End D
Band E	Tolerance (film resistors)	

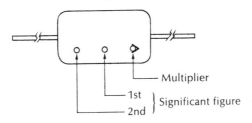

Capacitors. Fixed capacitors are either marked with a color code or have the value and tolerance printed on the body. The colors from Table A-2 are used according to one of the following standards:

1. Old three-dot code, mica capacitor:

All rated at 500 WVDC
Capacitance tolerance ± 20%

TABLE A-3 PREFERRED VALUES FOR RESISTORS

20% Tolerance	10% Tolerance	5% Tolerance	1% Tolerance		
10 ohms	10 ohms	10 ohms	10.0	21.5	46.4
		11	10.2	22.1	47.5
	12	12	10.5	22.6	48.7
		13	10.7	23.2	49.9
15	15	15	11.0	23.7	51.1
		16	11.3	24.3	52.3
	18	18	11.5	24.9	53.6
		20	11.8	25.5	54.9
22	22	22	12.1	26.1	56.2
		24	12.4	26.7	57.6
	27	27	12.7	27.4	59.0
		30	13.0	28.0	60.4
33	33	33	13.3	28.7	61.9
		36	13.7	29.4	63.4
	39	39	14.0	30.1	64.9
		43	14.3	30.9	66.5
47	47	47	14.7	31.6	68.1
		51	15.0	32.4	69.8
	56	56	15.4	33.2	71.5
		62	15.8	34.0	73.2
68	68	68	16.2	34.8	75.0
		75	16.5	35.7	76.8
	82	82	16.9	36.5	78.7
		91	17.4	37.4	80.6
			17.8	38.3	82.5
			18.2	39.2	84.5
			18.7	40.2	86.6
			19.1	41.2	88.7
			19.6	42.2	90.9
			20.0	43.2	93.1
			20.5	44.2	95.3
			21.0	45.3	97.6

2. Old six-dot code, mica capacitor:

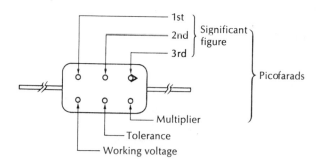

3. Current six-dot code, mica Capacitor:
    Note: EIA = Electronic Industry Association
    MIL = Military Standards
    AWS = American War Standards
    The sixth dot represents one of five classes, covering such characteristics as temperature coefficient, leakage resistance, or the like.

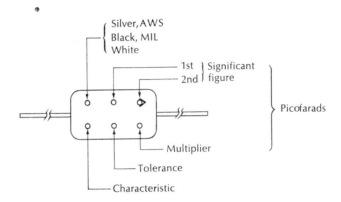

4. Ceramic Capacitors: These are generally tubular or disc type. They are either color coded or the value is printed on the device, as shown:

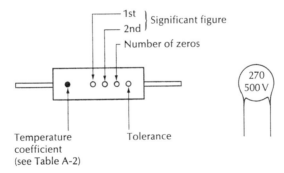

APPENDIX 3

TRIGONOMETRIC FUNCTIONS

These tables give the sine, cosine, and tangent for angles between 0 and 89.9 degrees in tenths of a degree.

0°–11.9°

Degs	Function	0.0°	0.1°	0.2°	0.3°	0.4°	0.5°	0.6°	0.7°	0.8°	0.9°
0	sin	0.0000	0.0017	0.0035	0.0052	0.0070	0.0087	0.0105	0.0122	0.0140	0.0157
	cos	1.0000	1.0000	1.0000	1.0000	1.0000	1.0000	0.9999	0.9999	0.9999	0.9999
	tan	0.0000	0.0017	0.0035	0.0052	0.0070	0.0087	0.0105	0.0122	0.0140	0.0157
1	sin	0.0175	0.0192	0.0209	0.0227	0.0244	0.0262	0.0279	0.0297	0.0314	0.0332
	cos	0.9998	0.9998	0.9998	0.9997	0.9997	0.9997	0.9996	0.9996	0.9995	0.9995
	tan	0.0175	0.0192	0.0209	0.0227	0.0244	0.0262	0.0279	0.0297	0.0314	0.0332
2	sin	0.0349	0.0366	0.0384	0.0401	0.0419	0.0436	0.0454	0.0471	0.0488	0.0506
	cos	0.9994	0.9993	0.9993	0.9992	0.9991	0.9990	0.9990	0.9989	0.9988	0.9987
	tan	0.0349	0.0367	0.0384	0.0402	0.0419	0.0437	0.0454	0.0472	0.0489	0.0507
3	sin	0.0523	0.0541	0.0558	0.0576	0.0593	0.0610	0.0628	0.0645	0.0663	0.0680
	cos	0.9986	0.9985	0.9984	0.9983	0.9982	0.9981	0.9980	0.9979	0.9978	0.9977
	tan	0.0524	0.0542	0.0559	0.0577	0.0594	0.0612	0.0629	0.0647	0.0664	0.0682
4	sin	0.0698	0.0715	0.0732	0.0750	0.0767	0.0785	0.0802	0.0819	0.0837	0.0854
	cos	0.9976	0.9974	0.9973	0.9972	0.9971	0.9969	0.9968	0.9966	0.9965	0.9963
	tan	0.0699	0.0717	0.0734	0.0752	0.0769	0.0787	0.0805	0.0822	0.0840	0.0857
5	sin	0.0872	0.0889	0.0906	0.0924	0.0941	0.0958	0.0976	0.0993	0.1011	0.1028
	cos	0.9962	0.9960	0.9959	0.9957	0.9956	0.9954	0.9952	0.9951	0.9949	0.9947
	tan	0.0875	0.0892	0.0910	0.0928	0.0945	0.0963	0.0981	0.0998	0.1016	0.1033
6	sin	0.1045	0.1063	0.1080	0.1097	0.1115	0.1132	0.1149	0.1167	0.1184	0.1201
	cos	0.9945	0.9943	0.9942	0.9940	0.9938	0.9936	0.9934	0.9932	0.9930	0.9928
	tan	0.1051	0.1069	0.1086	0.1104	0.1122	0.1139	0.1157	0.1175	0.1192	0.1210
7	sin	0.1219	0.1236	0.1253	0.1271	0.1288	0.1305	0.1323	0.1340	0.1357	0.1374
	cos	0.9925	0.9923	0.9921	0.9919	0.9917	0.9914	0.9912	0.9910	0.9907	0.9905
	tan	0.1228	0.1246	0.1263	0.1281	0.1299	0.1317	0.1334	0.1352	0.1370	0.1388
8	sin	0.1392	0.1409	0.1426	0.1444	0.1461	0.1478	0.1495	0.1513	0.1530	0.1547
	cos	0.9903	0.9900	0.9898	0.9895	0.9893	0.9890	0.9888	0.9885	0.9882	0.9880
	tan	0.1405	0.1423	0.1441	0.1459	0.1477	0.1495	0.1512	0.1530	0.1548	0.1566
9	sin	0.1564	0.1582	0.1599	0.1616	0.1633	0.1650	0.1668	0.1685	0.1702	0.1719
	cos	0.9877	0.9874	0.9871	0.9869	0.9866	0.9863	0.9860	0.9857	0.9854	0.9851
	tan	0.1584	0.1602	0.1620	0.1638	0.1655	0.1673	0.1691	0.1709	0.1727	0.1745
10	sin	0.1736	0.1754	0.1771	0.1788	0.1805	0.1822	0.1840	0.1857	0.1874	0.1891
	cos	0.9848	0.9845	0.9842	0.9839	0.9836	0.9833	0.9829	0.9826	0.9823	0.9820
	tan	0.1763	0.1781	0.1799	0.1817	0.1835	0.1853	0.1871	0.1890	0.1908	0.1926
11	sin	0.1908	0.1925	0.1942	0.1959	0.1977	0.1994	0.2011	0.2028	0.2045	0.2062
	cos	0.9816	0.9813	0.9810	0.9806	0.9803	0.9799	0.9796	0.9792	0.9789	0.9785
	tan	0.1944	0.1962	0.1980	0.1998	0.2016	0.2035	0.2053	0.2071	0.2089	0.2107

Trigonometric Functions

12°–28.9°

Degs	Function	0.0°	0.1°	0.2°	0.3°	0.4°	0.5°	0.6°	0.7°	0.8°	0.9°
12	sin	0.2079	0.2096	0.2113	0.2130	0.2147	0.2164	0.2181	0.2198	0.2215	0.2232
	cos	0.9781	0.9778	0.9774	0.9770	0.9767	0.9763	0.9759	0.9755	0.9751	0.9748
	tan	0.2126	0.2144	0.2162	0.2180	0.2199	0.2217	0.2235	0.2254	0.2272	0.2290
13	sin	0.2250	0.2267	0.2284	0.2300	0.2318	0.2334	0.2351	0.2368	0.2385	0.2402
	cos	0.9744	0.9740	0.9736	0.9732	0.9728	0.9724	0.9720	0.9715	0.9711	0.9707
	tan	0.2309	0.2327	0.2345	0.2364	0.2382	0.2401	0.2419	0.2438	0.2456	0.2475
14	sin	0.2419	0.2436	0.2453	0.2470	0.2487	0.2504	0.2521	0.2538	0.2554	0.2571
	cos	0.9703	0.9699	0.9694	0.9690	0.9686	0.9681	0.9677	0.9673	0.9668	0.9664
	tan	0.2493	0.2512	0.2530	0.2549	0.2568	0.2586	0.2605	0.2623	0.2642	0.2661
15	sin	0.2588	0.2605	0.2622	0.2639	0.2656	0.2672	0.2689	0.2706	0.2723	0.2740
	cos	0.9659	0.9655	0.9650	0.9646	0.9641	0.9636	0.9632	0.9627	0.9622	0.9617
	tan	0.2679	0.2698	0.2717	0.2736	0.2754	0.2773	0.2792	0.2811	0.2830	0.2849
16	sin	0.2756	0.2773	0.2790	0.2807	0.2823	0.2840	0.2857	0.2874	0.2890	0.2907
	cos	0.9613	0.9608	0.9603	0.9598	0.9593	0.9588	0.9583	0.9578	0.9573	0.9568
	tan	0.2867	0.2886	0.2905	0.2924	0.2943	0.2962	0.2981	0.3000	0.3019	0.3038
17	sin	0.2924	0.2940	0.2957	0.2974	0.2990	0.3007	0.3024	0.3040	0.3057	0.3074
	cos	0.9563	0.9558	0.9553	0.9548	0.9542	0.9537	0.9532	0.9527	0.9521	0.9516
	tan	0.3057	0.3076	0.3096	0.3115	0.3134	0.3153	0.3172	0.3191	0.3211	0.3230
18	sin	0.3090	0.3107	0.3123	0.3140	0.3156	0.3173	0.3190	0.3206	0.3223	0.3239
	cos	0.9511	0.9505	0.9500	0.9494	0.9489	0.9483	0.9478	0.9472	0.9466	0.9461
	tan	0.3249	0.3269	0.3288	0.3307	0.3327	0.3346	0.3365	0.3385	0.3404	0.3424
19	sin	0.3256	0.3272	0.3289	0.3305	0.3322	0.3338	0.3355	0.3371	0.3387	0.3404
	cos	0.9455	0.9449	0.9444	0.9438	0.9432	0.9426	0.9421	0.9415	0.9409	0.9403
	tan	0.3443	0.3463	0.3482	0.3502	0.3522	0.3541	0.3561	0.3581	0.3600	0.3620
20	sin	0.3420	0.3437	0.3453	0.3469	0.3486	0.3502	0.3518	0.3535	0.3551	0.3567
	cos	0.9397	0.9391	0.9385	0.9379	0.9373	0.9367	0.9361	0.9354	0.9348	0.9342
	tan	0.3640	0.3659	0.3679	0.3699	0.3719	0.3739	0.3759	0.3779	0.3799	0.3819
21	sin	0.3584	0.3600	0.3616	0.3633	0.3649	0.3665	0.3681	0.3697	0.3714	0.3730
	cos	0.9336	0.9330	0.9323	0.9317	0.9311	0.9304	0.9298	0.9291	0.9285	0.9278
	tan	0.3839	0.3859	0.3879	0.3899	0.3919	0.3939	0.3959	0.3979	0.4000	0.4020
22	sin	0.3746	0.3762	0.3778	0.3795	0.3811	0.3827	0.3843	0.3859	0.3875	0.3891
	cos	0.9272	0.9265	0.9259	0.9252	0.9245	0.9239	0.9232	0.9225	0.9219	0.9212
	tan	0.4040	0.4061	0.4081	0.4101	0.4122	0.4142	0.4163	0.4183	0.4204	0.4224
23	sin	0.3907	0.3923	0.3939	0.3955	0.3971	0.3987	0.4003	0.4019	0.4035	0.4051
	cos	0.9205	0.9198	0.9191	0.9184	0.9178	0.9171	0.9164	0.9157	0.9150	0.9143
	tan	0.4245	0.4265	0.4286	0.4307	0.4327	0.4348	0.4369	0.4390	0.4411	0.4431
24	sin	0.4067	0.4083	0.4099	0.4115	0.4131	0.4147	0.4163	0.4179	0.4195	0.4210
	cos	0.9135	0.9128	0.9121	0.9114	0.9107	0.9100	0.9092	0.9085	0.9078	0.9070
	tan	0.4452	0.4473	0.4494	0.4515	0.4536	0.4557	0.4578	0.4599	0.4621	0.4642
25	sin	0.4226	0.4242	0.4258	0.4274	0.4289	0.4305	0.4321	0.4337	0.4352	0.4368
	cos	0.9063	0.9056	0.9048	0.9041	0.9033	0.9026	0.9018	0.9011	0.9003	0.8996
	tan	0.4663	0.4684	0.4706	0.4727	0.4748	0.4770	0.4791	0.4813	0.4834	0.4856
26	sin	0.4384	0.4399	0.4415	0.4431	0.4446	0.4462	0.4478	0.4493	0.4509	0.4524
	cos	0.8988	0.8980	0.8973	0.8965	0.8957	0.8949	0.8942	0.8934	0.8926	0.8918
	tan	0.4877	0.4899	0.4921	0.4942	0.4964	0.4986	0.5008	0.5029	0.5051	0.5073
27	sin	0.4540	0.4555	0.4571	0.4586	0.4602	0.4617	0.4633	0.4648	0.4664	0.4679
	cos	0.8910	0.8902	0.8894	0.8886	0.8878	0.8870	0.8862	0.8854	0.8846	0.8838
	tan	0.5095	0.5117	0.5139	0.5161	0.5184	0.5206	0.5228	0.5250	0.5272	0.5295
28	sin	0.4695	0.4710	0.4726	0.4741	0.4756	0.4772	0.4787	0.4802	0.4818	0.4833
	cos	0.8829	0.8821	0.8813	0.8805	0.8796	0.8788	0.8780	0.8771	0.8763	0.8755
	tan	0.5317	0.5340	0.5362	0.5384	0.5407	0.5430	0.5452	0.5475	0.5498	0.5520

29°–45.9°

Degs	Function	0.0°	0.1°	0.2°	0.3°	0.4°	0.5°	0.6°	0.7°	0.8°	0.9°
29	sin	0.4848	0.4863	0.4879	0.4894	0.4909	0.4924	0.4939	0.4955	0.4970	0.4985
	cos	0.8746	0.8738	0.8729	0.8721	0.8712	0.8704	0.8695	0.8686	0.8678	0.8669
	tan	0.5543	0.5566	0.5589	0.5612	0.5635	0.5658	0.5681	0.5704	0.5727	0.5750
30	sin	0.5000	0.5015	0.5030	0.5045	0.5060	0.5075	0.5090	0.5105	0.5120	0.5135
	cos	0.8660	0.8652	0.8643	0.8634	0.8625	0.8616	0.8607	0.8599	0.8590	0.8581
	tan	0.5774	0.5797	0.5820	0.5844	0.5867	0.5890	0.5914	0.5938	0.5961	0.5985
31	sin	0.5150	0.5165	0.5180	0.5195	0.5210	0.5225	0.5240	0.5255	0.5270	0.5284
	cos	0.8572	0.8563	0.8554	0.8545	0.8536	0.8526	0.8517	0.8508	0.8499	0.8490
	tan	0.6009	0.6032	0.6056	0.6080	0.6104	0.6128	0.6152	0.6176	0.6200	0.6224
32	sin	0.5299	0.5314	0.5329	0.5344	0.5358	0.5373	0.5388	0.5402	0.5417	0.5432
	cos	0.8480	0.8471	0.8462	0.8453	0.8443	0.8434	0.8425	0.8415	0.8406	0.8396
	tan	0.6249	0.6273	0.6297	0.6322	0.6346	0.6371	0.6395	0.6420	0.6445	0.6469
33	sin	0.5446	0.5461	0.5476	0.5490	0.5505	0.5519	0.5534	0.5548	0.5563	0.5577
	cos	0.8387	0.8377	0.8368	0.8358	0.8348	0.8339	0.8329	0.8320	0.8310	0.8300
	tan	0.6494	0.6519	0.6544	0.6569	0.6594	0.6619	0.6644	0.6669	0.6694	0.6720
34	sin	0.5592	0.5606	0.5621	0.5635	0.5650	0.5664	0.5678	0.5693	0.5707	0.5721
	cos	0.8290	0.8281	0.8271	0.8261	0.8251	0.8241	0.8231	0.8221	0.8211	0.8202
	tan	0.6745	0.6771	0.6796	0.6822	0.6847	0.6873	0.6899	0.6924	0.6950	0.6976
35	sin	0.5736	0.5750	0.5764	0.5779	0.5793	0.5807	0.5821	0.5835	0.5850	0.5864
	cos	0.8192	0.8181	0.8171	0.8161	0.8151	0.8141	0.8131	0.8121	0.8111	0.8100
	tan	0.7002	0.7028	0.7054	0.7080	0.7107	0.7133	0.7159	0.7186	0.7212	0.7239
36	sin	0.5878	0.5892	0.5906	0.5920	0.5934	0.5948	0.5962	0.5976	0.5990	0.6004
	cos	0.8090	0.8080	0.8070	0.8059	0.8049	0.8039	0.8028	0.8018	0.8007	0.7997
	tan	0.7265	0.7292	0.7319	0.7346	0.7373	0.7400	0.7427	0.7454	0.7481	0.7508
37	sin	0.6018	0.6032	0.6046	0.6060	0.6074	0.6088	0.6101	0.6115	0.6129	0.6143
	cos	0.7986	0.7976	0.7965	0.7955	0.7944	0.7934	0.7923	0.7912	0.7902	0.7891
	tan	0.7536	0.7563	0.7590	0.7618	0.7646	0.7673	0.7701	0.7729	0.7757	0.7785
38	sin	0.6157	0.6170	0.6184	0.6198	0.6211	0.6225	0.6239	0.6252	0.6266	0.6280
	cos	0.7880	0.7869	0.7859	0.7848	0.7837	0.7826	0.7815	0.7804	0.7793	0.7782
	tan	0.7813	0.7841	0.7869	0.7898	0.7926	0.7954	0.7983	0.8012	0.8040	0.8069
39	sin	0.6293	0.6307	0.6320	0.6334	0.6347	0.6361	0.6374	0.6388	0.6401	0.6414
	cos	0.7771	0.7760	0.7749	0.7738	0.7727	0.7716	0.7705	0.7694	0.7683	0.7672
	tan	0.8098	0.8127	0.8156	0.8185	0.8214	0.8243	0.8273	0.8302	0.8332	0.8361
40	sin	0.6428	0.6441	0.6455	0.6468	0.6481	0.6494	0.6508	0.6521	0.6534	0.6547
	cos	0.7660	0.7649	0.7638	0.7627	0.7615	0.7604	0.7593	0.7581	0.7570	0.7559
	tan	0.8391	0.8421	0.8451	0.8481	0.8511	0.8541	0.8571	0.8601	0.8632	0.8662
41	sin	0.6561	0.6574	0.6587	0.6600	0.6613	0.6626	0.6639	0.6652	0.6665	0.6678
	cos	0.7547	0.7536	0.7524	0.7513	0.7501	0.7490	0.7478	0.7466	0.7455	0.7443
	tan	0.8693	0.8724	0.8754	0.8785	0.8816	0.8847	0.8878	0.8910	0.8941	0.8972
42	sin	0.6691	0.6704	0.6717	0.6730	0.6743	0.6756	0.6769	0.6782	0.6794	0.6807
	cos	0.7431	0.7420	0.7408	0.7396	0.7385	0.7373	0.7361	0.7349	0.7337	0.7325
	tan	0.9004	0.9036	0.9067	0.9099	0.9131	0.9163	0.9195	0.9228	0.9260	0.9293
43	sin	0.6820	0.6833	0.6845	0.6858	0.6871	0.6884	0.6896	0.6909	0.6921	0.6934
	cos	0.7314	0.7302	0.7290	0.7278	0.7266	0.7254	0.7242	0.7230	0.7218	0.7206
	tan	0.9325	0.9358	0.9391	0.9424	0.9457	0.9490	0.9523	0.9556	0.9590	0.9623
44	sin	0.6947	0.6959	0.6972	0.6984	0.6997	0.7009	0.7022	0.7034	0.7046	0.7059
	cos	0.7193	0.7181	0.7169	0.7157	0.7145	0.7133	0.7120	0.7108	0.7096	0.7083
	tan	0.9657	0.9691	0.9725	0.9759	0.9793	0.9827	0.9861	0.9896	0.9930	0.9965
45	sin	0.7071	0.7083	0.7096	0.7108	0.7120	0.7133	0.7145	0.7157	0.7169	0.7181
	cos	0.7071	0.7059	0.7046	0.7034	0.7022	0.7009	0.6997	0.6984	0.6972	0.6959
	tan	1.0000	1.0035	1.0070	1.0105	1.0141	1.0176	1.0212	1.0247	1.0283	1.0319

46°–62.9°

Degs	Function	0.0°	0.1°	0.2°	0.3°	0.4°	0.5°	0.6°	0.7°	0.8°	0.9°
46	sin	0.7193	0.7206	0.7218	0.7230	0.7242	0.7254	0.7266	0.7278	0.7290	0.7302
	cos	0.6947	0.6934	0.6921	0.6909	0.6896	0.6884	0.6871	0.6858	0.6845	0.6833
	tan	1.0355	1.0392	1.0428	1.0464	1.0501	1.0538	1.0575	1.0612	1.0649	1.0686
47	sin	0.7314	0.7325	0.7337	0.7349	0.7361	0.7373	0.7385	0.7396	0.7408	0.7420
	cos	0.6820	0.6807	0.6794	0.6782	0.6769	0.6756	0.6743	0.6730	0.6717	0.6704
	tan	1.0724	1.0761	1.0799	1.0837	1.0875	1.0913	1.0951	1.0990	1.1028	1.1067
48	sin	0.7431	0.7443	0.7455	0.7466	0.7478	0.7490	0.7501	0.7513	0.7524	0.7536
	cos	0.6691	0.6678	0.6665	0.6652	0.6639	0.6626	0.6613	0.6600	0.6587	0.6574
	tan	1.1106	1.1145	1.1184	1.1224	1.1263	1.1303	1.1343	1.1383	1.1423	1.1463
49	sin	0.7547	0.7559	0.7570	0.7581	0.7593	0.7604	0.7615	0.7627	0.7638	0.7649
	cos	0.6561	0.6547	0.6534	0.6521	0.6508	0.6494	0.6481	0.6468	0.6455	0.6441
	tan	1.1504	1.1544	1.1585	1.1626	1.1667	1.1708	1.1750	1.1792	1.1833	1.1875
50	sin	0.7660	0.7672	0.7683	0.7694	0.7705	0.7716	0.7727	0.7738	0.7749	0.7760
	cos	0.6428	0.6414	0.6401	0.6388	0.6374	0.6361	0.6347	0.6334	0.6320	0.6307
	tan	1.1918	1.1960	1.2002	1.2045	1.2088	1.2131	1.2174	1.2218	1.2261	1.2305
51	sin	0.7771	0.7782	0.7793	0.7804	0.7815	0.7826	0.7837	0.7848	0.7859	0.7869
	cos	0.6293	0.6280	0.6266	0.6252	0.6239	0.6225	0.6211	0.6198	0.6184	0.6170
	tan	1.2349	1.2393	1.2437	1.2482	1.2527	1.2572	1.2617	1.2662	1.2708	1.2753
52	sin	0.7880	0.7891	0.7902	0.7912	0.7923	0.7934	0.7944	0.7955	0.7965	0.7976
	cos	0.6157	0.6143	0.6129	0.6115	0.6101	0.6088	0.6074	0.6060	0.6046	0.6032
	tan	1.2799	1.2846	1.2892	1.2938	1.2985	1.3032	1.3079	1.3127	1.3175	1.3222
53	sin	0.7986	0.7997	0.8007	0.8018	0.8028	0.8039	0.8049	0.8059	0.8070	0.8080
	cos	0.6018	0.6004	0.5990	0.5976	0.5962	0.5948	0.5934	0.5920	0.5906	0.5892
	tan	1.3270	1.3319	1.3367	1.3416	1.3465	1.3514	1.3564	1.3613	1.3663	1.3713
54	sin	0.8090	0.8100	0.8111	0.8121	0.8131	0.8141	0.8151	0.8161	0.8171	0.8181
	cos	0.5878	0.5864	0.5850	0.5835	0.5821	0.5807	0.5793	0.5779	0.5764	0.5750
	tan	1.3764	1.3814	1.3865	1.3916	1.3968	1.4019	1.4071	1.4124	1.4176	1.4229
55	sin	0.8192	0.8202	0.8211	0.8221	0.8231	0.8241	0.8251	0.8261	0.8271	0.8281
	cos	0.5736	0.5721	0.5707	0.5693	0.5678	0.5664	0.5650	0.5635	0.5621	0.5606
	tan	1.4281	1.4335	1.4388	1.4442	1.4496	1.4550	1.4605	1.4659	1.4715	1.4770
56	sin	0.8290	0.8300	0.8310	0.8320	0.8329	0.8339	0.8348	0.8358	0.8368	0.8377
	cos	0.5592	0.5577	0.5563	0.5548	0.5534	0.5519	0.5505	0.5490	0.5476	0.5461
	tan	1.4826	1.4882	1.4938	1.4994	1.5051	1.5108	1.5166	1.5224	1.5282	1.5340
57	sin	0.8387	0.8396	0.8406	0.8415	0.8425	0.8434	0.8443	0.8453	0.8462	0.8471
	cos	0.5446	0.5432	0.5417	0.5402	0.5388	0.5373	0.5358	0.5344	0.5329	0.5314
	tan	1.5399	1.5458	1.5517	1.5577	1.5637	1.5697	1.5757	1.5818	1.5880	1.5941
58	sin	0.8480	0.8490	0.8499	0.8508	0.8517	0.8526	0.8536	0.8545	0.8554	0.8563
	cos	0.5299	0.5284	0.5270	0.5255	0.5240	0.5225	0.5210	0.5195	0.5180	0.5165
	tan	1.6003	1.6066	1.6128	1.6191	1.6255	1.6319	1.6383	1.6447	1.6512	1.6577
59	sin	0.8572	0.8581	0.8590	0.8599	0.8607	0.8616	0.8625	0.8634	0.8643	0.8652
	cos	0.5150	0.5135	0.5120	0.5105	0.5090	0.5075	0.5060	0.5045	0.5030	0.5015
	tan	1.6643	1.6709	1.6775	1.6842	1.6909	1.6977	1.7045	1.7113	1.7182	1.7251
60	sin	0.8660	0.8669	0.8678	0.8686	0.8695	0.8704	0.8712	0.8721	0.8729	0.8738
	cos	0.5000	0.4985	0.4970	0.4955	0.4939	0.4924	0.4909	0.4894	0.4879	0.4863
	tan	1.7321	1.7391	1.7461	1.7532	1.7603	1.7675	1.7747	1.7820	1.7893	1.7966
61	sin	0.8746	0.8755	0.8763	0.8771	0.8780	0.8788	0.8796	0.8805	0.8813	0.8821
	cos	0.4848	0.4833	0.4818	0.4802	0.4787	0.4772	0.4756	0.4741	0.4726	0.4710
	tan	1.8040	1.8115	1.8190	1.8265	1.8341	1.8418	1.8495	1.8572	1.8650	1.8728
62	sin	0.8829	0.8838	0.8846	0.8854	0.8862	0.8870	0.8878	0.8886	0.8894	0.8902
	cos	0.4695	0.4679	0.4664	0.4648	0.4633	0.4617	0.4602	0.4586	0.4571	0.4555
	tan	1.8807	1.8887	1.8967	1.9047	1.9128	1.9210	1.9292	1.9375	1.9458	1.9542

63°–79.9°

Degs	Function	0.0°	0.1°	0.2°	0.3°	0.4°	0.5°	0.6°	0.7°	0.8°	0.9°
63	sin cos tan	0.8910 0.4540 1.9626	0.8918 0.4524 1.9711	0.8926 0.4509 1.9797	0.8934 0.4493 1.9883	0.8942 0.4478 1.9970	0.8949 0.4462 2.0057	0.8957 0.4446 2.0145	0.8965 0.4431 2.0233	0.8973 0.4415 2.0323	0.8980 0.4399 2.0413
64	sin cos tan	0.8988 0.4384 2.0503	0.8996 0.4368 2.0594	0.9003 0.4352 2.0686	0.9011 0.4337 2.0778	0.9018 0.4321 2.0872	0.9026 0.4305 2.0965	0.9033 0.4289 2.1060	0.9041 0.4274 2.1155	0.9048 0.4258 2.1251	0.9056 0.4242 2.1348
65	sin cos tan	0.9063 0.4226 2.1445	0.9070 0.4210 2.1543	0.9078 0.4195 2.1642	0.9085 0.4179 2.1742	0.9092 0.4163 2.1842	0.9100 0.4147 2.1943	0.9107 0.4131 2.2045	0.9114 0.4115 2.2148	0.9121 0.4099 2.2251	0.9128 0.4083 2.2355
66	sin cos tan	0.9135 0.4067 2.2460	0.9143 0.4051 2.2566	0.9150 0.4035 2.2673	0.9157 0.4019 2.2781	0.9164 0.4003 2.2889	0.9171 0.3987 2.2998	0.9178 0.3971 2.3109	0.9184 0.3955 2.3220	0.9191 0.3939 2.3332	0.9198 0.3923 2.3445
67	sin cos tan	0.9205 0.3907 2.3559	0.9212 0.3891 2.3673	0.9219 0.3875 2.3789	0.9225 0.3859 2.3906	0.9232 0.3843 2.4023	0.9239 0.3827 2.4142	0.9245 0.3811 2.4262	0.9252 0.3795 2.4383	0.9259 0.3778 2.4504	0.9265 0.3762 2.4627
68	sin cos tan	0.9272 0.3746 2.4751	0.9278 0.3730 2.4876	0.9285 0.3714 2.5002	0.9291 0.3697 2.5129	0.9298 0.3681 2.5257	0.9304 0.3665 2.5386	0.9311 0.3649 2.5517	0.9317 0.3633 2.5649	0.9323 0.3616 2.5782	0.9330 0.3600 2.5916
69	sin cos tan	0.9336 0.3584 2.6051	0.9342 0.3567 2.6187	0.9348 0.3551 2.6325	0.9354 0.3535 2.6464	0.9361 0.3518 2.6605	0.9367 0.3502 2.6746	0.9373 0.3486 2.6889	0.9379 0.3469 2.7034	0.9385 0.3453 2.7179	0.9391 0.3437 2.7326
70	sin cos tan	0.9397 0.3420 2.7475	0.9403 0.3404 2.7625	0.9409 0.3387 2.7776	0.9415 0.3371 2.7929	0.9421 0.3355 2.8083	0.9426 0.3338 2.8239	0.9432 0.3322 2.8397	0.9438 0.3305 2.8556	0.9444 0.3289 2.8716	0.9449 0.3272 2.8878
71	sin cos tan	0.9455 0.3256 2.9042	0.9461 0.3239 2.9208	0.9466 0.3223 2.9375	0.9472 0.3206 2.9544	0.9478 0.3190 2.9714	0.9483 0.3173 2.9887	0.9489 0.3156 3.0061	0.9494 0.3140 3.0237	0.9500 0.3123 3.0415	0.9505 0.3107 3.0595
72	sin cos tan	0.9511 0.3090 3.0777	0.9516 0.3074 3.0961	0.9521 0.3057 3.1146	0.9527 0.3040 3.1334	0.9532 0.3024 3.1524	0.9537 0.3007 3.1716	0.9542 0.2990 3.1910	0.9548 0.2974 3.2106	0.9553 0.2957 3.2305	0.9558 0.2940 3.2506
73	sin cos tan	0.9563 0.2924 3.2709	0.9568 0.2907 3.2914	0.9573 0.2890 3.3122	0.9578 0.2874 3.3332	0.9583 0.2857 3.3544	0.9588 0.2840 3.3759	0.9593 0.2823 3.3977	0.9598 0.2807 3.4197	0.9603 0.2790 3.4420	0.9608 0.2773 3.4646
74	sin cos tan	0.9613 0.2756 3.4874	0.9617 0.2740 3.5105	0.9622 0.2723 3.5339	0.9627 0.2706 3.5576	0.9632 0.2689 3.5816	0.9636 0.2672 3.6059	0.9641 0.2656 3.6305	0.9646 0.2639 3.6554	0.9650 0.2622 3.6806	0.9655 0.2605 3.7062
75	sin cos tan	0.9659 0.2588 3.7321	0.9664 0.2571 3.7583	0.9668 0.2554 3.7848	0.9673 0.2538 3.8118	0.9677 0.2521 3.8391	0.9681 0.2504 3.8667	0.9686 0.2487 3.8947	0.9690 0.2470 3.9232	0.9694 0.2453 3.9520	0.9699 0.2436 3.9812
76	sin cos tan	0.9703 0.2419 4.0108	0.9707 0.2402 4.0408	0.9711 0.2385 4.0713	0.9715 0.2368 4.1022	0.9720 0.2351 4.1335	0.9724 0.2334 4.1653	0.9728 0.2317 4.1976	0.9732 0.2300 4.2303	0.9736 0.2284 4.2635	0.9740 0.2267 4.2972
77	sin cos tan	0.9744 0.2250 4.3315	0.9748 0.2232 4.3662	0.9751 0.2215 4.4015	0.9755 0.2198 4.4374	0.9759 0.2181 4.4737	0.9763 0.2164 4.5107	0.9767 0.2147 4.5483	0.9770 0.2130 4.5864	0.9774 0.2113 4.6252	0.9778 0.2096 4.6646
78	sin cos tan	0.9781 0.2079 4.7046	0.9785 0.2062 4.7453	0.9789 0.2045 4.7867	0.9792 0.2028 4.8288	0.9796 0.2011 4.8716	0.9799 0.1994 4.9152	0.9803 0.1977 4.9594	0.9806 0.1959 5.0045	0.9810 0.1942 5.0504	0.9813 0.1925 5.0970
79	sin cos tan	0.9816 0.1908 5.1446	0.9820 0.1891 5.1929	0.9823 0.1874 5.2422	0.9826 0.1857 5.2924	0.9829 0.1840 5.3435	0.9833 0.1822 5.3955	0.9836 0.1805 5.4486	0.9839 0.1788 5.5026	0.9842 0.1771 5.5578	0.9845 0.1754 5.6140

80°–89.9°

Degs	Function	0.0°	0.1°	0.2°	0.3°	0.4°	0.5°	0.6°	0.7°	0.8°	0.9°
80	sin	0.9848	0.9851	0.9854	0.9857	0.9860	0.9863	0.9866	0.9869	0.9871	0.9874
	cos	0.1736	0.1719	0.1702	0.1685	0.1668	0.1650	0.1633	0.1616	0.1599	0.1582
	tan	5.6713	5.7297	5.7894	5.8502	5.9124	5.9758	6.0405	6.1066	6.1742	6.2432
81	sin	0.9877	0.9880	0.9882	0.9885	0.9888	0.9890	0.9893	0.9895	0.9898	0.9900
	cos	0.1564	0.1547	0.1530	0.1513	0.1495	0.1478	0.1461	0.1444	0.1426	0.1409
	tan	6.3138	6.3859	6.4596	6.5350	6.6122	6.6912	6.7720	6.8548	6.9395	7.0264
82	sin	0.9903	0.9905	0.9907	0.9910	0.9912	0.9914	0.9917	0.9919	0.9921	0.9923
	cos	0.1392	0.1374	0.1357	0.1340	0.1323	0.1305	0.1288	0.1271	0.1253	0.1236
	tan	7.1154	7.2066	7.3002	7.3962	7.4947	7.5958	7.6996	7.8062	7.9158	8.0285
83	sin	0.9925	0.9928	0.9930	0.9932	0.9934	0.9936	0.9938	0.9940	0.9942	0.9943
	cos	0.1219	0.1201	0.1184	0.1167	0.1149	0.1132	0.1115	0.1097	0.1080	0.1063
	tan	8.1443	8.2636	8.3863	8.5126	8.6427	8.7769	8.9152	9.0579	9.2052	9.3572
84	sin	0.9945	0.9947	0.9949	0.9951	0.9952	0.9954	0.9956	0.9957	0.9959	0.9960
	cos	0.1045	0.1028	0.1011	0.0993	0.0976	0.0958	0.0941	0.0924	0.0906	0.0889
	tan	9.5144	9.6768	9.8448	10.02	10.20	10.39	10.58	10.78	10.99	11.20
85	sin	0.9962	0.9963	0.9965	0.9966	0.9968	0.9969	0.9971	0.9972	0.9973	0.9974
	cos	0.0872	0.0854	0.0837	0.0819	0.0802	0.0785	0.0767	0.0750	0.0732	0.0715
	tan	11.43	11.66	11.91	12.16	12.43	12.71	13.00	13.30	13.62	13.95
86	sin	0.9976	0.9977	0.9978	0.9979	0.9980	0.9981	0.9982	0.9983	0.9984	0.9985
	cos	0.0698	0.0680	0.0663	0.0645	0.0628	0.0610	0.0593	0.0576	0.0558	0.0541
	tan	14.30	14.67	15.06	15.46	15.89	16.35	16.83	17.34	17.89	18.46
87	sin	0.9986	0.9987	0.9988	0.9989	0.9990	0.9990	0.9991	0.9992	0.9993	0.9993
	cos	0.0523	0.0506	0.0488	0.0471	0.0454	0.0436	0.0419	0.0401	0.0384	0.0366
	tan	19.08	19.74	20.45	21.20	22.02	22.90	23.86	24.90	26.03	27.27
88	sin	0.9994	0.9995	0.9995	0.9996	0.9996	0.9997	0.9997	0.9997	0.9998	0.9998
	cos	0.0349	0.0332	0.0314	0.0297	0.0279	0.0262	0.0244	0.0227	0.0209	0.0192
	tan	28.64	30.14	31.82	33.69	35.80	38.19	40.92	44.07	47.74	52.08
89	sin	0.9998	0.9999	0.9999	0.9999	0.9999	1.000	1.000	1.000	1.000	1.000
	cos	0.0175	0.0157	0.0140	0.0122	0.0105	0.0087	0.0070	0.0052	0.0035	0.0017
	tan	57.29	63.66	71.62	81.85	95.49	114.6	143.2	191.0	286.5	573.0

APPENDIX 4

COMMON LOGARITHMIC TABLES

In the following log tables, the listed four-digit numbers are the *mantissa* of the complete logarithm, and in practice are preceded by a decimal point. The *characteristic* is then placed to the left of the decimal.

Characteristic	Decimal Point	Mantissa
4		8451

Hence, the logarithm is written 4.8451.

The mantissa is read from the table for the number (N) desired. For example, find the logarithm (log) of 7. In the N column, find 70 and read in the 0 column 8451, which is the mantissa for 7, 70, 700, 7000, and so on. Now, the characteristic has no purpose except to place the decimal point in the original number, and is *one less than the number of digits to the left* of the decimal point for any number greater than 1. Hence, since 7.0 has one digit to the left of the decimal, the complete logarithm for 7 is 0.8451, and is written

$$\log_{10} 7 = 0.8451.$$

Also, by this procedure,

$$\log_{10} 70 = 1.8451$$
$$\log_{10} 700 = 2.8451$$
$$\log_{10} 7000 = 3.8451$$

If the original number is 70.5, then read opposite 70, but down the 5 column:

$$\log_{10} 70.5 = 1.8482$$
$$\log_{10} 705 = 2.8482.$$

For numbers less than one, the characteristic is *negative*, and one less than the number of places from the decimal point to the first significant digit. Furthermore, the mantissa must be subtracted from one. For example, what is the common log of 0.7?

$$\text{Mantissa for } 7 = 0.8451$$
$$\text{and } 1 - 0.8451 = 0.1549;$$

there is one decimal place to the first digit (0.7), and one less than one is zero. Therefore,

$$\log_{10} 0.7 = -0.1549.$$

In the same way,

$$\log_{10} 0.07 = -1.1549$$
$$\log_{10} 0.007 = -2.1549$$
$$\log_{10} 0.0007 = -3.1549.$$

To find the *antilog* of positive logarithms, find the mantissa in the table, and read the numbers. Fix the decimal at *one more* than the characteristic:

Find the antilog of $0.8451 = 7.0$.

Find the antilog of $1.8451 = 70$.

Find the antilog of $2.8541 = 700$.

Find the antilog of $3.8541 = 7000$.

To find the antilog of a negative logarithm, subtract the mantissa from 1. Find the difference in the table and read the digits. Place the decimal point at *one more* place left than the characteristic indicates:

Find the antilog of $-0.1549 = 0.7$.

Find the antilog of $-1.1549 = 0.07$.

Find the antilog of $-2.1549 = 0.007$.

Find the antilog of $-3.1549 = 0.0007$.

To use the table to determine natural logs, use the following relationship:

$$\ln N = \log_e N = 2.3026 \times \log_{10} N$$
$$\log_{10} N = 0.4343 \times \log_e N.$$

That is, to find the $\log_e$ of a number, simply find the $\log_{10}$ and multiply it by 2.3026; and if the $\log_e$ is given, multiply it by 0.4343 to convert to $\log_{10}$.

COMMON LOGARITHM MANTISSAS ($\log_{10}$)

N	0	1	2	3	4	5	6	7	8	9	N
10	0000	0043	0086	0128	0170	0212	0253	0294	0334	0374	10
11	0414	0453	0492	0531	0569	0607	0645	0682	0719	0755	11
12	0792	0828	0864	0899	0934	0969	1004	1038	1072	1106	12
13	1139	1173	1206	1239	1271	1303	1335	1367	1399	1430	13
14	1461	1492	1523	1553	1584	1614	1644	1673	1703	1732	14
15	1761	1790	1818	1847	1875	1903	1931	1959	1987	2014	15
16	2041	2068	2095	2122	2148	2175	2201	2227	2253	2279	16
17	2304	2330	2355	2380	2405	2430	2455	2480	2504	2529	17
18	2553	2577	2601	2625	2648	2672	2695	2718	2742	2765	18
19	2788	2810	2833	2856	2878	2900	2923	2945	2967	2989	19
20	3010	3032	3054	3075	3096	3118	3139	3160	3181	3201	20
21	3222	3243	3263	3284	3304	3324	3345	3365	3385	3404	21
22	3424	3444	3464	3483	3502	3522	3541	3560	3579	3598	22
23	3617	3636	3655	3674	3692	3711	3729	3747	3766	3784	23
24	3802	3820	3838	3865	3874	3892	3909	3927	3945	3962	24
25	3979	3997	4014	4031	4048	4065	4082	4099	4116	4133	25
26	4150	4166	4183	4200	4216	4232	4249	4265	4281	4298	26
27	4314	4330	4346	4362	4378	4393	4409	4425	4440	4456	27
28	4472	4487	4502	4518	4533	4548	4564	4579	4594	4609	28
29	4624	4639	4654	4669	4683	4698	4713	4728	4742	4757	29
30	4771	4786	4800	4814	4829	4843	4857	4871	4886	4900	30
31	4914	4928	4942	4955	4969	4983	4997	5011	5024	5038	31
32	5051	5065	5079	5092	5105	5119	5132	5145	5159	5172	32
33	5185	5198	5211	5224	5237	5250	5263	5276	5289	5302	33
34	5315	5328	5340	5353	5366	5378	5391	5403	5416	5428	34
35	5441	5453	5465	5478	5490	5502	5514	5527	5539	5551	35
36	5563	5575	5587	5599	5611	5623	5635	5647	5658	5670	36
37	5682	5694	5705	5717	5729	5740	5752	5763	5775	5786	37
38	5798	5809	5821	5832	5843	5855	5866	5877	5888	5899	38
39	5911	5922	5933	5944	5955	5966	5977	5988	5999	6010	39
40	6021	6031	6042	6053	6064	6075	6085	6096	6107	6117	40
41	6128	6138	6149	6160	6170	6180	6191	6201	6212	6222	41
42	6232	6243	6253	6263	6274	6284	6294	6304	6314	6325	42
43	6335	6345	6355	6365	6375	6385	6395	6405	6415	6425	43
44	6435	6444	6454	6464	6474	6484	6493	6503	6513	6522	44
N	0	1	2	3	4	5	6	7	8	9	N

COMMON LOGARITHM MANTISSAS ($\log_{10}$)

N	0	1	2	3	4	5	6	7	8	9	N
45	6532	6542	6551	6561	6571	6580	6590	6599	6609	6618	45
46	6628	6637	6646	6656	6665	6675	6684	6693	6702	6712	46
47	6721	6730	6739	6749	6758	6767	6776	6785	6794	6803	47
48	6812	6821	6830	6839	6848	6857	6866	6875	6884	6893	48
49	6902	6911	6920	6928	6937	6946	6955	6964	6972	6981	49
50	6990	6998	7007	7016	7024	7033	7042	7050	7059	7067	50
51	7076	7084	7093	7101	7110	7118	7126	7135	7143	7152	51
52	7160	7168	7177	7185	7193	7202	7210	7218	7226	7235	52
53	7243	7251	7259	7267	7275	7284	7292	7300	7308	7316	53
54	7324	7332	7340	7348	7356	7364	7372	7380	7388	7396	54
55	7404	7412	7419	7427	7435	7443	7451	7459	7466	7474	55
56	7482	7490	7497	7505	7513	7520	7528	7536	7543	7551	56
57	7559	7566	7574	7582	7589	7597	7604	7612	7619	7627	57
58	7634	7642	7649	7657	7664	7672	7679	7686	7694	7701	58
59	7709	7716	7723	7731	7738	7745	7752	7760	7767	7774	59
60	7782	7789	7796	7803	7810	7818	7825	7832	7839	7846	60
61	7853	7860	7868	7875	7882	7889	7896	7903	7910	7917	61
62	7924	7931	7938	7945	7952	7959	7966	7973	7980	7987	62
63	7993	8000	8007	8014	8021	8028	8035	8041	8048	8055	63
64	8062	8069	8075	8082	8089	8096	8102	8109	8116	8122	64
65	8129	8136	8142	8149	8156	8162	8169	8176	8182	8189	65
66	8195	8202	8209	8215	8222	8228	8235	8241	8248	8254	66
67	8261	8267	8274	8280	8287	8293	8299	8306	8312	8319	67
68	8325	8331	8338	8344	8351	8357	8363	8370	8376	8382	68
69	8388	8395	8401	8407	8414	8420	8426	8432	8439	8445	69
70	8451	8457	8463	8470	8476	8482	8488	8494	8500	8506	70
71	8513	8519	8525	8531	8537	8543	8549	8555	8561	8567	71
72	8573	8579	8585	8591	8597	8603	8609	8615	8621	8627	72
73	8633	8639	8645	8651	8657	8663	8669	8675	8681	8686	73
74	8692	8698	8704	8710	8716	8722	8727	8733	8739	8745	74
75	8751	8756	8762	8768	8774	8779	8785	8791	8797	8802	75
76	8808	8814	8820	8825	8831	8837	8842	8848	8854	8859	76
77	8865	8871	8876	8882	8887	8893	8899	8904	8910	8915	77
78	8921	8927	8932	8938	8943	8949	8954	8960	8965	8971	78
79	8976	8982	8987	8993	8998	9004	9009	9015	9020	9025	79
N	0	1	2	3	4	5	6	7	8	9	N

COMMON LOGARITHM MANTISSAS ($\log_{10}$)

N	0	1	2	3	4	5	6	7	8	9	N
80	9031	9036	9042	9047	9053	9058	9063	9069	9074	9079	80
81	9085	9090	9096	9101	9106	9112	9117	9122	9128	9133	81
82	9138	9143	9149	9154	9159	9165	9170	9175	9180	9186	82
83	9191	9196	9201	9206	9212	9217	9222	9227	9232	9238	83
84	9243	9248	9253	9258	9263	9269	9274	9279	9284	9289	84
85	9294	9299	9304	9309	9315	9320	9325	9330	9335	9340	85
86	9345	9350	9355	9360	9365	9370	9375	9380	9385	9390	86
87	9395	9400	9405	9410	9415	9420	9425	9430	9435	9440	87
88	9445	9450	9455	9460	9465	9469	9474	9479	9484	9489	88
89	9494	9499	9504	9509	9513	9518	9523	9528	9533	9538	89
90	9542	9547	9552	9557	9562	9566	9571	9576	9581	9586	90
91	9590	9595	9600	9605	9609	9614	9619	9624	9628	9633	91
92	9638	9643	9647	9652	9657	9661	9666	9671	9675	9680	92
93	9685	9689	9694	9699	9703	9708	9713	9717	9722	9727	93
94	9731	9736	9741	9745	9750	9754	9759	9763	9768	9773	94
95	9777	9782	9786	9791	9795	9800	9805	9809	9814	9818	95
96	9823	9827	9832	9836	9841	9845	9850	9854	9859	9863	96
97	9868	9872	9877	9881	9886	9890	9894	9899	9903	9908	97
98	9912	9917	9921	9926	9930	9934	9939	9943	9948	9952	98
99	9956	9961	9965	9969	9974	9978	9983	9987	9991	9996	99
N	0	1	2	3	4	5	6	7	8	9	N

APPENDIX 5

FREQUENCY BANDS AND ALLOCATIONS

TABLE A-4 BAND CHARACTERISTICS

Band Number	Abbreviation		Frequency Range
2	ELF	(Extremely Low Frequency)	30 to 300 Hz
3	VF	(Voice Frequency)	300 to 3000 Hz
4	VLF	(Very Low Frequency)	3 to 30 kHz
5	LF	(Low Frequency)	30 to 300 kHz
6	MF	(Medium Frequency)	300 to 3000 kHz
7	HF	(High Frequency)	3 to 30 MHz
8	VHF	(Very High Frequency)	30 to 300 MHz
9	UHF	(Ultra High Frequency)	300 to 3000 MHz
10	SHF	(Super High Frequency)	3 to 30 GHz
11	EHF	(Extremely High Frequency)	30 to 300 GHz

Listed below are a few of the many services occupying the radio frequency bands. These services are typical, and this listing is not complete.

Commercial broadcasting:

	Radio	AM	535.00 to 1605.00 kHz (MF)
		FM	88.00 to 108.00 MHz (VHF)
	TV	2–4	54.00 to 72.00 MHz (VHF)
		5–6	76.00 to 88.00 MHz (VHF)
		7–13	174.00 to 216.00 MHz (VHF)
		14–83	470.00 to 890.00 MHz (UHF)

Maritime communications, navigation and aeronautical radio navigation
30 to 535 kHz (LF–MF)

Amateur radio, loran (long range navigation) international broadcasting, industrial broadcasting, CB radio
1605 kHz to 30 MHz (MF–HF–VHF)

Amateur radio
50 to 54 MHz (VHF)

Aeronautical navigation	108 to 122 MHz (VHF)
Studio-transmitter relay, government and private fixed and mobile communication	890 to 3000 MHz (UHF)
Satellite communications	8400 to 8500 MHz (EHF)

APPENDIX 6

THE INTERNATIONAL SYSTEM OF UNITS

Several measurement systems are used in the world today. Among these are the U.S. "English" (or conventional or customary) system (feet, pounds, gallons) and two varieties of the metric system, the CGS (centimeter, gram, second) and the MKS (meter, kilogram, second). Converting a measurement from one system to another—or sometimes even converting from one unit to another unit of the same system—often requires the use of awkward conversion factors, which can lead to serious errors. Increasing worldwide research and communication have increased the danger that errors will be made.

For centuries, workers in scientific and technical areas have sought a simple, efficient system of measure. France adopted a metric system in 1790. The authors of the U.S. Constitution recognized the importance of the regulation of weights and measures. Since 1893, the international meter and kilogram have been the standards of length and mass in the U.S., both for conventional and metric measurement.

Since 1870, a series of international meetings has been held for the purpose of establishing a standard measuring system. The Bureau of Standards represents the U.S. at these meetings. At a meeting in Paris in 1954, a standard system was worked out. This system is called *Le Système International d'Unités* (The International System of Units), or simply SI. Many countries have already adopted SI, and the system is now the official system of the United States. We are now in a "transition period," from customary units to SI.

The advantage of SI is that it is coherent—an idea we shall come back to shortly. The units themselves are not new; they have been chosen from the older, less precise metric system.

SI UNITS AND SYMBOLS

SI units are divided into three classes:

    Base units
    Supplementary units
    Derived units

Base Units

SI is based on seven units:

Quantity	Unit	Symbol
Length	meter	m
Mass	kilogram	kg
Time	second	s
Electric current	ampere	A
Thermodynamic temperature	kelvin	K
Amount of substance	mole	mol
Luminous intensity	candela	cd

Supplementary Units

The two units listed below may be considered either base units or derived units.

Quantity	Unit	Symbol
plane angle	radian	rad
solid angle	steradian	sr

Derived Units

Derived units are formed by combining base units, supplementary units and other derived units according to the algebraic relationships that link the quantities. The symbols for derived units are formed by means of mathematical signs. For instance, the SI unit for velocity is the meter per second; its symbol is m/s. Some derived units have been given special names. Some of these units are:

Quantity	Unit	Symbol	Formula
Frequency (of a periodic phenomenon)	hertz	Hz	1/s
Force	newton	N	$kg \cdot m/s^2$
Energy, work	joule	J	$N \cdot m$
Power	watt	W	J/s
Quantity of electricity, electric charge	coulomb	C	$A \cdot s$
Electric potential, potential difference, electromotive force	volt	V	W/A
Capacitance	farad	F	C/V
Electric Resistance	ohm	Ω	V/A
Conductance	siemens	S	A/V
Magnetic flux	weber	Wb	$V \cdot s$
Magnetic flux density	tesla	T	Wb/m^2
Inductance	henry	H	Wb/A

We said before that SI is a *coherent* system. This means that all basic units are related to one another by the factor 1. For instance, a force of 1 newton exerted through a length of 1 meter produces 1 joule of energy. (From the preceding list, we see that energy in joules is calculated by the formula $N \cdot m$, and $1 \cdot 1 = 1$. Furthermore if this 1 joule of energy is produced in 1 second, the result is 1 watt of power ($W = J/s = 1/1 = 1$). You can appreciate that this simplifies many calculations and eliminates the need for confusing decimal conversion factors. Furthermore, multiple and submultiple prefixes make it easy to express very large or very small values.) Common prefixes and their values are:

Multiplication Factor	Prefix	Symbol
1 000 000 000 000 000 000 = 10^{18}	exa	E
1 000 000 000 000 000 = 10^{15}	peta	P
1 000 000 000 000 = 10^{12}	tera	T
1 000 000 000 = 10^{9}	giga	G
1 000 000 = 10^{6}	mega	M
1 000 = 10^{3}	kilo	k
0.001 = 10^{-3}	milli	m
0.000 001 = 10^{-6}	micro	μ
0.000 000 001 = 10^{-9}	nano	n
0.000 000 000 001 = 10^{-12}	pico	p
0.000 000 000 000 001 = 10^{-15}	femto	f
0.000 000 000 000 000 001 = 10^{-18}	atto	a

To get a clearer idea of how SI "works," examine the following examples of derived units:

Quantity	SI Unit	Symbol
Area	square meter	m^2
Volume	cubic meter	m^3
Acceleration	meter per second squared	m/s^2
Wave number	1 per meter	m^{-1}
Current density	ampere per square meter	A/m^2
Magnetic field strength	ampere per meter	A/m

Finally, there are several symbols which are not part of SI, but which are used with it. These special symbols are:

Name	Symbol	Value in SI Units
minute (time)	m	1 m = 60 s
hour	h	1 h = 3600 s
day	d	1 d = 86,400 s
degree (angle)	°	1 ° = (π/180) rad
minute (angle)	'	1 ' = (π/10,800) rad
second (angle)	"	1 " = (π/648,000) rad
liter	L	1 L = 10^{-3} m^3
metric ton	t	1 t = 10^3 kg
hectare (land or water areas)	ha	1 ha = 10^4 m^2

This has been a brief and simplified survey of an important and complex subject. Fortunately, most electrical units are the same in conventional and SI; in physics and many other fields of study, however, there are many changed units. If you want additional information on the history and application of SI, check your school library, or write to either of the following:

American Society for Testing and Materials
1916 Race Street
Philadelphia, PA 19103

Metric Education
National Bureau of Standards
Department of Commerce
Washington, D.C.

INDEX

A

Abbreviations, unit, 44
Ac circuit, LC, 414–16
　measuring of, 427
　RC, 389–96
　RL, 352–57
　RLC, 418–22
　R only, 413–14
　three phase, 441–44
Admittance, 426
Alnico, 239
Alternation, 295
Alternating current, 278, 290–309
　average value, 293
　circuits, 290–309
　effective value, 292
　frequency, 294
　meters, 427–36
　nonsinusoidal, 479–502
　peak to peak, 292
　peak value, 292
　period, 294
　phase, 303–4
　rms value, 292
　sine wave, 291
　three phase, 441–44
　wavelength, 302
American wire gauge, 157
Ampere, 32
Ampere turn, 250
Amplifier, 439
Ammeter, 55
Anode, diode, 182
Apparent power, 349
Armature, 281
Atom, 23–25
Atom, helium, 25
　hydrogen, 25
　nucleus, 25
　number, 26
　orbit, (shell), 25
　structure of, 24
　weight, 26
Audiotape recorder, 286
Autotransformer, 335

B

Bandpass, 390, 401–2
Bandwidth, 465–67
　half-power points, 466
　value of, 466
Barrier region, 180–82
Base-line shift, 492
Batteries,
　alkaline, 218
　carbon-zinc, 214
　charging, 223–25
　dry cell, 214–18
　internal resistance, 227
　fuel cell, 226–27
　lead acid, 221
　lithium, 227
　nickel cadmium, 224
　parallel, 213
　polarization, 211
　primary cell, 212
　secondary cell, 212
　series, 213
　silver oxide, 226
　specific gravity, 224
　storage cell, 222
B-H curve, 255
Bleeder resistor, 445
Bridge circuit, balanced, 117
　unbalanced, 119
Blocking capacitor, 398
Brushes, motor, 269
Bus, 73

C

Cables, 159
Capacitor, 364–85
　ceramic, 374
　charge, discharge, 366, 402–7
　electrolytic, 371
　filter, 467–71
　mica, 374
　opens and shorts in, 384
　tantalum, 373
　tubular, 369
　variable, 370
Capacitor, voltage-variable, 473–75
Cathode-ray oscilloscope, 433–36
Cathode-ray tube, 433
CEMF, 278
Choke coil, *See* Inductor, iron core
Circuit, dc, 33
　open, parallel, 96
　open, series, 77

589

590 Index

Circuit, dc (*continued*)
  short, parallel, 96
  short, series, 77
  series parallel, 100
    opens and shorts, 112
Circular-mill area, 157
Coercive force, 258
Coil, ignition, 286
  iron-core, 275
Color code, capacitor, 380
  resistor, 191, 196–98
Complex numbers, 343–46
Compass, 239
Compound, 24
Conductance, 90, 426
Conductor, 27, 157, 162–63
Core, ferrite, 334
  powdered iron, 334
Corona discharge, 166
Common reference, 73
Constant-current source, 230
Constant-voltage source, 230
Core saturation, 334
Cosine of angle, 319
Coulomb, 29
Counterforce, 270
CRT, *see* Cathode-ray tube
Curie point, 244, 259, 260
Current carrier, 40
Current divider, 91, 512–14
Current, induced, 269
Current, spark discharge, 29
Current, tank, 462
Curve, exponential, 339
Cycle, ac, 294

D

dBm, 440
Decibel (dB), 439–41
Delta-wye transformation, 536
Density, flux, 245
Diamagnetism, 239
Dielectric, 168
Dielectric constant, 368
Dielectric strength, 369
Differentiator, 490
Diode, barrier, 180
  characteristics, 182
  forward bias, 181
  junction, 179
  PIV, 184
  reverse bias, 181
  voltage drop, 183
Diode, varactor, 473
Dipole, magnetic, 244
DMM, 150
Domain, magnetic, 243
Doorbell, 281

E

Electrical pressure (volts), 31
  current flow, 32
  resistance, 35
Electrical symbols, 34, Appendix 1
Electromagnet, 250
Electromotive force (EMF), 31
Electron, 24
Element, 23
Equivalent circuit, 101
ESL, 377
ESR, 377
Exponent, 44

F

f_1-f_2, 466
Farad, unit of, 367
Faraday's law, 272
Ferromagnetism, 239
FET-VOM, 150
Filter, 445
Filter circuit, bandpass, 469
  band-reject, 469
  high-pass, 468
  low-pass, 468
Frequency, 294
  audio, 301
  compensation, 494
  fundamental, 483
  harmonic, 483
  radio (RF), 301
  resonant, 453
  response, 398–402
Full-scale current (I_{fs}), 126
Flux, magnetic, 240, 245
Fuse, 40

G

Gases, 23
Generator, ac, 278–82
  dc, 269
Gilbert, 252
Ground, 73
  intermediate, 115

H

Half-power point, 466
Harmonic, 483
Henry, 313
Hole, 175
Hypotenuse, 317
Hysteresis, magnetic, 257

I

Impedance, RC circuit, 390
  RL circuit, 320
  RLC circuit, 419

Impedance matching, 231, 336
Inductance, 312
Inductor, air core, 358
　faults, 361
　ferrite core, 358
　iron core, 359
Integrator, 491
Insulator, 27, 168
Ion, 30
　current, 165

J

Joule, 337

K

Kirchhoff's laws, 518

L

Lamination, core, 332
Lenz's law, 270
Line, magnetic, 237
Liquid, 22
Loadstone, 237
Loss, hysteresis, 333
Loudspeaker, 284

M

Magnet, 236–64
　bar, 237
　horeshoe, 242
　pot-core, 242
Magnetic induction, 241
Magneticmotive force, 251
Magnetite, 237
Matter, 22
Meter, full-scale current, 126
　loading, 138
　multi-, 148
　multiplier, 134
　ohm's-per-volt, 135
　shunt, 129
　zero-ohm's adjust, 142
Meter movement, D'Arsonval, 126
　taut-band, 127
Meter scales, 128
Meter sensitivity, 127
Meter, type of,
　dynamometer, 432
　iron-vane, 432
　thermocouple, 432
　watt meter, 433
Millman's theorem, 533
MKS units, 261, Appendix 6
Motor, dc, 268
Molecule, 23

N

Neutron, 24
Norton's theorem, 528

Numbers, imaginary, 344
　positive, 343
　negative, 343

O

Oersted, 252
Ohm, 35
Ohmmeter, 55
Ohm's law, 40

P

Paramagnetism, 239
Passband, 466
PC boards, 161
Permeability, 244, 252
Period, 294
Permeance, 253
Phase angle, 303–4
　capacitive circuits, 379, 340
　inductive circuits, 314–17
Phasor, 305
Plates, capacitor, 364
Pole, magnetic, 237
Power factor, capacitive, 376–77
　inductive, 351
Power, ac circuit, 437–38
　dc circuit, 36, 52–54
Power of ten, 44
　adding, 46
　dividing, 48
　extracting root, 49
　multiplying, 48
　raising to power, 48
　subtracting, 47
Power, parallel circuit, 92
　series circuit, 36
Probe, attenuator, 498
Proton, 24
Polarity, voltage drop, 68
Pulsating dc, 282, 446
Pulse, 479–98
　duration, 481
　duty period, 482
　fall time, 481
　frequency, 482
　repetition rate, 482
　rest period, 482
　period, 481

Q

Q, 459
　high, 460
　low, 461

R

Radian, 304
Ratio and proportion, 509–12
RC coupling, 396, 488

592 Index

Reactance, capacitive, 365, 378–79
Rectifier, 428
  full-wave bridge, 448
  half wave, 448
  power supply, 444
Ripple voltage, 448
Reactance, capacitive, 365, 378–79
  inductive, 320
Relay, 284
Reluctance, 245, 253
Residual magnetism, 245
Resistor, 35
  parallel connection, 86
  series connection, 63
Resistor, color code, 196
  fixed, 35
  IC type, 200
  potentiometer, 191
  taper, 205
  temperature coefficient, 198
  tolerance, 198
  voltage rating, 202
  wire-wound, 192
Retentivity, 245

S

Semiconductor, 28
  crystalline structure of, 171
  intrinsic, 172
  lattice structure of, 172
  N doping, 173
  passive, 170
  P doping, 175
Saturation, iron core, 256
Separation of charge, 29
Series circuit, 39
Shell, 27
SI units, 260, Appendix 6
Siemens, 426
Signalling, in-band, 471
Sine of angle, 319
Sine wave, average value, 294
  peak to peak, 292
  peak value, 289, 292
  rms value, 292
Slip rings, 279
Solenoid, 287
Solids, 22
Stray capacitance, 414
  inductance, 414
Superposition theorem, 529
Susceptance, 426

T

Tangent of angle, 319
Temperature coefficient, 164, 376
Time constant (t_c), RC, 403
  RL, 338

Thevenin's theorem, 522
Tolerance, capacitor, 376
  resistor, 196–98
Transducer, 301
Transformer, 291, 328
  efficiency, 332
  ferrite core, 500
  pot core, 500
  primary winding, 329
  pulse, 500
  secondary winding, 329
  turns ratio, 330
Trap, filter, 471
Triangle, right, 316
Trigonometry, basic concepts, 317–22
True power, 349

V

Valence, 170
Varactor diode, 473–75
Vector, 305
Videotape recorder, 287
Volt, 31
Voltage divider, 69, 407, 507–13
  capacitive, 407
  compensated, 494
  inductive, 325–28
  resistive, 69, 507–13
Voltage drop, 65
Voltage, gain of, 459
Voltage, induced, 270
Voltage reference, 69, 72
  series aiding, 71
  series opposing, 71
Voltage, signal, 300
Voltage, working, capacitance, 376
Voltmeter, 55
VOM, 56, 125

W

Waveform,
  dc component of, 485–88
  differentiated, 309, 480
  integrated, 309, 480
  pulse, 307
  ramp, 480
  rectangular, 482
  sawtooth, 309, 480
  sinusoid, 291–95
  square wave, 308, 479
  staircase, 309
Watt, 36, 52
Wavelength, 302
Wavetrap, 464
  parallel, 465
  series, 465